Bioremediation of Inorganic Compounds

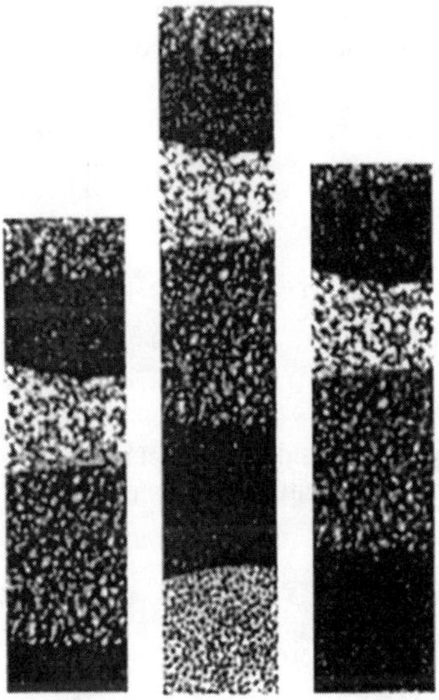

Editors

Andrea Leeson, Brent M. Peyton, Jeffrey L. Means, and Victor S. Magar

The Sixth International In Situ and On-Site Bioremediation Symposium

San Diego, California, June 4–7, 2001

BATTELLE PRESS
Columbus • Richland

Library of Congress Cataloging-in-Publication Data

International In Situ and On-Site Bioremediation Symposium (6th : 2001 : San Diego, Calif.)
Bioremediation of inorganic compounds : the Sixth International In Situ and On-Site Bioremediation Symposium : San Diego, California, June 4-7, 2001 / editors, A. Leeson … [et al.].
p. cm. -- (The Sixth International In Situ and On-Site Bioremediation Symposium ; 9)
Includes bibliographical references and index.
ISBN 1-57477-119-1 (hc. : alk. paper)
1. Metal wastes--Biodegradation--Congresses. 2. Inorganic compounds--Biodegradation--Congresses. 3. Bioremediation--Congresses. I. Leeson, Andrea, 1962- . II. Title. III. Series: International In Situ and On-Site Bioremediation Symposium (6th : 2001 : San Diego, Calif.). Sixth International In Situ and On-Site Bioremediation Symposium ; 9.
TD192.5.I56 2001 vol. 9
[TD196.M4]
628.5 s--dc21
[628.5'2]

2001041197

Printed in the United States of America

Battelle Press
505 King Avenue
Columbus, Ohio 43201, USA
614-424-6393 or 1-800-451-3543
Fax: 1-614-424-3819
Internet: press@battelle.org
Website: www.battelle.org/bookstore

For information on future environmental conferences, write to:
Battelle
Environmental Restoration Department, Room 10-123B
505 King Avenue
Columbus, Ohio 43201-2693
Phone: 614-424-7604
Fax: 614-424-3667
Website: www.battelle.org/conferences

CONTENTS

Foreword vii

Acid Mine Drainage

Treatment of Acid Mine Drainage Using Membrane Bioreactors. *P. Rao, R. Govind, and H.H. Tabak* 1

Biorecovery of Metals From Acid Mine Drainage. *R.A. Scharp, F. Kawahara, J. Burckle, J. Allen, and R. Govind* 9

Effect of Metal Ions on Acetate-Utilizing Mixed Culture of Sulfate-Reducing Bacteria (SRB). *V.P. Utgikar, H.H. Tabak, J.R. Haines, S.M. Harmon, and R. Govind* 17

Statistical Modeling of Sulfate Removal in Mine Drainage Treatment. *I.A. Cocos, G.J. Zagury, B. Clement, and R. Samson* 27

Status and Performance of Engineered SRB Reactors for Acid Mine Drainage Control. *M.H. Zaluski, J.M. Trudnowski, M.C. Canty, and M.A. Harrington-Baker* 35

Neutralization of Acidic Mining Lakes Via In Situ Stimulation of Bacteria. *R. Frommichen, M. Koschorreck, K. Wendt-Potthoff, and K. Friese* 43

Bioremediation of Metals from an Acid Mine Drainage at Cane Creek, Coal Valley Site. *V.M. Ibeanusi, E. Archibold, L. Hannon, E. Garry, A. Hines, and A. Sola* 53

Heavy Metals

Removal of Heavy Metals and Organic Compounds from Anaerobic Sediments. *J.H.P. Watson and D.C. Ellwood* 61

In Situ Treatment of Metals-Contaminated Groundwater Using Permeable Reactive Barriers. *D.J.A. Smyth, D.W. Blowes, S.G. Benner, and C.J. Ptacek* 71

Removal of Heavy Metals Using a Rotating Biological Contactor. *S.C. Costley, F.M. Wallis, and M.D. Laing* 79

Heavy Metals In Situ Bioprecipitation and Adsorption on a Manufacturing Site (Belgium). *D. Nuyens, L. Bastiaens, J. Vos, J. Gemoets, and L. Diels* 87

In Situ Bioremediation of a Soil Contaminated with Heavy Metals and Arsenic. *S.N. Groudev, P.S. Georgiev, K. Komnitsas, I.I. Spasova, and I. Paspaliaris* 97

In Situ Bioremediation of Metals-Contaminated Groundwater Using Sulfate-Reducing Bacteria: A Case History. *J.A. Saunders, M.-K. Lee, J.M. Whitmer, and R.C. Thomas* 105

Soil Heavy Metal Contamination and Practical Approaches to Remediation in Some Parts of China. *Q. Wang, Y. Dong, Y. Cui, and X. Liu* 113

Treatment of Metal-Contaminated Soil and Water by Sulphate Reducing Bacteria. *T. Hakansson and B. Mattiasson* 123

Metal Speciation and Toxicity in Chromated Copper Arsenate-Contaminated Soils. *C.F. Balasoiu, G.J. Zagury, and L. Deschenes* 129

Naphthalene Degradation and Concurrent Cr(VI) Reduction by *Pseudomonas putida* ATCC17484. *S. Ghoshal, A. Al-Hakak, and J. Hawari* 139

Selection and Development of Cadmium Resistance in Bacterial Consortia. *H.M. Knotek-Smith, L.A. Deobald, M. Ederer, and D.L. Crawford* 147

Uranium Wastes

Microbially Mediated Reduction and Immobilization of Uranium in Groundwater at Konigstein. *W. Lutze, Z. Chen, D. Diehl, W. Gong, H.E. Nuttall, and G. Kieszig* 155

Uranium Sequestration by Microbially Induced Phosphorus Bioavailability. *L.G. Powers, H.J. Mills, A.V. Palumbo, C.L. Zhang, and P.A. Sobecky* 165

Nitrogen Removal and Nitrogen-Enhanced Biodegradation

Bioremediation of Nitrate-Contaminated Wastewater. *P.C. Mishra and N. Behera* 173

Natural Attenuation of Nitrate in the Big Ditch Watershed, Illinois. *S. Shiffer, R. Sanford, T. Matos, E. Mehnert, D.A. Keefer, W.S. Dey, and T.R. Holm* 179

In Situ Evaluation of Embedded Carrier in Soil to Reduce Nitrate Leaching from Cropland. *T. Higashi, T. Oshio, and T. Kawakami* 187

Autotrophic Denitrification of Bank Filtrate Using Elemental Sulfur. *H.S. Moon, K. Nam, and J.Y. Kim* 195

High-Rate Denitrification in Biofilm-Electrode Reactor Combined with Microfiltration. *M. Prosnansky and Y. Sakakibara* 201

Nitrite May Accumulate in Denitrifying Walls When Phosphate is Limiting. *W.J. Hunter* 209

Cost-Effective Biological/Chemical Treatment of Ammonia/Ammonium Impacted Media. *W.D. Brady, D. Graves, J. Klens, and D. Strybel* 215

Enhanced In Situ Biological Denitrification: Comparing Two Amendment Delivery Systems. *C. Jones, H.E. Nuttall, and B. Faris* 223

Hydrocarbon Biodegradation Rates and Water Potential in Nitrogen Augmented Desert Soils. *C. Walecka-Hutchison and J.L. Walworth* 231

Electrokinetic Movement of Biological Amendments Through Natural Soils to Enhance In Situ Bioremediation. *D. Gent, R.M. Bricka, L.D. Hansen, D.D. Truax, and M.E. Zappi* 241

Perchlorate

Bioremediation of Perchlorate in Groundwater and Reverse Osmosis Rejectates. *M.E. Losi, V. Hosangadi, D. Tietje, T. Giblin, and W.T. Frankenberger* 249

Evaluation of In Situ Biodegradation of Perchlorate in a Contaminated Site. *Z. Zhang, T. Else, P. Amy, and J. Batista* 257

Biological Perchlorate Removal from Drinking Waters Incorporating Microporous Membranes. *J.R. Batista and J. Liu* 265

Enhanced Natural Attenuation of Perchlorate in Soils Using Electrokinetic Injection. *W.A. Jackson, M.-A. Jeon, J.H. Pardue, and T. Anderson* 273

Case Study of Ex-Situ Biological Treatment of Perchlorate-Contaminated Groundwater. *A.P. Togna, W.J. Guarini, S. Frisch, M. Del Vecchio, J. Polk, C. Murray, and D.E. Tolbert* 281

In Situ Bioremediation of Perchlorate-Contaminated Soils. *J.R. Kastner, K.C. Das, V.A. Nzengung, J. Dowd, and J. Fields* 289

Successful Field Demonstration of In Situ Bioremediation of Perchlorate in Groundwater. *M.L. McMaster, E.C. Cox, S.L. Neville, and L.T. Bonsack* 297

Perchlorate Degradation in Bench- and Pilot-Scale Ex Situ Bioreactors. *B.E. Logan, K. Kim, and S. Price* 303

In Situ Removal of Perchlorate From Groundwater. *W.J. Hunter* 309

In Situ Biotreatment of Perchlorate and Chromium in Groundwater. *M.W. Perlmutter, R. Britto, J.D. Cowan, and A.K. Jacobs* 315

On-Site Evaluation of Selenium Treatment Technologies. *J.C. Cherry and J. Saran* 323

Pilot-Scale Nitrate, Selenium, and Cyanide Removal. *D.J. Adams and T. Pickett* 331

Author Index 339

Keyword Index 367

FOREWORD

The papers in this volume correspond to presentations made at the Sixth International In Situ and On-Site Bioremediation Symposium (San Diego, California, June 4-7 2001). The program included approximately 600 presentations in 50 sessions on a variety of bioremediation and supporting technologies used for a wide range of contaminants.

This volume focuses on *Bioremediation of Inorganic Compounds*. This important subject area represents a diverse group of compounds including metals, perchlorate, arsenic, and nitrogen in a variety of environmental media, including acid mine drainage, soils, sediments, and groundwater. The application of bioremediation to such diverse conditions confirms its immense value in the area of hazardous waste site remediation.

The author of each presentation accepted for the symposium program was invited to prepare an eight-page paper. According to its topic, each paper received was tentatively assigned to one of ten volumes and subsequently was reviewed by the editors of that volume and by the Symposium chairs. We appreciate the significant commitment of time by the volume editors, each of whom reviewed as many as 40 papers. The result of the review was that 352 papers were accepted for publication and assembled into the following ten volumes:

Bioremediation of MTBE, Alcohols, and Ethers — 6(1). Eds: Victor S. Magar, James T. Gibbs, Kirk T. O'Reilly, Michael R. Hyman, and Andrea Leeson.

Natural Attenuation of Environmental Contaminants — 6(2). Eds: Andrea Leeson, Mark E. Kelley, Hanadi S. Rifai, and Victor S. Magar.

Bioremediation of Energetics, Phenolics, and Polycyclic Aromatic Hydrocarbons — 6(3). Eds: Victor S. Magar, Glenn Johnson, Say Kee Ong, and Andrea Leeson.

Innovative Methods in Support of Bioremediation — 6(4). Eds: Victor S. Magar, Timothy M. Vogel, C. Marjorie Aelion, and Andrea Leeson.

Phytoremediation, Wetlands, and Sediments — 6(5). Eds: Andrea Leeson, Eric A. Foote, M. Katherine Banks, and Victor S. Magar.

Ex Situ Biological Treatment Technologies — 6(6). Eds: Victor S. Magar, F. Michael von Fahnestock, and Andrea Leeson.

Anaerobic Degradation of Chlorinated Solvents— 6(7). Eds: Victor S. Magar, Donna E. Fennell, Jeffrey J. Morse, Bruce C. Alleman, and Andrea Leeson.

Bioaugmentation, Biobarriers, and Biogeochemistry — 6(8). Eds: Andrea Leeson, Bruce C. Alleman, Pedro J. Alvarez, and Victor S. Magar.

Bioremediation of Inorganic Compounds — 6(9). Eds: Andrea Leeson, Brent M. Peyton, Jeffrey L. Means, and Victor S. Magar.

In Situ Aeration and Aerobic Remediation — 6(10). Eds: Andrea Leeson, Paul C. Johnson, Robert E. Hinchee, Lewis Semprini, and Victor S. Magar.

In addition to the volume editors, we would like to thank the Battelle staff who assembled the ten volumes and prepared them for printing: Lori Helsel, Carol Young, Loretta Bahn, Regina Lynch, and Gina Melaragno. Joseph Sheldrick, manager of Battelle Press, provided valuable production-planning advice and coordinated with the printer; he and Gar Dingess designed the covers.

The Bioremediation Symposium is sponsored and organized by Battelle Memorial Institute, with the assistance of a number of environmental remediation organizations. In 2001, the following co-sponsors made financial contributions toward the Symposium:

Geomatrix Consultants, Inc.
The IT Group, Inc.
Parsons
Regenesis
U.S. Air Force Center for Environmental Excellence (AFCEE)
U.S. Naval Facilities Engineering Command (NAVFAC)

Additional participating organizations assisted with distribution of information about the Symposium:

Ajou University, College of Engineering
American Petroleum Institute
Asian Institute of Technology
National Center for Integrated Bioremediation Research & Development (University of Michigan)
U.S. Air Force Research Laboratory, Air Expeditionary Forces Technologies Division
U.S. Environmental Protection Agency
Western Region Hazardous Substance Research Center (Stanford University and Oregon State University)

Although the technical review provided guidance to the authors to help clarify their presentations, the materials in these volumes ultimately represent the authors' results and interpretations. The support provided to the Symposium by Battelle, the co-sponsors, and the participating organizations should not be construed as their endorsement of the content of these volumes.

Andrea Leeson & Victor Magar, Battelle
2001 Bioremediation Symposium Co-Chairs

TREATMENT OF ACID MINE DRAINAGE USING MEMBRANE BIOREACTORS

Rakesh Govind and Prasanna Rao (Department of Chemical Engineering, University of Cincinnati, Cincinnati, OH)
Henry H. Tabak (U.S. EPA, NRMRL, Cincinnati, OH)

ABSTRACT: Acid mine drainage is a severe water pollution problem attributed to past mining activities. The exposure of the post-mining mineral residuals to water and air results in a series of chemical and biological oxidation reactions, that produce an effluent which is highly acidic and contains high concentrations of various metal sulfates. Several treatment techniques utilizing sulfate reducing bacteria have been proposed in the past; however few of them have been practically applied to treat acid mine drainage. This research deals with membrane reactor studies to treat the acid mine drainage water from Berkeley Pit in Butte, Montana using hydrogen-consuming sulfate reducing bacteria. Eventually, the membrane reactor system can be applied towards the treatment of acid mine drainage to produce usable water.

INTRODUCTION

Metals are an integral part of the United States economy, and this is also true throughout the world. The residual effects of metals and their use, particularly in aqueous streams, continues to be a problem for metal producers and users, as well as federal and state regulators. Innovative and alternative techniques which allow for the economic control/recovery of metals is one alternative that lends itself not only to human health and environmental protection, but also to resource conservation and reuse of valuable commodities.

Acid mine drainage (AMD) is a common problem for the mining and smelting industries throughout the world. These drainages typically contain dissolved metals of high concentration and more than 3 g/L sulfate. Low pH and presence of heavy metals makes AMD treatment a major concern because of the possible deleterious effects of the effluent on the environment. There are over 40,000 remote abandoned mines in the state of Montana alone, and several thousand such mines in other states such as Pennsylvania, Ohio and West Virginia. Acid mine water, upwelling from these remote mines, mainly during the spring season, results in massive destruction of surrounding vegetation.

Conventional AMD treatments use lime to precipitate metals. These treatments present some serious limitations in terms of application and effectiveness. They usually result in production of mixture of unstable metal hydroxides which also lead to a greater disposal expense. In recent years, the use of sulfate reducing bacteria (SRB) to biodegrade sulfate and precipitate metals in acid mine drainage has been proposed as an alternative to hydroxide precipitation.

One problem with using a biomass is that the precipitated metals are lost with the wasted bacteria. Govind (1997) separated the biological stage from the precipitation stage for the treatment of acid mine drainage. The bioreactor converted sulfate to hydrogen sulfide gas that was then used in three precipitation units. The water treated was a synthetic mixture of similar composition to Berkeley Pit water. Not only were the metals able to be separated from the biomass, but they were also removed effectively. The copper concentration was reduced 99.9%, the manganese concentration was reduced 98.2%, the zinc concentration was reduced 99.9%, the arsenic concentration was reduced 97.1%, the cadmium concentration was reduced 94.6%, and lead was reduced 81.2%.

In this study, membrane reactor system is used to biologically reduce sulfate to hydrogen sulfide gas, which can be used to precipitate the metals from acid mine drainage. Membrane reactors have been used in a variety of applications, including waste water treatment, chemical processing and air pollution control (Govind and Itoh, 1989).

Traditional methods of using hydrogen and carbon dioxide gas mixture for sulfate reduction include the use of gas sparged reactors. The gas mixture is bubbled through the reactor liquid, with the liquid bubbles rising through the liquid containing active SRBs. The gases dissolve and diffuse to the active cells, resulting in the formation of sulfides. Since hydrogen is rather insoluble in water, the unreacted gases exiting the reactor are re-pressurized and recycled. The hydrogen sulfide gas formed and stripped from the liquid, needs to be separated from the recycle gas stream, to prevent its accumulation.

The main disadvantages of the sparged gas reactor system are as follows: (1) large hydrogen gas mixture recycle, since hydrogen gas has very low aqueous solubility; (2) substantial gas-phase pressure drop, which results in the use of large recycle gas compressors; (3) safety issues resulting from hydrogen gas compression for recycle; (4) low conversions of sulfate in the water to sulfides, due to poor mass transfer of the hydrogen gas to the active hydrogen-consuming SRBs;(5) difficulties in separating the hydrogen sulfide gas produced from the recycle hydrogen/carbon dioxide gas mixture; and (6) tall reactor sizes due to low conversions and washout of the active SRB population.

The use of a membrane reactor system, schematically shown in Figure 1, overcomes the problems of using gas sparged reactors. The main advantages of the membrane reactor system are as follows: (1) the microporous membrane surface presents a very large surface area to the liquid phase, resulting in high mass fluxes, compared to the surface area of the much larger rising gas bubbles in the sparged reactor system; (2) the hydrogen sulfide gas is formed outside the membrane and hence does not mix with the pressurized gas inside the hollow fibers (refer to Figure 1); (3) there is no requirement of a gas recycle compressor, which is major advantage especially with the safety issues concerned with hydrogen gas compression; (4) the membrane surface provides a suitable support surface for immobilization of active SRBs resulting in the formation of biofilms; the concentration of active SRBs (present as biofilms) is substantially greater than the concentration of SRBs that can be achieved in suspended culture gas-sparged reactors, resulting in substantially higher sulfate reduction rates; (5) formation of biofilms prevents washout problems associated with suspended culture reactors; and (6) the investment and operating cost of the proposed reactor are significantly lower than a tall liquid-phase sparged reactor system. Due to mass transfer limitations, sparged gas reactors will have a significantly higher volume than membrane reactors, and the operating costs of sparged reactors is higher compared to membrane systems mainly due to gas recompression and recycle costs.

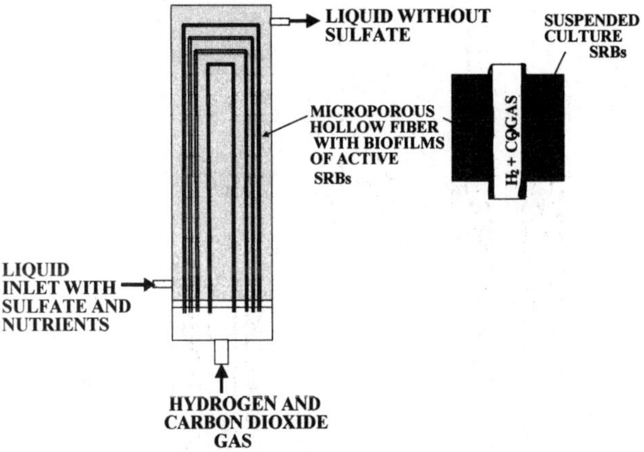

FIGURE 1. Schematic of the membrane reactor system using hydrogen-consuming sulfate reducing bacterial biofilms outside the membrane hollow fibers.

Objective. The main objective of this research was to operate the membrane reactor system to determine the rate of sulfate reduction using hydrogen-consuming SRBs. Acid mine water from the Berkeley Pit, Butte, MT, was obtained to conduct this study. The average composition of the acid mine water is summarized in Table 1.

TABLE 1. Dissolved Metal Concentrations in Berkeley Pit Mine Water.

Compound	Concentration (mg/L)
Al^{+3}	293
Cu^{+2}	223
Mn^{+2}	223
Fe^{+2}	514
Zn^{+2}	630
Cd^{+2}	1.38
Ni^{+2}	2.14
As^{+3}	0.512
Co^{+2}	1.23
$SO4^{2-}$	2,400
Cl^{-}	16
Na^{+1}	213

MATERIALS AND METHODS

Precipitation of Metals. A large batch of acid mine water was mixed with calcium hydroxide and hydrogen sulfide gas mixture (50% hydrogen sulfide and 50% carbon dioxide) was bubbled to precipitate the metal sulfides and hydroxides as well as calcium sulfate, resulting in acid mine water at pH = 8.0. At this pH of 8.0, the majority of copper, zinc, aluminum, iron, manganese and other metals were precipitated as either sulfides and/or hydroxides, resulting in water with high sulfate content. At the end of the batch precipitation process, after a pH of 8.0 has been attained, the water was filtered to separate the precipitates.

Development of Master Culture. Sludge was withdrawn from a local municipal plant's anaerobic digester. Approximately 200 g of wet filtered sludge was mixed with 1.5 L of a basal medium, which had the following composition (per liter of distilled water): 5.3 g ammonium sulfate, 2 g sodium acetate, 0.5 g potassium phosphate, 1 g sodium chloride, 0.2 g magnesium sulfate, 0.1 g calcium chloride, 1 mL reazurin solution (0.2% in water) and 10 mL of trace element solution. The trace element solution contained per liter distilled water, the following chemicals: 12.8 g nitrilotriacetic acid neutralized to pH 6.5 with sodium hydroxide, 300 mg ferrous chloride, 20 g cuprous chloride, 100 mg manganese chloride, 170 mg cobalt chloride, 100 mg zinc chloride, 10 mg boric acid and 10 mg of sodium molybdate. The medium was autoclaved, cooled and the following components were added from sterile stock solution: 50 mL of 8% sodium carbonate, 5.5 mL 25% hydrochloric acid and 1 mL of sodium thiosulfate solution (0.5M in water). The pH of the basal medium was 7.2 after preparation.

The gas phase used was 80 % hydrogen/20 % carbon dioxide or nitrogen gas. The pH of the culture was maintained by adding sulfuric acid. After the sediment culture had been grown for four weeks, the solution was filtered through a coarse filter to separate the sediment coarse particles. The solution was decanted to prevent all the sediment particles from being filtered. The turbid solution was then added as inoculum to the SRB master culture reactors.

Three master culture reactors were set up to grow sulfate-reducing bacteria for operating the membrane reactors. The volume of each reactor was 2 L and the working volume of the culture was about 1800 mL. The reactors were gas-tight so that there was no oxygen contamination. The original bacterial culture for these reactors was taken from the anaerobic digester sludge from the Wastewater Treatment Plant of City of Cincinnati. 200 gm of wet sludge was mixed with 1.5 L of basal medium and about 65 mL of nutrient solution. All three reactors were operated in the incubator at a temperature of 30°C and the contents of the reactors were stirred using the magnetic stirrers. Hydrogen and carbon dioxide gas mixture in a 50/50 ratio was bubbled through the reactors. The gas exiting the reactor was bubbled through zinc acetate solution and thus the production of hydrogen sulfide was measured. The pH of the culture was measured periodically and was found to vary between 7 and 8.

Batch Studies to Determine Kinetics of Sulfate Reduction. A master culture reactor was spiked with known sulfate concentration of 3200 ppm. The sulfate concentrations were measured after a regular time interval of three hours using a Dionex Ion Chromatograph. The amount of biomass was measured by drying a known volume of sample and measuring the dry weight of the cells.

Operation of the Membrane Reactor Process. The acid mine water obtained after filtration from the batch precipitation step, was pumped through the membrane module (refer to Figure 2). The membrane module consisted of a FibreFlo hollow fiber capsule filter with a surface area of 0.6 ft^2. The membrane was made of polypropylene, which was found to have a least uptake of metals. Three membrane modules were initially loaded with the hydrogen-consuming sulfate reducing bacterial culture by pumping the water containing culture from the master culture reactor through the shell-side of the membrane module for several hours. As the culture flowed through the shell side of the membrane module, sufficient amount of biomass was attached to the outside surface of the hollow membrane fibers.

The metal-free, filtered acid mine water was placed in three 500 mL reservoirs. Nutrient solution was added to this water. After the culture was attached to the outside of the membrane hollow fibers, the filtered acid mine water was pumped through the shell-side of the membrane module using a multi-channel peristaltic pump. The hydrogen-carbon dioxide gas mixture, used for culturing the inoculum in the master culture reactors was used to pressurize the inside of the membrane hollow fibers. The bubble pressure inside the membrane module was 30 psig. The gas mixture at a pressure of 24 psig was used to pressurize the insides of the membrane, so the biomass on the outer side can avail it. A 0.2-micron inline filter was used to prevent the biomass from the membrane reactors from contaminating the AMD water in the reservoirs. Three such membrane reactors were operated. Samples from the reservoirs were taken and analyzed for sulfate concentrations using an ion chromatograph.

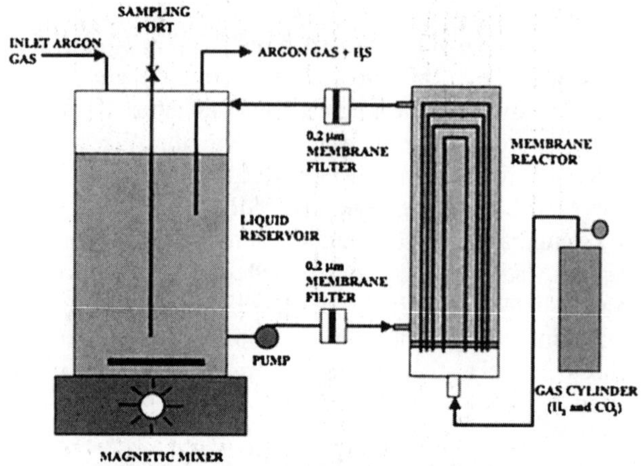

FIGURE 2. Schematic of the membrane reactor experimental set-up.

RESULTS AND DISCUSSION

Culture Growth Characteristics. Growth of SRBs on hydrogen and carbon dioxide alone was not successful. However, growth occurred when acetate was present in the basal medium. Similar results have also been obtained by previous investigators. Acetate is mainly used for culture growth and hydrogen is consumed to reduce sulfate to sulfide. Results of culture growth are shown in Figure 3. Based on the experimental data, the yield of hydrogen consuming SRBs (g cells, dry weight) per unit weight of sulfate consumed (Y_{SO4}) was calculated to be 0.04 + 0.003. Literature yield have varied in the range of 0.03 – 0.06 g/g. The value of yield is about 10% of the yield for acetate-consuming SRB mixed cultures, which demonstrates that the SRBs in the master culture reactors were hydrogen-consuming SRBs rather than acetate-consuming SRBs.

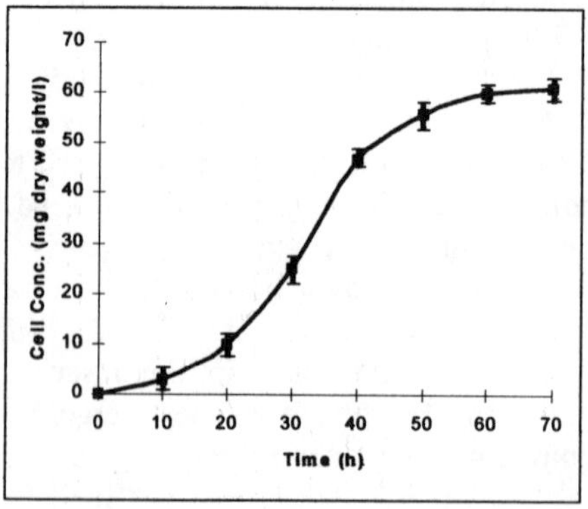

FIGURE 3. Growth of cells as a function of time.

Results from Batch Test Studies in Master Culture Reactors. In the batch test studies, sulfate concentration in the master culture reactor decreased with time as shown in Figure 4.

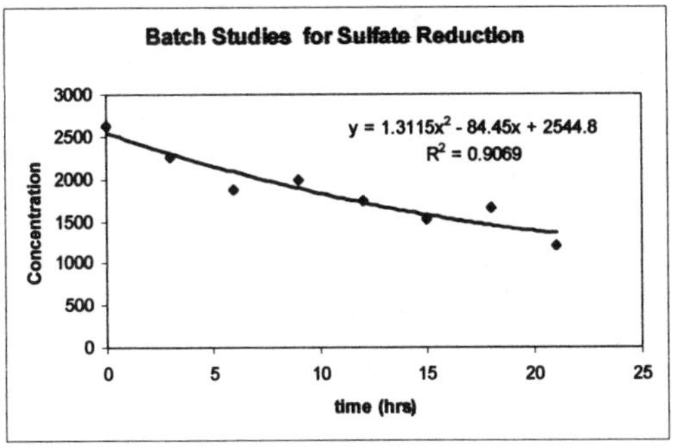

FIGURE 4. Sulfate reduction in batch tests in the master culture reactor.

The sulfate reduction data was analyzed using the Monod model:

$$-\frac{dC_{SO4}}{dt} = \frac{\mu_{SO4} C_{SO4} X_{SRB}}{Y_{SO4}(K_{SO4} + C_{SO4})} \qquad (1)$$

where C_{SO4} is the sulfate concentration in the batch reactor
μ_{SO4} is the maximum specific growth rate for the SRB culture
Y_{SO4} is the yield of the SRB culture
K_{SO4} is the half-saturation constant for the SRB culture
X_{SO4} is the dry weight concentration of SRB culture in the reactor

Using the above model, the following values of the biokinetic parameters were determined with the dry weight SRB biomass concentration in the reactor to be 1.11×10^5 ppm.

$K_{SO4} = 4{,}695 \pm 23$ ppm and $(Y_{SO4}/\mu_{SO4}X_{SRB}) = 0.0121 \pm 0.002$ hr/ppm

Results from Membrane Reactor Studies. Membrane reactor studies were conducted by sampling the liquid from the reservoir periodically and determining the sulfate concentration. The hydrogen and carbon dioxide gas mixture was supplied to the inside of the hollow fibers during this study. As sulfate reduction occurred due to biological reduction by the hydrogen-consuming SRBs, liquid samples were withdrawn and analyzed for sulfate using the EPA standard method 4110. The results of this study are shown in Figure 5.

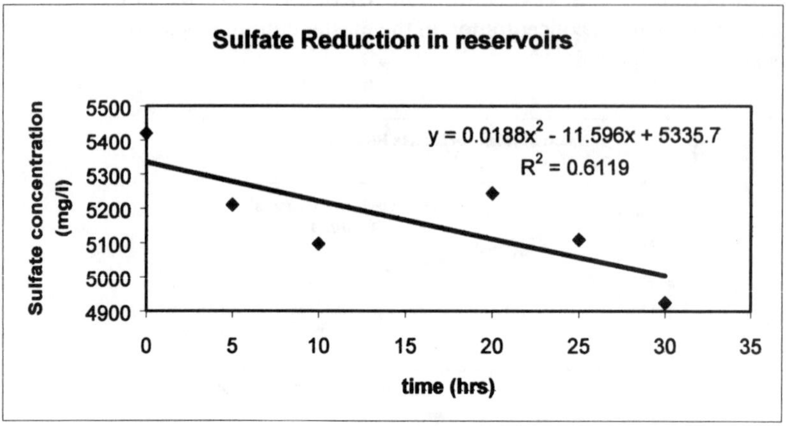

FIGURE 5. Sulfate reduction as a function of time in the reservoir of the membrane reactor experimental system.

Sulfate Reduction at Higher Flowrates. The membrane modules were operated at flowrates of 25, 50, 75 and 100 mL/min using a peristaltic pump. The percent sulfate reduction in the reservoirs was found approximately 13, 16.6, 14.5 and 10.5%. All the runs were conducted at an ambient temperature of 25^0 C. The reduction in the rate at higher flowrate was due to the detachment of the biomass from the outside surface of the hollow fibers.

CONCLUSIONS

Experimental studies were conducted on achieving sulfate reduction using polypropylene hollow fiber membrane reactor system using hydrogen-consuming SRBs. Master Culture Reactor studies showed that hydrogen-consuming SRBs could be cultured from anaerobic digester sludges. The nutrient media used was adequate for growing hydrogen consuming SRBs and biokinetic studies showed that the yield of the bacterial culture was very low. Membrane reactor studies conducted using the hydrogen-consuming SRBs showed that the reactor is capable of reducing sulfate efficiently in a short residence time.

REFERENCES

Govind, R. and N. Itoh. 1989. *Membrane Reactor **Technology***, AIChE Symposium Series, American Institute of Chemical Engineers, New York, NY.

Govind, R., U. Kumar, R. Puligadda, J. Antia, and H. Tabak. 1997. "Biorecovery of Metals from Acid Mine Drainage." *Emerging Technologies in Hazardous Waste Management*, 7, 91-101.

BIORECOVERY OF METALS FROM ACID MINE DRAINAGE

Richard A. Scharp and Fred Kawahara (NRMRL, U.S. EPA, Cincinnati, OH)
John Burckle (Grantee, U.S. EPA/ NRMRL) Jefferey Allen and Rakesh Govind
(Department of Chemical Engineering, University of Cincinnati, Cincinnati, OH)

ABSTRACT: Acid mine water is an acidic, metal-bearing wastewater generated by the oxidation of metallic sulfides by certain bacteria in both active and abandoned mining operations. The wastewaters contain substantial quantities of dissolved solids with the particular pollutants dependant upon the mineralization occurring at the mined faces. Those usually encountered and considered of concern for risk assessment are: arsenic, cadmium, copper, iron, lead, manganese, zinc and sulfate. Occasionally, other minor elements are encountered that must be considered. The pollution generated by abandoned mining activities in the area of Butte, Montana has resulted in the designation of the Silver Bow Creek – Butte Area as the largest Superfund (National Priorities List) site in the U.S. This paper reports the preliminary results of bench-scale studies conducted to develop a resource recovery based remediation process for the clean up of the Berkeley Pit. The process utilizes selective, sequential precipitation (SSP) of metal hydroxides and sulfides, such as copper, zinc, aluminum, iron and manganese, from the Berkeley Pit AMD for their removal from the water in a form suitable for additional processing into marketable precipitates and pigments.

INTRODUCTION

Acid mine drainage is a major environmental issue in the United States and other countries wherever mining has been practiced on a large scale. Acid mine drainage (AMD) causes billions of dollars of damage to natural vegetation, silvaculture, rivers, natural habitats and aquatic life. Innovative and alternative techniques that facilitate the economic control and recovery of metal values is one alternative that lends itself not only to protection of human health and the environment, but also to the recovery of valuable commodities and resource conservation. Previous studies have attempted to precipitate metals from acid mine drainage using sulfate reducing bacteria (SRBs) (Bhattacharyya *et al.* 1979, 1981; Maree and Strydom, 1987; Allen *et al.*, 1999).

A major problem associated with use of sulfate reducing bacteria (SRBs) for precipitating metal sulfides is metal toxicity to active SRB cultures. Govind et al. (1997) proposed a two-stage process in which the metal precipitation step was separated from the SRB bioreactor system. Hydrogen sulfide produced by the SRBs in the bioreactor system was used in the precipitation step to form insoluble metal sulfides. In this study, the precipitation process developed and tested is designed to use biogenic hydrogen sulfide gas generated from the wastewater; hence the precipitation process is referred to as a biorecovery process. This arrangement is shown in Figure 1, below, and the corresponding paper presented at this conference (Rao *et al.*, 2001).

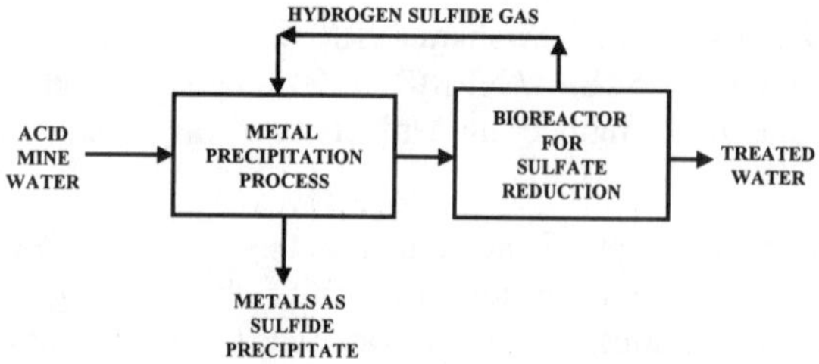

FIGURE 1. Schematic of the sequential treatment system wherein the metal precipitation step is separated from the biological sulfate reduction step.

Objective. The main objective of this study was to use sulfide and hydroxide precipitation to separate metals from acid mine drainage to achieve high recoveries and metal precipitate purities, while producing a water suitable for discharge to the environment. Table 1 shows the average composition of acid mine drainage (pH = 2.2) from the Berkeley Pit. Metals that could potentially be marketed in sulfide form to the smelters, such as copper, zinc and manganese are present in appreciable concentrations as dissolved sulfates.

TABLE 1. Average Concentrations of Metals in the Acid Mine Drainage Treatment Plant Feedwater

Compound	Concentration (mg/L)
Al^{+3}	293
Cu^{+2}	223
Mn^{+2}	223
Fe^{+2}	514
Zn^{+2}	630
Cd^{+2}	1.38
Ni^{+2}	2.14
As^{+3}	0.512
Co^{+2}	1.23
$SO4^{2-}$	2,400
Cl^{-}	16
Na^{+1}	213

MATERIALS AND METHODS

Metal Concentrations. The concentration of metals including Fe^{2+}, Cu^{2+}, Zn^{2+}, Al^{3+}, Mn^{2+}, As^{2+}, Cd^{2+}, Co^{2+}, Pb^{2+}, Ni^{2+} in the acid mine water was analyzed using an ICAP 61E Plasma Emission Spectrometer (Thermo Jarrell Ash Corporation) using EPA Method 6010. Liquid samples were filtered through a

0.22 μm membrane filter (Cole Parmer Company) to remove solids and diluted with 2% nitric acid solution to avoid precipitation of metals due to changes in pH. Dried precipitate samples were redissolved in concentrated nitric acid solution, and then diluted for analysis.

Precipitation Process. The precipitation process studied is shown in Figure 2. Each precipitator unit is equipped with pH probes, feed lines for pH adjustment using hydroxide, and H_2S/CO_2 gas feed lines, where needed for metal sulfide precipitation. The baffles in the settling tanks reduce turbulence and enhance precipitate settling.

The precipitation process consists of four separate Precipitator-Settler units. Argon gas was used to blanket the feed tank to prevent oxidation of the ferrous sulfate present in the acid mine water. A synthetic gas mixture (50 mole % hydrogen sulfide and 50 mole% carbon dioxide) was bubbled into the precipitator using a fine sparger. The first precipitator-settler unit, in which copper and zinc were co-precipitated for operational convenience, was operated at pH of 3.0. The effluent from the first unit was sparged with nitrogen gas to remove any dissolved sulfide and pumped into the second precipitator-settler unit. The pH in the second precipitator-settler unit was adjusted by addition of 12N potassium hydroxide solution.

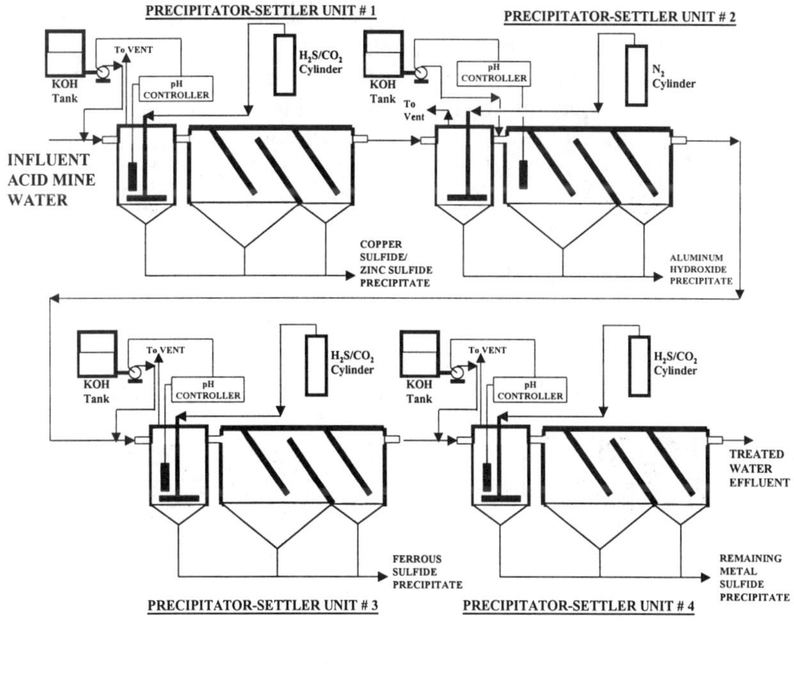

FIGURE 2. Schematic of the four step precipitation process studied for recovery of metals from acid mine drainage.

The effluent water from the second precipitator-settler unit was pumped into the third precipitator-settler unit, in which the pH was maintained by continuous and automatic addition of potassium hydroxide solution. A mixture of 50% hydrogen sulfide and 50% carbon dioxide gas was sparged into this third precipitator-settler unit, to precipitate Ferrous Sulfide from the acid mine water.

The water from the third precipitator-settler unit was pumped into the fourth and last precipitator-settler unit, where the 50% hydrogen sulfide and 50% carbon dioxide gas mixture was sparged at pH above 8.0 to precipitate the remaining metals as sulfide precipitates. Previous studies had shown that, above pH of 8.0, all remaining metal sulfides (mainly manganese sulfide) will precipitate.

Defining Recovery and Purity. Recovery is a measure of how well the precipitator-settler is able to precipitate the desired metal(s). Mathematically, recovery is defined as follows:

$$R = \frac{F_i x_i + B x_B - B y_o - F_i y_o}{F_i x_i + B x_B} \times 100\% \qquad (1)$$

where: F_i = Liquid stream into the reactor (ml/min or ml)
x_i = Composition of metal "i" in liquid stream into the reactor (ppm)
B = Base stream into the reactor (ml/min or ml)
x_B = Composition of metal "i" in base stream into the reactor (ppm)
G_i = Gas stream into the reactor (SCFH)
x_G = Composition of metal "i" in gas stream into the reactor (ppm)
G_o = Gas stream leaving the reactor (SCFH)
y_G = Composition of metal "i" in gas stream leaving the reactor (ppm)
F_o = Liquid stream leaving the reactor (ml/min or ml)
y_o = Composition of metal "i" in liquid stream leaving the reactor (ppm)
S = Solid product (g/min or g)
y_S = Mole fraction of metal "i" in solid product
ρ_i = Density of the liquid stream into the reactor (g/ml)
ρ_B = Density of the base stream into the reactor (g/ml)
ρ_o = Density of the liquid stream leaving the reactor (g/ml)

Purity is the amount of desired metal precipitated from a reactor divided by the sum of all metals precipitated.

$$P = \frac{F_i (x_i - x_o) + B(x_B - x_o)}{\sum_{\text{all metals}} F_i (x_i - x_o) + B(x_B - x_o)} \times 100\% \qquad (2)$$

Metal removal is computed as the difference between initial metal concentration in the feed and the metal concentration in the effluent leaving the system. It is then normalized by dividing by the initial concentration.

$$\text{Removal} = \frac{(x_i)_{\text{first stage}} - (x_o)_{\text{last stage}}}{(x_i)_{\text{first stage}}} \times 100\% \quad (3)$$

RESULTS AND DISCUSSION

Table 2 shows the results of metal recoveries for the four stage precipitation-settling process. The results show that the first reactor was able to remove over 99% of the copper and zinc as a combined precipitate. After obtaining only a 90% recovery during the first set of experiments, the pH set-point for the pH controller was increased and aluminum recovery rose to 94%. Iron and manganese were both recovered at 99% and 95%, respectively, in some experiments. Thus, each step in the process was successful in removing the desired metal.

TABLE 2. Metal Recoveries for each metal for the four stage precipitation process operating on Berkeley Pit acid mine drainage.

Experiment Number	Precipitator-Settler Unit # 1		Precipitator-Settler # 2	Precipitator-Settler # 3	Precipitator-Settler # 4
	Cu	Zn	Al	Fe	Mn
1	100%	99.9%	97.0%	100%	
2	100%	100%	96.8%	100%	
3	100%	99.9%		97.6%	
4	100%	100%	82.1%	100%	
5	100%	100%	86.7%	100%	100%
6	100%	100%	88.8%	100%	100%
7	100%	100%	95.2%	100%	
8	100%	100%	89.9%	100%	100%
9			84.6%		
10	100%	99.9%			
11	99.9%	99.9%			
12	100%	99.9%	92.0%	100%	100%
13	100%	99.9%	96.1%	99.6%	99.0%
14	100%	99.5%	96.1%	99.8%	95.1%
15	100%	99.8%	97.1%	99.9%	99.1%
16	100%	99.8%	92.9%	99.6%	96.3%
17	99.9%	99.9%	97.1%	99.8%	100%
18	99.9%	99.9%	99.2%	100%	99.0%
19	100%	99.8%	77.1%	100%	99.3%
20	99.8%	99.7%	92.8%	99.7%	100%
21	99.9%	99.9%	98.5%	99.6%	89.2%

To determine whether or not each step was able to remove only the desired metal, the purity of the resulting precipitate was measured. The combined copper and zinc purity in the first precipitator was consistently over 95% purity for each metal. The aluminum hydroxide was typically over 90% pure. The third precipitator product, iron sulfide, was not consistently pure. The product obtained during the earlier portion of the tests indicate achievement of purities above 90%.

The data collected during the later period was consistently around 60% purity mainly due to co-precipitation of manganese.

TABLE 3. Metal purities calculated using mass balance analysis for the precipitation process for the Berkeley Pit acid mine water.

Experiment Number	Copper/Zinc Sulfide Meas.	Copper/Zinc Sulfide Calc.	Aluminum Hydroxide Meas.	Aluminum Hydroxide Calc.	Ferrous Sulfide Meas.	Ferrous Sulfide Calc.	Manganese Sulfide Meas.	Manganese Sulfide Calc.
1		98.2%		97.0%	93.9%	92.0%		
2		99.5%		96.8%		79.7%		
3		99.7%			94.2%	92.4%		
4	96.8%	95.8%	66.1%*	82.1%		93.9%		
5	99.2%	97.1%	68.2%*	86.7%	54.4%	66.7%	33.1%	28.8%
6	98.3%	98.0%		88.8%	78.3%*	34.3%*	51.0%	65.6%
7		98.6%		95.2%		35.2%*		
8		98.2%	96.8%	89.9%	94.9%	92.4%	35.3%	50.9%
9	97.4%			84.6%				
10	99.2%	99.8%						
11		79.6%						
12	98.6%	98.1%	94.3%	92.0%		55.5%		7.2%
13		81.7%		96.1%		67.5%		10.1%
14	98.7%	99.7%	93.2%	96.1%	66.8%	59.6%	14.0%	9.2%
15		99.7%		97.1%		57.4%		8.1%
16	98.9%	99.5%	95.2%	92.9%	58.4%	61.9%	8.0%	12.0%
17		77.2%		97.1%		57.4%		34.3%
18	98.7%	99.7%	89.1%	99.2%	63.3%	58.1%	9.2%	8.9%
19		99.8%		77.1%		61.6%		33.7%

*Denotes abnormal point

The poor results for iron and manganese purity are attributed to poor pH control and the presence of oxygen in these stages.

The overall removal efficiency for each metal was calculated from the mass balance for the entire process. Because metal removal indicates the entire process's ability to remove metals, this value measures the overall treatment capability. Thus, the process is able to remove 95% or more of aluminum, cobalt, copper, iron, manganese, nickel and zinc. The process removed some of the calcium, chromium, and magnesium. Sodium and potassium were not removed from Berkeley Pit water. The percentage removals for magnesium varied from 9 – 49% and that for calcium were in the range of 16 – 53%. Chromium concentrations in the feed acid mine drainage were about 0.11 ppm and since the overall removal of chromium was based on the exit concentration being below the detection limit of the analytical instrument, chromium removals were calculated using the instrument detection limit. The high degree of metal removal from the wastewater is confirmed by the analytical measurement results of the metal concentration in treated wastewater effluent, as given in Table 4.

TABLE 4. Analysis of effluent water from the last precipitator-settler unit.
(as measured by ICAP)

Compound	Concentration (mg/L)
Al^{+3}	0.50
Ca^{+2}	291.9
Cr^{+3}	BDL
Cu^{+2}	BDL
Mn^{+2}	0.40
Fe^{+2}	0.10
Zn^{+2}	BDL
Cd^{+2}	BDL
Ni^{+2}	BDL
As^{+3}	Not measured
Co^{+2}	BDL
$SO4^{2-}$	2400
NO_3^{-1}	0.32
Cl^{-1}	16.0
Na^{+1}	213
K^{+1}	3532

ECONOMICS

An engineering cost analysis of the process was performed to provide an idea of the potential economic impact/benefit of the process if developed and implemented. The estimated cost savings and economic value for converting the Berkeley Pit clean-up to a profitable commercial resource recovery operation is shown in Table 5.

TABLE 5. Potential economic impact/benefit.

Process	Capital Cost (MM$)	Cash Flow per year (MM$/year)	Cost Savings per year (MM$/year)	Payout time (years)
1. ROD 2 stage lime precipitation	9.4	– 4.5	0	infinity
2. SSP Copper and zinc only	14.4	– 2.1	2.4	2.1
3. SSP Copper, zinc, aluminum, iron and manganese recovery	19.5	10.5 to 24.8	15.0 to 26.9	to 0.6

1. The ROD process consists of a 2 stage lime precipitation plant with secured subtitle D disposal of the wastes.
2. SSP with recovery of copper and zinc only , with secured subtitle D disposal of the other metals wastes.
3. SSP Copper, zinc, aluminum, iron, and manganese recovery

CONCLUSIONS

The four stage precipitation process was able to separate the metal sulfides and hydroxides at reasonable high recoveries and precipitate purities. Several problems were overcome in terms of metal separation and ability to precipitate the fine particles using the inclined plate precipitators. Purities for iron sulfide were not consistently high, mainly due to lack of proper mixing of the hydroxide solution, which resulted in localized regions of high pH and consequently co-precipitation of manganese sulfide. The presence of oxygen in these stages also adversely affected the ability to achieve pure sulfide precipitates. However, the precipitation process was able to demonstrate high overall metal removals and produced final effluent water which could be considered as a valuable product.

REFERENCES

Allen, J., R. Govind, R. Scharp, H. Tabak, and F. Bishop. 1999. Paper presented at the Annual AIChE Meeting, November 1999.

Bhattacharyya, D., A. B. Jumawan, G. Sun, C. Sund-Hagelberg, and K. Schwitzgebel. 1981. "Precipitation of Heavy Metals with Sodium Sulfide: Bench-Scale and Full-Scale Experimental Results." *AIChE Symposium Series*, *77* (209): 31-38.

Bhattacharyya, D., A. B. Jumawan, Jr., and R. B. Grieves. 1979. "Separation of Toxic Heavy Metals by Sulfide Precipitation." *Separation Science and Technology*. *14* (5): 441-452.

Govind, R., U. Kumar, R. Puligadda, J. Antia, and H. Tabak. 1997. "Biorecovery of Metals from Acid Mine Drainage." *Emerging Technologies in Hazardous Waste Management*, *7*: 91-101.

Maree, J. P. and W. F. Strydom. 1987. "Biological Sulphate Removal from Industrial Effluent in an Upflow Packed Bed Reactor." *Water Research*, *21* (2): 141-146.

Rao, P., R. Govind, and H. Tabak. 2001. "Treatment of Acid Mine Drainage using Membrane Bioreactors." Paper to be presented at The Sixth International Symposium on *In Situ and On Site Bioremediation*, San Diego, CA, June 4-7, 2001.

EFFECT OF METAL IONS ON ACETATE-UTILIZING MIXED CULTURE OF SULFATE-REDUCING BACTERIA (SRB)

Vivek P. Utgikar (National Research Council, U.S. EPA, Cincinnati, Ohio)
Henry H. Tabak, John R. Haines and Stephen M. Harmon, (U.S. Environmental Protection Agency, Cincinnati, Ohio)
Rakesh Govind (University of Cincinnati, Cincinnati, Ohio)

ABSTRACT: The inhibitory and toxic impact of metal ions needs to be quantified for maintaining a stable process for the treatment of metal-bearing sulfate-rich wastewaters using sulfate-reducing bacteria (SRB). The ultimate toxic concentrations of zinc and copper (lowest concentration at which no sulfate reduction activity was detected) were found to be 20 mg/L and 12 mg/L respectively for an acetate utilizing mixed culture of SRB. The EC_{50} (initial metal ion concentration at which the activity of the culture is reduced by 50%) values were obtained from the ionic concentration measurements and found to be 16.5 mg/L and 10.5 mg/L for zinc and copper respectively. The metal sulfides formed due to reaction between metal ions in the solution and biogenic sulfide inhibited the sulfate reducing activity of the SRB. The effect of metal ions on an active, growing culture of SRB was investigated by spiking copper/zinc ion into the batch reactors containing acetate-utilizing SRB. The bioactivity was monitored through physico-chemical measurements of pH, oxidation-reduction potential (ORP), ionic concentrations in the reactor and hydrogen sulfide concentration in the headspace. The actual concentrations of the metal ions after spiking were much lower than calculated concentrations and ranged from 5-10 mg/L indicating a significant level of sulfide protection in the reactors. The sulfate reduction activity of the reactors gradually resumed as could be seen from the increase in the pH and a decrease in the ORP. The reactors did not change over to methanogenic mode based on the analysis of the headspace gas sample. The microbial population measurements using an MPN technique indicated that the active SRB population decreased after exposure to metal ion.

INTRODUCTION

Acid mine drainage (AMD) streams at low pH (ca. 2) containing high concentrations of sulfate and metal ions can contaminate the freshwater sources (Wildeman et al., 1991). Biological remediation of AMD using sulfate-reducing bacteria (SRB) involves reduction of sulfate ion to sulfide and concomitant increase in the pH. Heavy metal ions can also be removed from the AMD in this process by precipitating them as insoluble metal sulfides by reaction with the biogenic sulfide ion. (Barton and Tomei, 1995; Utgikar et al., 2000). The presence of heavy metal ions can be detrimental to the process as many heavy metals are toxic to microorganisms. Heavy metal ions can deactivate enzymes by reacting with their functional groups, denature proteins and compete with essential cations (Mazidji et al., 1992; Mosey and Hughes, 1975). The net effect of metal ions on a mixed culture may be a reduction in total number or species diversity (Babich and Stotzky, 1985; Gadd and Griffiths,

1978; White et al., 1997). Exposure to heavy metals can result in an operational shift from sulfidogenic to methanogenic conditions in a mixed culture bioreactor (Capone et al., 1983). Quantification of the toxic and/or inhibitory impact of the heavy metal ions on SRB is critical for maintaining a stable operating process based on SRB metabolism.

Toxic concentrations of heavy metals (to sulfate reducers) have been reported to range from a few ppm (mg/L) to as much as 100 ppm (Booth and Mercer, 1963; Hao et al., 1994; Loka Bharathi et al., 1990; Poulson et al., 1997; Saleh et al., 1964; Temple and Le Roux, 1964; Utgikar et al., 2001). Miller (1950) has reported a stimulatory response of a *Desulfovibrio* strain to trace quantities of heavy metals. Biogenic sulfide can insulate the mixed SRB cultures from the toxic effects of the metal ions by precipitating them (Temple and Le Roux, 1964). However, Undissociated hydrogen sulfide is also a toxic/inhibitory agent for the SRB (Kalyuzhnyi et al., 1998). Metal sulfide/hydroxide/phosphate precipitation, biosorption, complexation and chelation with organic compounds, biotransformations and ionic interactions are the confounding factors in metal ion toxicity determination. An extensive study on the effect of metal ions on acetate-utilizing mixed cultures of SRB was conducted at the National Risk Management Research Laboratory of the U.S. Environmental Protection Agency in Cincinnati, Ohio. The results of the study are presented in this paper.

Objective. The objective of this study was to determine the toxic and inhibitory effect of copper(II) and zinc(II) on the metabolic activity of a mixed culture of sulfate-reducing bacteria. The ultimate toxic concentrations (the lowest initial metal ion concentration at which no sulfate reduction occurs) were determined for the two metals using Hungate tubes. The initial metal ion concentrations at which the activity of the culture is 50% of that of control [no Cu(II) or Zn(II)] were also determined and designated as EC_{50}. The effect of metal ions on active cultures of SRB was also determined through studies on a batch reactor. The study also examined the effect of metal sulfides on the bioactivity.

MATERIALS AND METHODS

Mixed SRB Culture. Anaerobic digester sludge obtained from a domestic wastewater treatment plant in Cincinnati, Ohio, was seeded in a 20-L glass master culture reactor and sulfidogenic conditions were provided in the by maintaining an excess of sulfate over acetate. This master culture reactor was maintained at 35°C and a pH of 7.5 ± 0.5. A weekly drain-and-fill schedule for the master culture involved replacing 20% of the volume by fresh Postgate's C medium (Atlas, 1993) modified to contain acetate instead of lactate. Aliquots of the master culture washed thoroughly with deionized water to remove traces of dissolved sulfide were used in the studies.

Nutrient Medium Design. The pH of the medium was adjusted to 6.60 ± 0.05 to minimize the precipitation of metal ions as hydroxides, and the phosphate concentration was lowered to 50 mg/L, to preclude precipitation as phosphates. The nutrient medium contained 6 g/L sodium acetate instead of lactate. The rest of the constituents were same as Postgate medium "C" for sulfate reducers (Atlas, 1993). The pH and C/P ratios are still within the range of optimum conditions determined for some sulfate-reducers (Okabe and Characklis, 1992; Reis et al., 1992).

Toxicity Testing using Hungate Tubes. Metal ion (copper or zinc) containing solutions were prepared by adding stock metal sulfate (copper or zinc) solutions to the nutrient medium. The pH of the spiked solutions was adjusted to 6.60 $\pm$ 0.05. The nutrient medium without copper or zinc was designated as the no-metal control. The no-metal control and metal-containing media were dispensed into 15 mL anaerobic screw-cap butyl rubber stoppered Hungate Tubes, autoclaved and seeded with mixed SRB culture upon cooling. The inoculation was carried out in an anaerobic chamber (Model 855-AC, Plas-Labs, Lansing, Michigan). The inoculated tubes were incubated at 35°C for a period of 7-14 days and observed for blackening. Concentrations of dissolved metals (copper, zinc, iron) in the tubes were determined at the end of incubation period. The quantities of metals present in the sulfide precipitate were determined by redissolution of the precipitate in concentrated hydrochloric acid-nitric acid mixture followed by the metal ion analysis of solution. Turbidity of the tubes was measured by inserting the Hungate Tubes in the sample cell of the turbidimeter.

Serum Bottle Studies. Nutrient media spiked with copper or zinc as described above were dispensed into 125 mL pyrex glass serum bottles. The bottles were capped with a butyl rubber stopper-aluminum crimp seal and autoclaved for 15 minutes at 121°C. The bottles were seeded after cooling with the mixed SRB culture in the anaerobic chamber, and incubated at 35°C. The bottles were sampled for the analysis of dissolved ions as a function of time. Samples of biomass were also examined under the scanning electron microscope.

Batch Reactor Studies. One liter jacketed glass batch reactors were filled with the nutrient medium and inoculated with the mixed SRB culture. Nitrogen gas was purged through the reactor to strip off the hydrogen sulfide formed due to sulfate-reduction. Purge gas was bubbled through a zinc acetate solution to trap the hydrogen sulfide. Formation of the white zinc sulfide precipitate in the acetate trap served as an indicator of the SRB activity. Samples were drawn from the sample point for the analysis of sulfide ion concentration. The pH and the ORP of the reactor were also monitored using the respective probes. The purge gas was analyzed for the presence of hydrogen sulfide and methane. Copper (or zinc) ion stock solution was spiked into the reactor four days after the start-up and samples were withdrawn immediately following the spiking and at regular intervals afterwards for the determination of actual ionic concentrations. The bacterial population was estimated by a most probable number technique.

Analytical Methods. The metal ion concentrations were determined by an inductively coupled argon plasma emission spectroscopy [Method 3120, APHA (1998)] on a Perkin-Elmer Optima 3300 DV instrument. The pH and ORP of the solutions was measured by a Digi-sense Digital pH/mV/ORP meter or an Accumet 25 pH/mV/ISE meter. Sulfide ion concentration was measured by a Sulfide Ion Selective

Electrode according to the Standard Method 4500G (APHA, 1998). The purge gas was analyzed on a Fisher Model 1200 Gas Partitioner. The nephelometric turbidity measurements of the samples were conducted on a Hach 2100N Turbidimeter

A Most Probable Number (MPN) Technique for Enumeration of SRB. Semisolid MPN medium was prepared by adding 1.5 g/L of Difco anaerobic agar to the Postgate B medium modified by replacing sodium lactate by 3 g/L of sodium acetate. The pH adjusted to 7.0 $\pm$ 0.2 to maintain the similarity between the constituents of the media used in toxicity studies and MPN tests. 9 mL of the medium was dispensed into anaerobic aluminum crimp seal, butyl-rubber stoppered Hungate tubes each (#2048, Bellco Glass, Vineland, NJ) and autoclaved at 121°C for 15 minutes. Serial ten-fold dilutions of SRB samples were prepared inside the anaerobic chamber using deaerated, deionized water.1 mL of the appropriate dilution was inoculated into the MPN medium tube. 6-8 serial dilutions were incubated at 35°C in a constant temperature room, with 5 tubes at each dilution and observed for blackening (ferrous sulfide precipitate) for the detection of SRB activity. MPN were calculated using the most probable number calculator program developed by Klee (1996).

Scanning Electron Micrographs. The biomass samples were fixed by immersion in a 3% glutaraldehyde solution for 2 hours (Millonig, 1976), dehydrated using ethanol displacement of water (Hayat, 1989), and dried in a critical point dryer (Autosamdri 814B, Tousimis, Rockville, MD). The dried samples were coated with carbon and examined under a JSM 5800LV scanning electron microscope (JEOL, Tokyo, Japan).

RESULTS AND DISCUSSION

Ultimate Toxic Concentrations. SRB systems are unique with respect to the toxicity of the metal ions due to the fact that in an active culture the metal ion concentrations will decrease with time due to precipitation by reaction with the biogenic sulfide. So the initial metal ion concentration is the highest concentration to which the SRB culture is exposed (Poulson et al., 1997). The confounding effects of metal hydroxide and phosphate precipitation were avoided through the proper design of the nutrient medium. The extent of biosorption was found to be negligible and has been reported elsewhere (Utgikar et al., 2001). The nutrient medium used in the study contained ferrous ions that are precipitated as ferrous sulfide in an active culture. The resultant blackening of the Hungate tubes is the indicator of the sulfate reduction activity. The lowest initial metal ion concentration that prevented the initiation of the sulfate reduction activity of the culture was defined as the ultimate toxic concentration for the metal. The lack of sulfate reduction activity was inferred from the absence of blackening of the tubes. The ultimate toxic concentrations of Cu(II) and Zn(II) were found to be 12 mg/L and 20 mg/L respectively in these studies.

EC_{50}. EC_{50} was defined as the initial metal ion concentration at which the activity of the mixed culture was 50% of that of the no-metal control. Different metal ions are expected to impact the SRB to a different extent, and EC_{50} can serve as an indicator of the relative impact of the various metal ions. It can also be used to compare the toxicity of a toxicant (metal ions in this case) to SRB to the toxicity of the same toxicant to other microorganisms. For example, the EC_{50} of a toxicant in the Microtox® acute toxicity test (Azur Environmental, Carlsbad, CA) is defined on the basis of 50% extinction in the light output from the luminescent bacteria *Vibrio fischeri.* Various physico-chemical properties measured in this study (pH, MPN, turbidity and ionic concentrations) were evaluated for use as the indicator of the SRB activity. The metal ions present in the system are progressively depleted due to the formation of insoluble sulfides as the biological sulfate reduction proceeds. The difference between the initial and final total metal ion concentrations was the amount of metal precipitated that served as a measure of the SRB activity. Table 1 shows the metal ion removal for SRB cultures exposed to copper, zinc and the no-metal control. The amount shown for the no-metal control is the amount of iron (Fe(II)) precipitated out of the solution.

TABLE 1. Total Metal Ion Removal by the SRB activity

Metal Ion	Initial Metal (Cu/Zn) Ion Concentration, mg/L	Metal Removed, mmol x 10^3
No-metal Control	0	9.1 ± 1.0
Zn	2.7	10.2 ± 0.3
	5.2	9.7 ± 0.1
	9.8	9.1 ± 0.2
	14.5	6.5 ± 0.9
	19.4	1.8 ± 0.8
Cu	2.1	8.9 ± 0.1
	4.1	9.4 ± 1.6
	8.2	9.7 ± 1.4
	12.0	1.0 ± 0.13
	16.0	1.3 ± 0.4

EC_{50} values based on the removal of metal were 16.5 mg/L and 10.5 mg/L for Zn(II) and Cu(II), respectively. These values indicated that copper was more toxic to the SRB than zinc. These measurements were conducted at the end of 7-day incubation period for the Hungate tubes and are qualified as the 7-day EC_{50} values. These values may be different if some other time period is chosen for the study. It should be noted that EC_{50} values are time-dependent in the Microtox® acute toxicity test as well. Measurements of other physico-chemical properties – pH, turbidity, metals recovered in the sulfide precipitate and bacterial population – provided supplementary data in support of the EC_{50} values stated above. Ionic concentration measurements provide a rapid method for the estimation of the effect of metal ions on sulfate reduction activity and a basis for comparing the toxicity of different metal ions.

Serum Bottle Studies - Metal Sulfide Inhibition. As mentioned above, zinc and copper present in the system will be depleted progressively in an active culture. This expected concentration decrease was indeed observed in the serum bottles seeded with copper at the initial concentrations of ca. 6 and 10 mg/L, and the zinc at the initial concentrations of ca. 6, 12 and 19 mg/L. This indicated that the exposure to the metal ions had not proved to be ultimately toxic to the mixed SRB cultures. The concentration of the metal ions was expected to reach zero after a certain period in these serum bottles. This was observed for the serum bottles spiked with ca. 6 mg/L copper or zinc. However, complete elimination of the metal ions was not observed in the other serum bottles. The concentrations tended to level off after ca. 50 hours for copper and 70 hours for zinc. The residual metal ion concentrations after 180 hours ranged from 30% - 50% of the initial concentrations stated above. It was concluded that the activity of the SRB culture was inhibited because the metal ion concentrations would decrease in an uninhibited system. The concentration of the metal ions present in the system was lower than the initial concentration, and hence was not the reason for the inhibition. This inhibition of the SRB activity was attributed to formation of metal sulfides. The mechanism of inhibition was considered to be physical in nature. In essence, insoluble metal sulfides acted as a barrier that prevented the access of the electron donor-acceptor pair (organic molecule-sulfate) to the active bacterial site or enzyme that catalyzed the sulfate reduction. Figure 1 shows the scanning electron micrograph of the culture taken from the serum bottle spiked with copper along with the elemental distribution obtained by energy dispersive x-ray analysis. It can be seen that copper and sulfur profiles are identical in shape to the bacterial cell, lending supporting evidence for the proposed hypothesis for the observed inhibition. Similar effect was also observed for the samples withdrawn from other serum bottles containing copper or zinc but not with the no-metal control serum bottles containing iron.

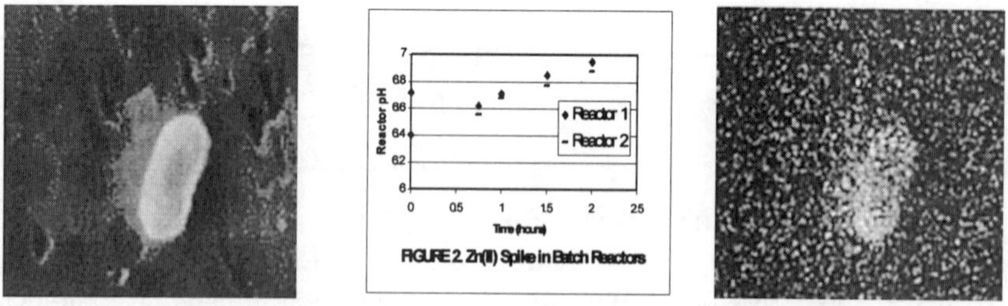

(a) (b) (c)

FIGURE 1. Scanning Electron Micrograph and Energy-Dispersive X-Ray Analysis of the SRB Culture Exposed to Copper(II): (a) Scanning Electron Micrograph (x20000) (b) Copper Distribution (c) Sulfur Distribution

Batch Reactor Studies. Figure 2 shows the effect of zinc spike in the batch reactors. Spiking the reactors with zinc sulfate stock solution resulted in the instantaneous lowering of the pH from 6.7 to 6.4. The ORP of the reactors (not shown in the figure) increased from ca. –400 mV to –150 mV. However, the pHs of both the reactors increased subsequently as shown in the Figure. This increase in pH was an indication of the biological sulfate reduction, as sulfate reduction is accompanied by an increase in pH. A concomitant decrease in the ORP was also observed, ultimately reaching below –300 mV for both the reactors. The zinc stock solution was spiked in the amount calculated to result in 100 mg/L of Zn(II) in the bioreactors. However, the actual concentrations in the two reactors after spiking were only ca. 30 mg/L. The dissolved sulfide ion concentrations in the reactors also decreased from an average of 50 mg/L to 10 mg/L, indicating that the biogenic sulfide was reducing the concentration of the spiked zinc ions. The pHs increased to 7.90 and zinc concentration reduced to < 5 mg/L in both the reactors within 48 hours after spiking.
The bacterial populations (estimated by the MPN method) did decrease by >90% in both the reactors upon exposure to zinc ion.

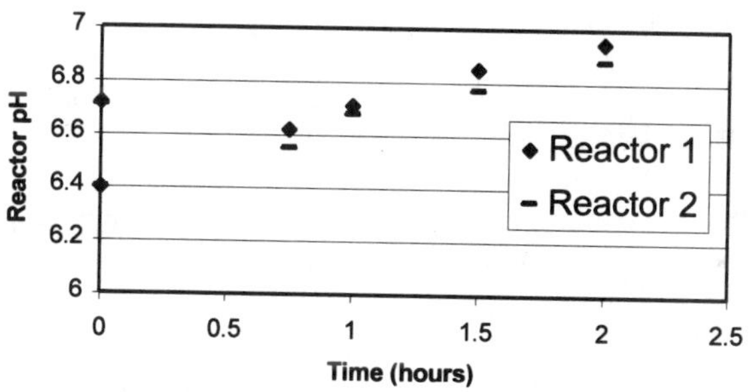

FIGURE 2. Zn(II) Spike in Batch Reactors

Data obtained from the experiments involving spiking of copper in the batch reactors showed similar trends. The stock solution was spiked to result in a concentration of 25 mg/L of Cu(II) in the reactors. However, the actual concentrations were <10 mg/L in both the reactors. A similar increase in pH, and decreases in ORP, bacterial populations and Cu(II) concentrations were observed in the reactors. The analysis of the purge gas (for either copper or zinc spike) did not indicate any presence of methane, indicating that the reactor operation had not shifted from sulfidogenic to methanogenic mode.

CONCLUSIONS

The effect of copper and zinc ions and their sulfides on an acetate-utilizing mixed culture of SRB was investigated in this study. The ultimate toxic concentrations obtained from the Hungate Tube study were 20 mg/L and 12 mg/L for zinc and copper respectively. The corresponding EC_{50} values were 16.5 mg/L and

10.5 mg/L. It was found that the metal sulfide precipitates inhibited the sulfate reduction activity of the culture, possibly by creating a physical barrier around the bacterial cells. The energy dispersive x-ray analysis of the samples revealed elemental distribution of copper/zinc and sulfur identical in shape to the bacterial cells. The batch reactor studies indicated that a significant level of sulfide protection existed in the reactors that could reduce the concentration of the spiked metal and insulate the culture from the toxic effects of the metal ions.

ACKNOWLEDGEMENTS

The study was conducted while the first author (Vivek P. Utgikar) held a Research Associateship of the National Research Council. The technical support by Mr. Navendu Chaudhary and the analytical support by Dr. Neal Sellers and Mr. Michael Goss are gratefully appreciated.

REFERENCES

American Public Health Association (APHA), American Water Works Association (AWWA), Water Environment Federation (WEF). *Standard Methods for the Examination of Water and Wastewater*, 20th Ed.; American Public Health Association: Washington, DC, 1998.

Atlas, R.M. 1993. *Handbook of Microbiological Media*. Parks, L. C. (Ed.), CRC Press: Boca Raton, FL.

Babich, H., and G. Stotzky. 1985. "Heavy metal toxicity to microbe-mediated processes: a review and potential application to regulatory policies". *Environ. Res. 36*:111-137.

Barton, L.L., and F.A. Tomei. 1995. "Characteristics and activities of sulphate-reducing bacteria". In L.L. Barton (Ed.), *Sulphate Reducing Bacteria*, pp. 1-22. Plenum Press: New York, NY.

Booth, G.H., and S.J. Mercer. 1963. "Resistance of copper to some oxidizing and reducing bacteria". *Nature. 199*:622.

Capone, D.G., D.D. Reese, and R.P. Kiene. 1983. "Effects of metals on methanogenesis, sulfate reduction, carbon dioxide evolution and microbial biomass in anoxic salt water sediments". *Appl. Environ. Microbiol. 45*:1586-1591.

Gadd, G. M., and A. J. Griffiths. 1978. "Microorganisms and heavy metal toxicity". *Microbial Ecol. 4*:303-317.

Hao, O.J., L. Huang, J.M. Chen, and R.L. Buglass. 1994. "Effects of metal additions on sulfate reduction activity in wastewaters". *Toxicol. Environ. Chem. 46*:197-212.

Hayat, M. A. 1989. *Principles and Techniques of Electron Microscopy. Biological Applications*. 3rd Ed., CRC Press: Boca Raton, FL.

Kalyuzhnyi, S., V. Fedorovich, P. Lens, L. Hulshoff Pol, and G. Lettinga. 1998. "Mathematical modelling as a tool to study population dynamics between sulfate reducing and methanogenic bacteria". *Biodegradation. 9*:187-199.

Klee, A.J. 1996. *Most Probable Number Calculator, version 4.04*, National Risk Management Research Laboratory, U.S. Environmental Protection Agency, Cincinnati, OH.

Loka Bharathi, P.A., V. Sathe, and D. Chandramohan. 1990. "Effect of lead, mercury, cadmium on a sulphate-reducing bacterium". *Environ. Pollut. 67*:361-374.

Mazidji, C. N., B. Koopman, G. Bitton, and D. Neita. 1992. "Distinction between Heavy Metal and Organic Toxicity using EDTA Chelation and Microbial Assays". *Environ. Toxicol. Water Quality: An Int. J. 7*:339-353.

Miller, L.P. 1950. "Formation of Metal Sulfides through the Activities of Sulfate-Reducing Bacteria". *Contr. Boyce Thompson Inst. 16*:85-89.

Millonig, G. 1976. In M. Saviolo (Ed.) *Laboratory Manual of Biological Electron Microscopy*. Vercelli, Italy.

Mosey, F.E., and D.A. Hughes. 1975. "The Toxicity of Heavy Metal Ions to Anaerobic Digestions". *Water Pollut. Cont. 74*:18-39.

Okabe, S., and W.G. Characklis. 1992. "Effect of Temperature and Phosphorus Concentration on Microbial Sulfate Reduction by *Desulfovibrio desulfuricans*". *Biotechnol. Bioeng. 39*:1031-1042.

Poulson, S.R., P.J.S. Colberg, and J.I. Drever. 1997. "Toxicity of heavy metals (Ni, Zn) *to Desulfovibrio desulfuricans*". *Geomicrobiol. J. 14*:41-49.

Reis, M.A.M., J.S. Almeida, P. Lens, and M.J.T. Carrondo. 1992. "Effect of Hydrogen Sulfide on Growth of Sulfate-Reducing Bacteria". *Biotechnol. Bioeng. 40*:493-600.

Saleh, A.M., R. Macpherson, and J.D.A. Miller. 1964. "Effect of Inhibitors on Sulphate-Reducing Bacteria: A Compilation". *J. Appl. Bact. 27*(2):281-293.

Temple, K.L., and N.W. Le Roux. 1964. "Syngenesis of sulfide ores: sulfate-reducing bacteria and copper toxicity". *Econ. Geol. 59*:271-278.

Utgikar, V.P., B.–Y. Chen, H.H. Tabak, D.F. Bishop, and R. Govind. 2000. "Treatment of acid mine drainage: I. Equilibrium biosorption of zinc and copper on non-viable activated sludge". *Int. Biodeterior. Biodegrad. 46*:19-28.

Utgikar, V.P., B.–Y. Chen, N. Chaudhary, H.H. Tabak, J.R. Haines, and R. Govind, 2001. "Acute toxicity of acid mine water heavy metals to acetate-utilizing sulfate-reducing bacteria". Paper accepted for publication by *Environ. Toxicol. Chem.*

White, G., J.A. Sayer, and G.M. Gadd. 1997. "Microbial solubilization and immobilization of toxic metals: key biogeochemical processes for treatment of contaminaion". *FEMS Microbiol. Rev. 20*:503-516.

Wildeman, T., G.A. Brodie, and J.F. Gusek. 1991. *Draft Handbook for Constructed Wetlands receiving Acid Mine Drainage.* U.S. Environmental Protection Agency: Cincinnati, OH.

STATISTICAL MODELING OF SULFATE REMOVAL IN MINE DRAINAGE TREATMENT

Ioana A. Cocos, Gérald J. Zagury, Bernard Clément and Réjean Samson
(École Polytechnique de Montréal, Montréal, Québec, Canada)

ABSTRACT: Sulfate-reducing reactive walls installed in-situ in the path of acid mine drainage (AMD) contaminated groundwater, present a promising passive treatment technology. However, a rigorous and methodical selection of the most appropriate reactive mixture composition still needs to be investigated. The aim of this study was the modeling of the sulfate-reduction rate in order to assess the variables that significantly affect it. Reactivity of 17 mixtures was assessed in batch reactors (in duplicates) using a synthetic AMD. Results indicate that within 41 days, sulfate concentrations decreased from initial concentrations of 2000-3200 mg/L to final concentrations of < 90 mg/L, while metal removal efficiencies ranged between 51 - 84 % for Ni and 73 - 93 % for Zn. The generated sulfate-reduction rate predictive model, which had very satisfactory parameters ($R^2 = 0.90$, $F = 62.3$ (p-level $<10^{-13}$)) identified reactive mixture carbon sources as the critical variables for sulfate-reduction rate.

INTRODUCTION

Mine tailings are generated in large quantities and exposed to open air. Acid mine drainage (AMD) generation is associated with biological and physico-chemical oxidation of sulfide ores in contact with infiltrating water in tailings impoundment. These processes release sulfuric acid and heavy metals into the drainage waters, which can lead to acidification of recipient and to deterioration of aquatic ecosystems. Since AMD pollution is widespread and often intermittent, low maintenance and economic treatment processes are preferred.

Biological approach to AMD treatment, which is a possible alternative to chemical treatment, intend to induce sulfate-reducing conditions particularly by adding an organic waste product to stimulate the activity of sulfate reducing bacteria (SRB). Bacterial sulfate reduction has been identified as a potentially valuable process for removing contaminants from mine drainage (Dvorak et al., 1992; Blowes et al., 1995). Due to its ability to convert sulfate to H_2S, an anaerobic group of bacteria is used as an inexpensive sulfide source. SRB obtain growth energy by oxidation of organic substrates (CH_2O), removal of hydrogen atoms from the organic molecule, and use of sulfate as the terminal electron acceptor (Hao et al., 1996):

$$SO_4^{2-} + 2CH_2O \xrightarrow{SRB} H_2S + 2HCO_3^- \qquad (1)$$

Furthermore, the dissolved sulfide binds with most metals released from AMD to form insoluble metal sulfide according to the reaction:

$$H_2S + Me^{2+} \rightarrow MeS + 2H^+ \quad (2)$$

The remediation and prevention of acid mine drainage contaminated groundwater through the use of a sulfate-reducing reactive wall, is being more and more recognized as an effective alternative to conventional collection and treatment program (Benner et al., 1997; 1999). The reactive wall consists of a reactive cell installed below the ground surface in the path of AMD-contaminated ground waters. The composition of the reactive mixture is crucial for the efficiency of the treatment process (Kuyucak and St-Germain, 1994; Prasad et al., 1999; Waybrant et al., 1998). Waybrant et al. (1998) assessed the reactivity of eight organic-carbon reactive mixtures containing four components: an organic source, a bacterial source, a neutralizing agent, and a porous medium. They found that reactivity of the mixtures varied with those containing several organic sources being most reactive. However, a rigorous and statistically based selection of the most reactive mixture still needs to be investigated.

Therefore, the objective of this study was to generate a predictive model for sulfate reduction in order to assess the critical variable components on sulfate-reduction rate, and to quantify the influence of each component on reactivity. For this purpose, upper and lower constraints were settled for five components (three different organic substrates and two porous supports), while the other mixture components (bacterial source, neutralizing agent and urea) remained constant.

MATERIALS AND METHODS

Experimental Set-up. Laboratory batch experiments using 0.5 L glass reaction flasks were conducted at room temperature (22 ± 1°C), to assess the potential of different reactive mixtures to promote sulfate reduction. A synthetic acid drainage was used to ensure constant quality of water throughout the study. Simulated AMD composition (Table 1) was based on the chemical characterization of a groundwater plume discharging from the study mine site (Abitibi, Quebec).

TABLE 1. Composition of synthetic acid mine drainage.

Element	Desired concentration (mg/L)	Chemical used	Amount of chemical added per liter (mg)
Fe	800	$FeSO_4 \cdot 7H_2O$	3,971
Cu	0.08	$CuSO_4$	0.2
Zn	7.3	$ZnSO_4 \cdot 7H_2O$	32
Ni	0.8	$NiSO_4 \cdot 6H_2O$	3.5
Co	0.89	$CoSO_4 \cdot 7H_2O$	4.2
Mn	47	$MnSO_4 \cdot H_2O$	144.5
SO_4	2,940	Na_2SO_4	1,093
pH	5.5-6.0	-	-

Required quantities of sulfate salts of Fe, Cu, Ni, Zn, Co, Mn, and Na were added to a 1-liter volumetric flask and made up to volume with deionized water

(18.2 MΩ·cm). 6N NaOH was slowly added to buffer the solution until the final pH was around 5.5-6.0.

Reactive mixtures used in batch tests consisted of a bacterial source, a pH neutralizing material, urea, a carbon source, and a porous media, all proportions considered on a dry basis.

1. Constant Components. Urea was added to reach an appropriate C: N: P ratio of 110:7:1 to support high SRB activity (Gerhardt, 1981). The added percentage of urea was constant (3 wt %) for all mixtures.

The SRB source used was a creek sediment (the presence of SRB was suspected by the black color and the H_2S odor of the sediment) sampled from the anaerobic zone of an AMD-affected stream in an inactive mine area (Eastern Townships, Quebec). Enumeration of SRB in the sediment using the Most Probable Number (MPN) technique (ASTM, 1990) showed a SRB concentration of 30×10^3 cells/100 mL. A constant percentage of bacterial source (37 wt %), was added in all mixtures.

Limestone (2 wt %) was also added in order to maintain near neutral conditions (Blowes et al., 1995).

2. Variable Components. The organic components chosen for the experimental design were based on the experience gained from previous case studies (Kuyucak and St-Germain, 1994; Waybrant et al., 1998; Prasad et al., 1999).

Selection of alternative carbon sources was performed on two groups of waste materials: cellulosic wastes (wood chips, leaf compost) and organic wastes (poultry manure). Results of previous research indicated that cellulosic waste alone would not sustain SRB growth (Kuyucak and St-Germain, 1994). Consequently, a mixture of cellulosic wastes and poultry manure was considered. Wood chips, leaf compost and poultry manure were selected for their ability to sustain sulfate reduction. These substrates contain carbon with varying degrees of lability, either complex or simple organic compounds. Thus, reactive mixtures should have both long-term and ephemeral carbon sources, which should ensure shorter and longer periods of time for degradation. Wood chips and leaf compost were obtained from a local provider, while a local poultry farm provided the manure.

Two types of porous support (reactive or inert) were used in order to assess their different influence on the SRB activity, which strongly depends on the presence of a physical support (Lyew and Sheppard, 1999). A readily available porous support (i.e. oxidized tailings) was obtained from the previously mentioned study mine site (Abitibi, Quebec), while clean silica sand was obtained from a local stone quarry.

3. Statistical Design. Constraints were imposed for the chosen carbon sources, and for porous media. The upper and lower boundaries (wt %) based on previous studies (Waybrant et al., 1998; Prasad et al., 1999) were: 0 to 30 % for wood chips (X_1) and leaf compost (X_2), 10 to 20 % for poultry manure (X_3), and 0 to 8 % for both oxidized tailings (X_4) and silica sand (X_5). In addition, the following factor constraints (wt %) were imposed: $30 < X_1 + X_2 < 40$, $5 < X_4 + X_5 < 8$, and $X_1 + X_2 + X_3 + X_4 + X_5 = 58$. The remaining 42 % consisted of 3% urea, 37% creek sediment and 2% limestone.

Constraints on mixture proportions were processed using the Experimental Design operating module of STATISTICA software (StatSoft, Inc., 1995). The 5 factors mixture design generated 16 vertex and one centroid point to be tested (Table 2). Consequently, 17 reactors were run in duplicates in order to assess reactive mixture efficiencies in terms of sulfate reduction and heavy metal precipitation.

TABLE 2. Reactive mixture composition (dry wt %).

	R_1	R_2	R_3	R_4	R_5	R_6	R_7	R_8	R_9	R_{10}	R_{11}	R_{12}	R_{13}	R_{14}	R_{15}	R_{16}	R_{17}
X_1	30	30	0	0	30	10	30	10	30	30	3	3	30	30	10	10	17.87
X_2	0	0	30	30	10	30	10	30	3	3	30	30	10	10	30	30	17.87
X_3	20	20	20	20	10	10	10	10	20	20	20	20	13	13	13	13	15.75
X_4	8	0	8	0	8	8	0	0	5	0	5	0	5	0	5	0	3.25
X_5	0	8	0	8	0	0	8	8	0	5	0	5	0	5	0	5	3.25

After the mixtures components were added, all flasks were filled with 300 mL of deoxygenated synthetic mine drainage (by sparging with N_2), and the flasks were sealed.

Sampling and Analysis. All sampling sessions were conducted in an anaerobic glovebox under nitrogen atmosphere. Mine drainage water samples were removed with a syringe, by piercing the reaction flasks sampling ports fitted with Teflon-lined septa. Reactive mixtures were allowed to settle for approximately 12 h before sampling on day 0. Different volumes of water samples were also collected on days 5, 9, 13, 17, 21, 25, 33, and 41. Measurements of pH (ORION ROSS 8175 BN electrode), and oxido-reduction potential (ORP) (ACCUMET electrode), were performed in the sampling solution according to Standard Methods (APHA, 1998).

Filtered samples (0.45 μm cellulose acetate filter paper) were used for Ni, Zn, and SO_4^{2-} determination. Ni, and Zn concentrations were determined by atomic absorption spectrometry (AAS), and sulfate concentrations were determined by spectrophotometry using standardized substances and a Hach digital spectrophotometer DR/2010 (HACH Procedure Manual, 1998). Total organic carbon (TOC) was determined on a non-filtered sample using a Total Carbon Analyzer (Dorhmann DC-80) by ultra-violet promoted persulfate oxidation (APHA, 1998).

SRB were enumerated in the mixture suspension on day 0 and day 41 using the MPN technique (ASTM, 1990). Inoculated tubes were incubated at 30°C for 21 days. The growth of SRB was estimated by the formation of a black FeS precipitate.

RESULTS AND DISCUSSION

Sulfate-Reduction Rates. Batch experiments were run until depletion of AMD sulfate concentrations. After 41 days, sulfate concentrations decreased from 2020-3254 mg/L to < 90 mg/L depending on mixture composition. Sulfate-reducing conditions developed rapidly in all mixtures after an initial acclimation period, which ranged from 0 to 21 days. Significant populations of SRB were present throughout the experiment suggesting that sulfate removal occurred following bacterial activity. Average initial SRB concentration was $5.36 \pm 3.22 \times 10^3$ cells/100mL of mixture suspension. At the end of the experiment the average concentration was $2.54 \pm 2.22 \times 10^4$ cells/100mL.

Sulfate-removal rate per day ($T_{0\text{-}41}$) was calculated for each reactor using linear least-squares regression analysis for the entire period of study (Table 3). Sulfate-reduction rates varied between 35.6 mg/L-d and 87.6 mg/L-d, with an average value of 70.7 ± 12.1 mg/L-d for the 17 reactors. The coefficients of determination (R^2) ranged from 0.82 to 0.98 depending on mixture composition.

TABLE 3. Sulfate-reduction rates (mg/L-d) for duplicate (A, B) samples.

	#1	#2	#3	#4	#5	#6	#7	#8	#9	#10	#11	#12	#13	#14	#15	#16	#17
A	85.8	86.5	85.4	78.0	53.2	66.5	63.1	35.6	77.7	84.8	81.6	83.6	61.6	71.7	63.4	58.8	70.7
B	77.9	87.6	70.8	79.8	57.8	64.6	58.4	52.0	79.1	84.0	77.0	84.3	64.0	67.8	63.0	58.8	67.6

Mine Drainage Quality. Dramatic decreases were observed for Ni and Zn concentrations. Heavy metal removal efficiencies ranged between 51-84 % for nickel (with an average of 72%), and between 73-93 % for zinc (with an average of 88%) depending on the reactive mixture. Furthermore, results showed that around 75 % of total nickel removal, and around 85 % of total zinc removal was completed within 17 days. At this time, pH was stable near 8.07 ± 0.16 for all mixtures while average ORP was – 454 ± 15 mV. Because of the 2% $CaCO_3$ addition, average initial pH was 9.0 ± 0.2 in all reactors and then started to decrease. Therefore, part of Ni and Zn initial removal could be attributed to their low solubility at this pH value. Sulfate-reduction rates in addition with elevated SRB populations and ORP values, at which the SRB activity is optimum (Prasad et al., 1999), strongly suggest that sulfate reduction and at least part of metal removal was bacterially mediated. Average final TOC (3044 ± 693 mg/L) showed that all reactors had some carbon left.

Modeling of Sulfate-Reduction Rate. Statistical modeling of the sulfate-reduction rate must take into account the nature of the experimental space. Mixture designs were modeled with second-degree polynomials. Because of the linear dependence between the components, the polynomials were forced to go through the origin, that is with a zero intercept. Ratios of components were used in order to break the linear dependence. In this case, the polynomials were not subjected to have a zero intercept.

Sulfate-reduction rate (Y) was modeled using a response function (equation 3) with mixture component ratios ($Z_i = X_i/X_3$); X_3 variable was chosen as a denominator among the five variables because it varies between 10 and 20 (wt %), being the only one that is never zero.

$$Y = \beta_0 + \sum_{i=1}^{5} \beta_i Z_i + \sum_{i=1}^{5} \beta_{ii} Z_i^2 + \sum\sum_{i<j}^{5} \beta_{ij} Z_i Z_j ; \quad i, j = 1, 2, 4, 5 \qquad (3)$$

The fitted model was:

$$Y_1 = 79.08 - 9.34 \frac{X_1 X_2}{X_3^2} - 11.16 \frac{X_1 X_5}{X_3^2} + 23.36 \frac{X_5^2}{X_3^2} + 5.05 \frac{X_2 X_4}{X_3^2} \qquad (4)$$

This equation has a multiple determination coefficient (R^2) of 0.90, all five components (X_1 to X_5) being involved. Using a forward stepwise regression method, the first term selected ($R^2 = 0.76$) was a combination of the three carbon sources: wood chips (X_1), leaf compost (X_2) and poultry manure (X_3). Furthermore, considering this term's negative sign, the smallest its value the higher the sulfate-reduction rate. A small value implies either a high X_3 or a low X_1X_2 value. It should be noted that X_3 (poultry manure) is a second-degree term, its impact on sulfate-reduction rate being therefore much more important.

The addition of the other terms (involving the porous support) in this model increased the multiple determination coefficient R^2 by only 0.14 (from 0.76 to 0.90).

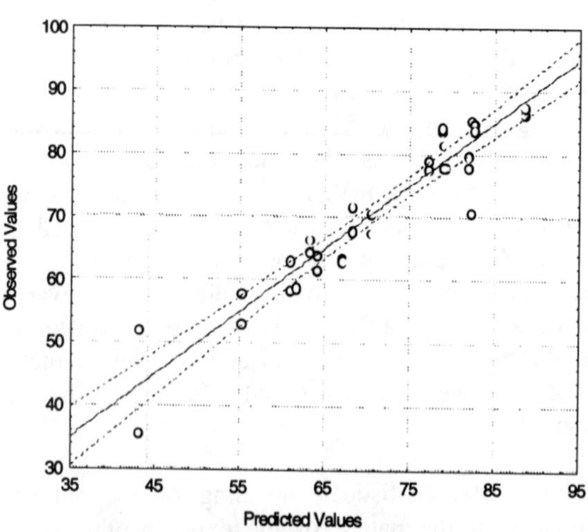

FIGURE 1. Predicted versus observed values.

According to the analysis of variance, the F-criterion, which is the sum of squares due to regression over the sum of squares for error divided by the respective degrees of freedom was highly significant (62.3, p-level $< 10^{-13}$). That means that the two groups (effects and residuals) are considerably different from each other.

Predicted values are in good agreement with observed values (Fig.1). Potential clusters of the model, which are not well predicted were also identified. The two important outliers are the first replicate of mixture R_8 and the second replicate of mixture R_3, for which the predicted $T_{0\text{-}41}$ values compared to observed values were 43.0 mg/L-d (compared to 35.6) and 82.1 mg/L-d (compared to 70.8).

CONCLUSIONS

- Final concentrations of sulfates in the synthetic AMD were lower than 90 mg/L and the average rate of sulfate reduction was 70.7 ± 12.1 mg/L-d. Sulfate-reduction rates varied between 35.6 mg/L-d and 87.6 mg/L-d indicating a strong influence of mixture composition on its reactivity.
- Heavy metal removal efficiencies ranged between 51 and 84 % for nickel and 73 and 93 % for zinc depending on the reactive mixture.
- Modeling of sulfate-reduction rate was carried out using a forward stepwise regression in order to generate a statistical model for sulfate-reduction rate and to identity the variables affecting it. The generated sulfate-reduction rate predictive model ($R^2 = 0.90$, $F = 62.3$ (p-level $< 10^{-13}$)), identified poultry manure and the two other carbon sources as the critical variables. More specifically, in the reactive mixture poultry manure proportion (X_3) should be higher, while a smaller proportion should characterize the product of the two more complex carbon sources. The results also showed a minimal influence of the tested porous supports (either oxidized tailings or silica sand) on sulfate-reduction rate.
- The sulfate-reduction rate predictive model generated in this study can be used to determine the initial barrier composition for further use in porous sulfate-reducing reactive walls. However, another important aspect, which should be considered for barrier reactive mixture efficiency is its longevity. Therefore, further research is needed to study the impact of mixture composition on long-term sulfate and metal removal.

ACKNOWLEDGEMENTS

The authors acknowledge the financial support of the Chair partners: Alcan, Bell Canada, Canadian Pacific Railway, Cambior, Centre d'expertise en analyse environnementale du Québec, Gaz de France/Électricité de France, Hydro-Québec, Ministère des Affaires Municipales et de la Métropole, Natural Science and Engineering Research Council (NSERC), Petro-Canada, Solvay (Belgium), Total Fina Elf (France), and Ville de Montréal.

REFERENCES

APHA. 1998. *Standard methods for the examination of water and wastewater*, 20th edition, American Public Health Association, Washington D. C.

ASTM. 1990. *Standard test method for sulfate reducing bacteria in water and water-formed deposit.* D4412-84, American Public Health Association, Washington D. C.

Benner, S.G., D.W. Blowes, and C.J. Ptacek. 1997. “A full-scale porous reactive wall for prevention of acid mine drainage.” *Ground Water Monit. Rem.* Fall 1997: 99-107.

Benner, S.G., D.W. Blowes, W.D. Gould, R.B. Herbert, and C.J. Ptacek. 1999. "Geochemistry of a permeable reactive barrier for metals and acid mine drainage." *Environ. Sci. Technol. 33*:2793-2799.

Blowes, D.W., C.J. Ptacek, and J.G. Bain. 1995. "Treatment of mine drainage water using *in situ* permeable reactive walls.", *Mining and the Environment*, Sudbury, pp. 979-987.

Dvorak, D.H., R.S. Hedin, H.M. Edenborn, and P. McIntire. 1992. "Treatment of metal contaminated water using bacterial sulfate reduction: results from pilot scale reactors." *Biotechnol. Bioeng. 40*:609-616.

Gerhardt, P. 1981. *Manual of methods for general microbiology*. Washington D.C. ASM Publication.

HACH Procedure Manual. 1998. *DR/2010 Spectrophotometer Handbook*. 4ed, HACH Company, USA.

Hao, O. J., J. M. Chen, L. Huang, and R. L. Buglass. 1996. "Sulfate reducing bacteria." *Critical Reviews in Environ. Sci. and Technol. 26*(1):155-187.

Kuyucak, N., and P.St-Germain. 1994. "In situ treatment of acid mine drainage by sulfate reducing bacteria in open pits: scale-up experiences." *International Land Reclamation and Mine Drainage Conference*, Pittsburgh, pp. 303-309.

Lyew, D., and J. Sheppard. 1999. "Sizing considerations for gravel beds treating acid mine drainage by sulfate reduction." *J Environ. Qual.. 28*:1025-1030.

Prasad, D., M. Wai, P. Bérubé, and J.G. Henry. 1999. "Evaluating substrates in the biological treatment of acid mine drainage." *Environ. Technol. 20*:449-458.

StatSoft Inc. 1995. *STATISTICA for Windows [Computer program manual]*. Tulsa, OK: StatSoft, Inc., 2300 East 14th Street, Tulsa, OK, 74104-4442.

Waybrant, K.R., D.W. Blowes, and C.J. Ptacek. 1998. "Selection of reactive mixtures for use in permeable reactive walls for treatment of mine drainage." *Environ. Sci. Technol. 32*:1972-1979.

STATUS AND PERFORMANCE OF ENGINEERED SRB REACTORS FOR ACID MINE DRAINAGE CONTROL

Marek H. Zaluski, John M. Trudnowski, Marietta C. Canty,
Mary Ann Harrington-Baker
(MSE Technology Applications, Inc., Butte, Montana, USA)

ABSTRACT: Acid mine drainage (AMD) emanates from many abandoned mine sites in the Western United States. Such drainage, having an elevated content of dissolved metals and low pH, presents an environmental problem that needs to be economically addressed. Sulfate-reducing bacteria (SRB) have the ability to immobilize dissolved metals, by precipitating them as sulfides, and increase pH provided that a favorable biochemical environment is created. To demonstrate this concept, three passive SRB bioreactors of different content and placement were constructed at an abandoned mine site near Butte, Montana. Performance of the reactors during the last 2 years indicates that the technology is effective in removing dissolved metals including copper, zinc, cadmium, and aluminum. There is also indication that the reactors' performance is sensitive to construction details including the amount of reactive media and its placement method. This demonstration is funded by the U.S. Environmental Protection Agency (EPA) and is jointly administered by the EPA and the U.S. Department of Energy (DOE). The project is implemented by MSE Technology Applications, Inc., Butte, Montana.

INTRODUCTION

Acid mine drainage (AMD) emanates from many abandoned mines in the Western United States, causing significant environmental problems by contaminating surface waters and groundwater with dissolved metals and raising their acidity. Conventional treatment of AMD is often not feasible due to the remoteness of the site, the lack of power, and limited site accessibility. Thus, for such sites, there is a need for a passive remedial technology to immobilize metals and increase the pH of the AMD.

Sulfate-reducing bacteria have the ability to increase pH and alkalinity of the water and to immobilize dissolved metals by precipitating them as sulfides, provided that a favorable biochemical environment is created. Such conditions were engineered within three field bioreactors (reactors) that were built at an abandoned mine site (Calliope Mine) in the vicinity of Butte, Montana.

PRINCIPLES OF THE SRB TECHNOLOGY

Acid mine drainage is a typical result of mining of sulfide-rich ore bodies. Acid mine water is formed when sulfide-bearing minerals, particularly pyrite [iron bisulfide (FeS_2)], are exposed to oxygen and water as described by the following overall reaction (Equation 1).

$$FeS_2 + 15/4\ O_2 + 7/2\ H_2O \dashrightarrow Fe(OH)_3 + 2SO_4^{2-} + 4H^+ \quad (1)$$

This reaction results in increased acidity of the water (lowered pH), increased metal mobility, and the formation of sulfate.

When provided with an organic carbon source, sulfate-reducing bacteria (SRB) are capable of reducing the sulfate to soluble sulfide by using sulfate as a terminal electron acceptor. Acetate, and bicarbonate ions are also produced. The soluble sulfide reacts with the metals in the AMD to form insoluble metal sulfides. The bicarbonate ions increase pH and alkalinity of the water (Equations 2 and 3).

$$SO_4^{2-} + 2CH_2O \text{ -------> } H_2S + 2HCO_3^- \quad (2)$$

$$H_2S + M^{+2} \text{ ---> } MS + 2H^+, \text{ where } M = \text{metal} \quad (3)$$

SITE FEATURES

Acid Mine Drainage Source. The abandoned Calliope mine site, located in Silver Bow County, Montana, includes a collapsed adit discharging water into a large (66,000 cubic yards) waste rock pile. This relatively good quality water flows over the top of the waste rock and accumulates in a small flow-through pond (Pond) at the toe of the pile. The AMD is mostly produced by atmospheric water that infiltrates the waste rock pile and reappears on the surface at the toe of the pile enriched in metals and with a pH of 2.6. This AMD also flows to the Pond where it mixes with good quality water and lowers its pH. A portion of the water that accumulates in the Pond has been diverted for treatment to three engineered bioreactors that were built at the site to demonstrate the SRB technology.

The quality of the Pond water and its pH is related to the amount of atmospheric water that infiltrates into the waste rock pile and leaches metals. With the exception of the first 8 months of the reactors' operation, atmospheric precipitation has been much below normal. Consequently, the quality of the Pond water significantly improved. Table 1 includes analytical information on the water quality of the reactors' influent.

TABLE 1. AMD analytical data and target concentrations for bioreactors.

Analyte	Concentration (μg/L)			Comments
	AMD (maximum)	AMD (minimum)	Target Effluent	
Aluminum	14,100	11.0	1,000	SMCL*=50 to 200 μg/L
Cadmium	41.9	3.7	5	MCL**=5 μg/L
Copper	3,050	7.2	100	MCL=1,300 μg/L
Iron	8,670	15.5	1,000	
Manganese	3,770	882	2,000	
Zinc	11,100	1,240	4,000	
Sulfate	229,000	69,800		
pH	7.52	3.29	6 to 8	

*SMCL = Suggested maximum contaminant level
**MCL = Maximum contaminant level

Bioreactors Layout. All three SRB reactors (denoted II, III, and IV) were designed and constructed (Zaluski et al., 1999) in parallel (Figure 1) downstream from the Pond. This allowed the AMD to be piped to and treated by the reactors using gravity flow. The reactors, constructed in the fall of 1998, were designed to evaluate the SRB technology applied in slightly different environmental conditions. Two reactors were placed below grade (Reactors II and III), and one was placed above grade (Reactor IV). The below-ground reactors were built to minimize temperature changes and prevent freezing. The above-ground reactor was built to evaluate the effect of the cold weather and freezing on the system. In addition, Reactors II and IV were built with a pretreatment section to evaluate the effect of inducing an optimal pH and oxidation-reduction potential (E_H) on the efficiency of the SRB.

Bioreactor Configuration. Each reactor was filled with a combination of organic carbon, crushed limestone, and cobbles (Figure 1) placed in three or four discrete chambers. The first two chambers of Reactors II and IV constitute the pretreatment section and include a chamber filled with organic matter and a chamber filled with crushed limestone. Following the pretreatment section is a primary treatment section that includes another chamber with organic matter and a chamber filled with cobbles. A pretreatment section was not included in Reactor III in order to evaluate its contribution to overall reactor efficiency.

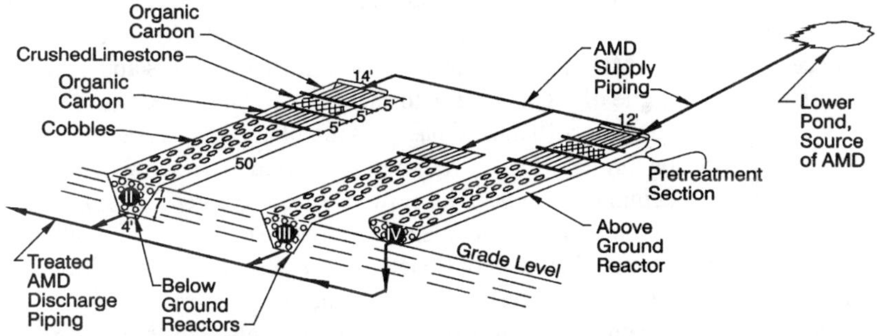

FIGURE 1. Layout of bioreactors.

Each media component is expected to play an important role in the treatment train: (a) organic carbon is the bacterial food supply and SRB source; (b) for the pretreatment section a chamber with organic carbon was included to lower the E_H of AMD; (c) crushed limestone provides buffering capacity to increase the alkalinity of AMD in the pretreatment section; and (d) cobbles placed in the last chamber of the reactors constitute stable substrate for bacterial growth.

Reactors II and III, 71.5 feet [21.8 meters (m)] and 61 feet (3.7 m) in length respectively, were constructed in 14-foot-wide (4.3 m) trapezoidal (4-foot-wide (1.2 m) bottom) trenches below grade; Reactor IV, 72.5 feet (22.1 m) in length, was constructed in a 12-foot-wide (3.7 m) metal half-culvert elevated above grade. The chambers filled with organic carbon or limestone are each 5 feet (1.5 m) in length, whereas the chambers filled with cobbles are 50 feet (15.2 m) in length.

OPERATION OF BIOREACTORS

Flow Rate. The reactors have been operated since December 1998. The two subsurface reactors (II and III) were designed to flow year-round. The above-ground reactor (IV) was designed to be shut down for winter to let it freeze full of AMD. Figure 2 shows the reactors flowed at a rate of 1 gpm (0.228 m^3/h) for the majority of time. For 4 months in the summer of 2000, the flow rate was doubled to nearly 2 gpm (0.456 m^3/h). Although flow for Reactor IV was shut down for the winter, a center portion of the reactor did not freeze.

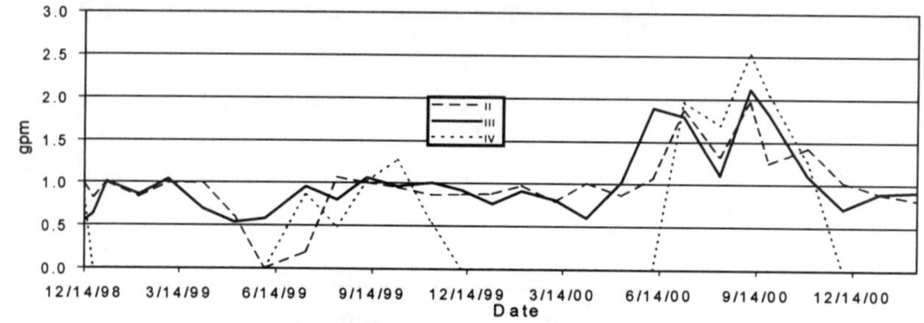

FIGURE 2. Reactor flow.

A flow rate of 1 gpm (0.228 m^3/h) corresponds to a 5-1/2 day residence time for the AMD in Reactors II and IV and a 4-1/2 day residence time in Reactor III. The residence time of the AMD in a single organic carbon chamber was approximately 10 hours for the flow rate of 1 gpm (0.228 m^3/h).

Flow through Reactors III and IV has been maintained as desired for most of the time. The flow rate through Reactor II, however, started to decrease in May 1999 and ceased at the beginning of June. The flow rate was restored in July after the upgradient cell with organic carbon was chemically treated to remove biofouling and associated plugging. Similar behavior of Reactor II was observed again in May 2000. This time, the permeability of the upgradient chamber was increased using an appropriate physical treatment. The repetitive plugging events of the most upgradient chamber in Reactor II seems to be attributed to a tighter packing of organic carbon in this chamber in comparison to other chambers with organic carbon.

Performance Monitoring. Performance of each reactor has been monitored by monthly sampling of the influent and effluent and continuous monitoring of selected parameters using appropriate sensors and data loggers. Aqueous samples have been analyzed for sulfate; alkalinity; SRB count[1]; heterotrophic bacteria count; dissolved oxygen (DO); E_H (with silver/silver chloride reference electrode); and metals including aluminum, zinc, cadmium, copper, iron, manganese, and cadmium. The results are presented in Figures 3 through 9.

[1] The SRB most probable number (MPN) was determined using American Institute Recommended Practice 38, First Ed., May 1959.

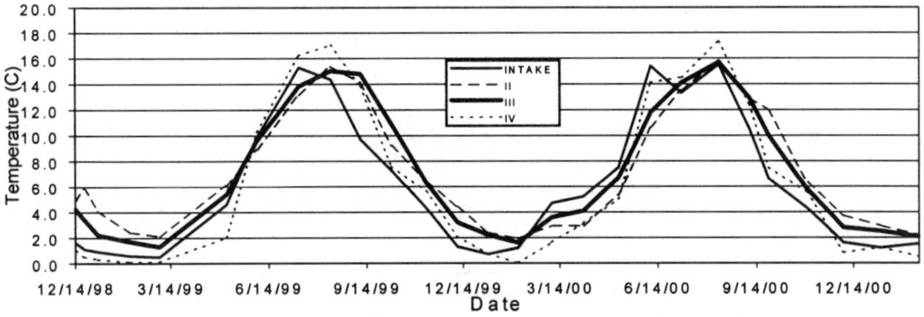

FIGURE 3. Reactors temperature.

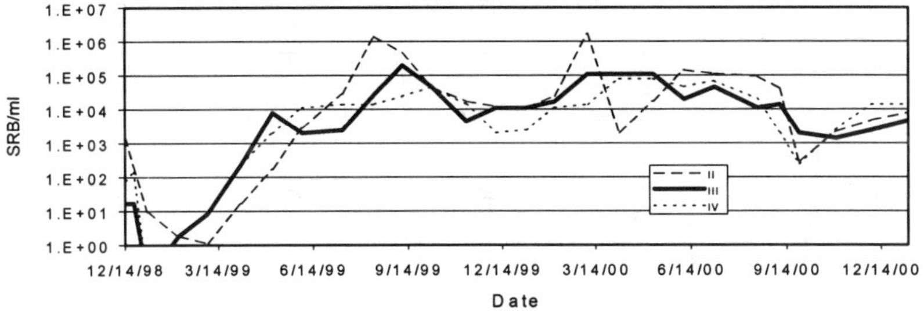

FIGURE 4. SRB population.

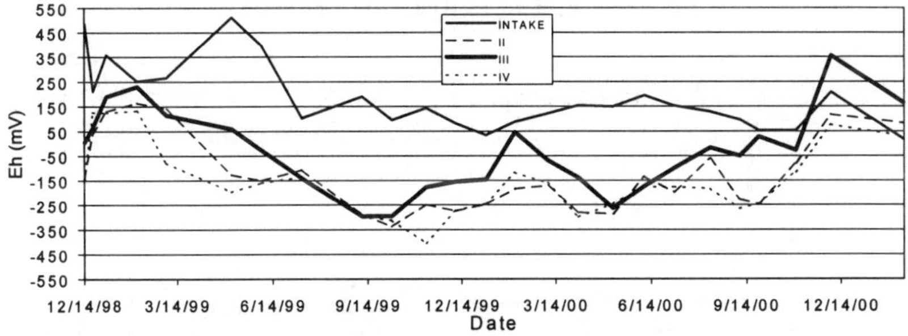

FIGURE 5. E_H trends.

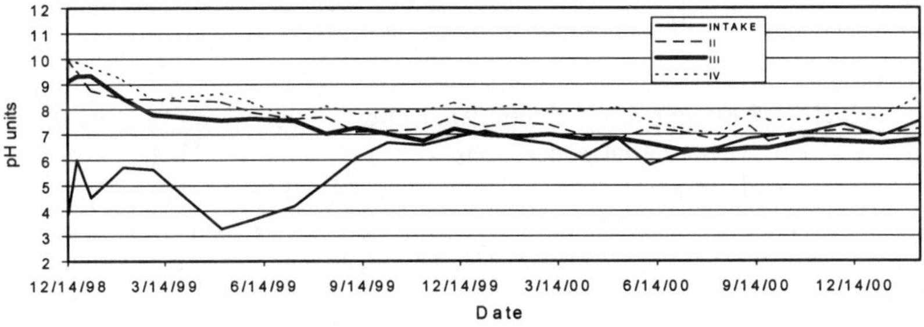

FIGURE 6. pH trends.

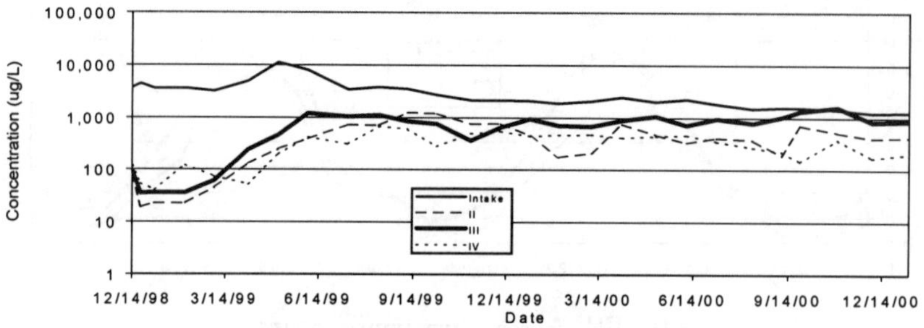

FIGURE 7. Concentration of zinc.

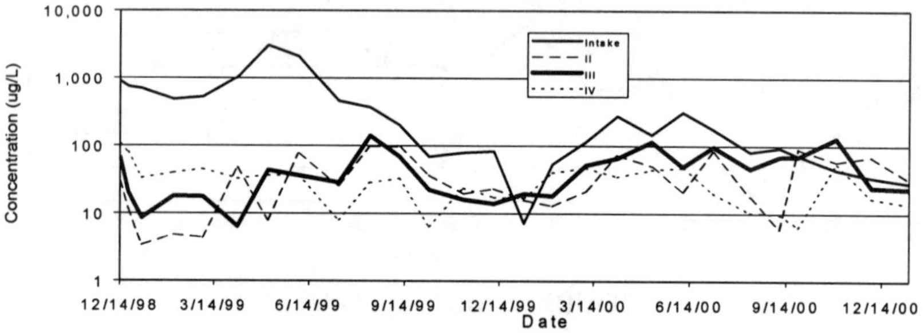

FIGURE 8. Concentration of copper.

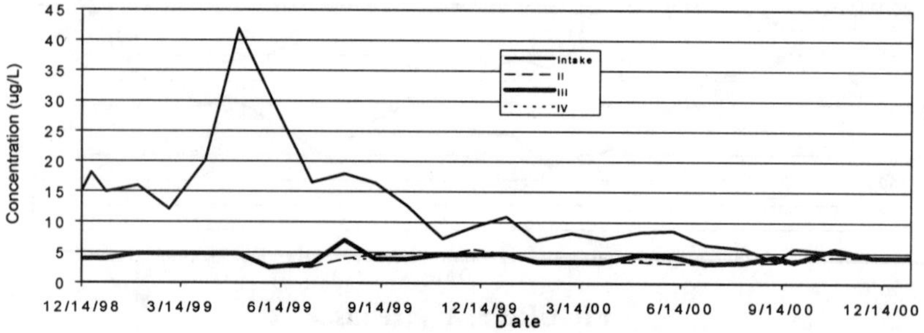

FIGURE 9. Concentration of cadmium.

RESULTS

The first 8 months of operation (Zaluski et al., 2000) can be described as period in which the microbial populations were becoming established within the reactors. It should be noted that the reactors were started in the winter when temperatures were not ideal for microbial growth. As the reactor temperatures (Figure 3) began to increase in April and May 1999, an increase in SRB populations (Figure 4) was also seen. During the second winter of operation, the well-established SRB population was not affected by the low temperatures.

The affect of the SRB population on E_H can be seen in Figure 5. The E_H decreased to -295 millivolts (mV) for Reactor III in September 1999 and -405 mV for Reactor IV in October 1999. During the same time, the SRB populations grew to a level of $2x10^5$ MPN for Reactor III. Sulfate-reducing bacteria population did not change significantly until September 2000, while the E_H level remained at -250 mV on average until October 2000. The subsequent correlation between the SRB population and E_H is not well defined, perhaps because of the impact of the flow rate that was nearly doubled during the summer of 2000.

The initial increase in effluent pH (Figure 6) can largely be attributed to alkalinity present within the organic substrate rather than to the presence of limestone. This is indicated by an insignificant pH difference in the effluent from Reactors II and III, the latter having no limestone pretreatment chamber. As the SRB became established and the effluent pH from each reactor dropped to the range of 7 to 8 for the period of March to October 1999, the pH differential between the influent and effluent can be attributed to SRB activity.

Beginning in October 1999, the quality of the influent AMD improved and until August 2000, its pH was only 1 unit lower than that of the effluent from the reactors. The slightly lower pH of the effluent from Reactor III during this period indicates the limestone chamber, which is present in Reactors II and IV, may have contributed to the increase of pH in these reactors. This effect on pH by the limestone chamber would have been in addition to the effects of the SRB.

Much of the metals removal observed during the first 7 months of operation can be attributed to adsorption. This effect was observed in previous SRB systems in which metals are initially removed by sorption onto the organic substrate (Canty, 1999). Once sorption sites fill and SRB populations become established, many metals, e.g., zinc, cadmium, and copper, are removed through SRB activity. This has also been proved at the Calliope site where despite relatively low concentrations, metals like zinc and copper [Figures 7 and 8 (in logarithmic scale)] and cadmium (Figure 9), were removed from the AMD to threshold levels that were approximately 500 micrograms per liter (μg/L) to 800 μg/L for zinc, 5 μg/L for cadmium, and 80 μg/L for copper.

The SRB activity is well proven by the removal of zinc, whose concentration in the effluent during the first 7 months was rising as the sorptive capacity of organic carbon was being filled. During this period, the percent of zinc removal was as high as 99%. Once the sorptive capacity was filled, zinc concentration stabilized at the threshold of approximately 500 μg/L for Reactors II and IV and 800 μg/L for Reactor III. These values correspond to an approximately 75% and 60% of zinc removal by SRB in the respective reactors. The slightly lower percent of removal in Reactor III is attributed to a smaller total supply of organic carbon, as this reactor has only one chamber with organic matter.

CONCLUSIONS

The reactors are scheduled to operate until June 2001, i.e., they will be operating for a total of 31 months. At that time, the reactors will be decommissioned, and the solid matrix of the reactors will be sampled to draw final conclusions on the appropriateness of the reactors design. At the present time, the following conclusions are deem justified:

- A drop in the SRB population in September 2000, nearly 3 months since the flow rates were doubled, may indicate flushing out of the bacteria at that flow velocity.
- A slightly lower metal removal efficiency of Reactor III that contained only one chamber with organic matter may indicate that for reactors constructed in cold-climate regions a residence time of 10 hours is close to minimal.
- It takes some time for SRB to be established in a new bioreactor. Once established and supplied with organic carbon, they maintain a population of 10^4 MPN or higher (in the aqueous medium) within a temperature range of 2 °C to 16 °C.
- Winter freezing of a well-established SRB population has little or no effect on their activity for the remainder of the year.
- Although it seems like a limestone chamber slightly increased effluent pH, its role was not dominant for the overall performance of the reactors.
- Observation of the reactors to date indicates that cobble sections are not important for the efficiency of SRB reactors.
- For the Calliope site climatic and hydrochemical conditions, the thresholds for zinc, cadmium, and copper removal are approximately 500 µg/L, 5 µg/L, and 80 µg/L, respectively.
- To avoid self-plugging (biofouling) of the reactor during its operation, a detailed specification of organic carbon placement is necessary.

ACKNOWLEDGMENTS

This work is funded by the U.S. Environmental Protection Agency (EPA) and is jointly administered by the EPA and the U.S. Department of Energy (DOE) through an Interagency Agreement and under DOE contract number DE-AC22-96EW96405. The project is implemented by MSE Technology Applications, Inc., Butte, Montana. Portions of this manuscript were presented at the Society for Mining Engineers ICARD 2000 Conference in May 2000.

REFERENCES

Canty, M. June 1999. "Overview of the Sulfate-Reducing Bacteria Demonstration Project under the Mine Waste Technology Program." *Mining Engineering*.

Zaluski, M., M. Foote, K. Manchester, M. Canty, M., Willis, J. Consort, J. Trudnowski, M. Johnson, and M. A. Harrington-Baker. April 19-22, 1999. "Design and Construction of Bioreactors with Sulfate-Reducing Bacteria for Acid Mine Drainage Control." *Proceedings of The Fifth International In Situ and On-Site Bioremediation Symposium*. Battelle Press, San Diego, CA.

Zaluski, M., J. Trudnowski, M. Canty, and M. A. Harrington-Baker. May 2000. "Performance of Field-Bioreactors with Sulfate-Reducing Bacteria to Control Acid Mine Drainage." *Proceedings of Society for Mining Engineers ICARD 2000 Conference*. Denver, CO.

NEUTRALIZATION OF ACIDIC MINING LAKES VIA *IN SITU* STIMULATION OF BACTERIA

René Frömmichen, Matthias Koschorreck, Katrin Wendt-Potthoff & Kurt Friese
(UFZ, Magdeburg, Germany)

ABSTRACT: Laboratory- and field-scale mesocosm studies were conducted from 1998 to 1999 to determine the efficacy of controlled eutrophication and saprobization for deacidification of extremely acidic mining lakes. Twelve to 17 months after treatment with the carbon source *Carbokalk*, a sugar industry by-product (solid precipitate of nonsugars after lime clarification of extracted sugar beet juice) and straw as the periphyton surface, acidity consumption was observed in the mesocosms. Via the accumulation of iron monosulfide, pyrite and elemental sulfur in the sediment of both the laboratory- and field-mesocosms, neutralization rates between 12 and 18 mol m^{-2} yr^{-1} were calculated. Nevertheless, in the water column of both mesocosms proton consumption differed by orders of magnitude. In the laboratory mesocosm the pH increased from 2.6 to around 6.5 throughout the 1.25 m water phase. In contrast, the pH in the 6.4 m water phase of the field mesocosm displayed no significant change.

INTRODUCTION

Intensive lignite mining activities have existed in Lusatia, Germany since the mid-19th century. As a result, the watertable has been drastically lowered over an area of approximately 2500 km^2 producing a volume deficit of 13 billion m^3. During open cast lignite mining, pyrite exposed to oxygen in the overburden resulted in microbial iron and sulfur oxidation and the production of sulfuric acid. After termination of mining activities in 1956, significant volumes of this acid were flushed into nearby developing lakes via rainfall runoff and/or ascending groundwater levels. As a direct consequence, more than 100 lakes in Lusatia are severely acidified and contain high concentrations of ferric iron and sulfate. At full capacity, these lakes will cover an area of 208 km^2 and hold approximately 4.5 billion m^3 of water. Approximately half of these lakes have pH-values in the range of 2.5 - 3 (Schultze et al., 1999). At these levels, no significant biological activity is present to neutralize the acidic lake conditions. Application of expensive methods such as iron precipitation used by commercial waterworks or the addition of limestone is not considered economically viable or environmentally conscious.

Objective. Researchers at the UFZ Environmental Research Centre developed an *in situ* deacidification strategy using dissimilatory iron- and sulfate reduction (Frömmichen et al., 2000a, b). In nature, *in situ* deacidification via sulfate- and iron reduction is possible only in those mining lakes in which stable stratification occurs or where deep water is significantly enriched with organic carbon. In most acidic mining lakes, the process of bacterial deacidification occurs at very slow rates (Peine, 1998).

The *in situ* deacidification strategy developed at the UFZ takes advantage of stable lake stratification situations. During remediation, the lake is considered to be a mixable reactor. The sediment is supplied with organic carbon sources and synthetic periphyton surfaces are placed upon the surface of the lake sediment. Both the periphyton surfaces and organic carbon stimulate anaerobic respiration of ferric iron and sulfate in the sediment in the deep watercolumn. Ferrous iron and sulfide are precipitated in the hypolimnion as iron monosulfides. Further diagenetic processes lead to more stable combinations, as pyrite and elemental sulfur. Under anoxic conditions these can be permanently stored in the sediment. With increased pH in the oxic epilimnion, the iron buffer determining acidity can be destroyed via precipitation of ferric iron as schwertmannite, jarosite or goethite.

Site Description. Sediment and lake water from the mining lake ML-111 of the Koyne-Plessa mining district in Lusatia (Germany) were used to construct mesocosms at the UFZ.

Open cast mining of lignite in the Koyne-Plessa district was operated betweem 1929 and 1956. Flooding of ML-111 was completed in 1969. Morphologically, the lake belongs to shallow water lakes of category 3 (Klapper, 1992) with an average water depth of 4.5 m and a comparatively small hypolimnion. The maximum depth is 10.2 m.

The photoautotrophic biomass is very limited in the epilimnion. Only in a few individual cases does the biovolume exceed 1 $mm^3\ L^{-1}$. The phytoplankton population is dominated by chrysophyceae of the *Ochromonas* species. Chlorophyceae of the *Chlamydomonas* species are also common. Zooplankton consists mainly of heliozoa, ciliates and rotifers (Packroff, 1998).

TABLE 1. Average concentrations of the most important compounds untreated water of the ML-111 mining lake

Compound	Concentration (mmol L^{-1})
Acidity	15.5
Ferric Iron	2.69
Aluminium	1.29
Sulfate	12.5
Calcium	5.74
Magnesium	1.23
Ammonia N	71.0 E-03
Total Phosphorus	0.32 E-03
Soluble Reactive Phosphorus	0.23 E-03

The chemical composition of the water column varied little during the investigation period between 1998 and 1999. The pH of ML-111 lake water is approximately 2.6, while electric conductivity ($\kappa_{25°C,\ ML\text{-}111}$) averages 2.5 mS cm^{-1}. Both total inorganic and organic carbon contents are less than 0.1 mmol L^{-1}.

Table 1 shows the average values for each of the most important chemical compounds found in the ML-111 water column. Divergence from mean values was only observed during times of temperature stratification in the water column during summer stagnation and winter ice cover (Herzsprung et al., 1998).

MATERIALS AND METHODS

The laboratory-scale mesocosms consisted of open Duran®-glass columns (Schott, Germany) 1.50 m in height. Untreated (control) and treated mesocosms each represented a lake section of an area of 0.071 m^2. The columns were filled with 14.8 ± 0.3 kg homogenized fresh lake sediment (0.20 m sediment layer in each column) and 88.0 ± 0.5 L of lake water (1.25 m water layer). The columns were equipped with sampling ports set at 0.10, 0.60 and 0.90 m from the top. For sediment sampling, the columns were divided 0.20 m above the column bottom over a detachable flanged system. Two separate thermostats were installed (Julabo Ltd.) for the purpose of creating a synthetic temperature stratification similar to that of a lake. The temperature in the bottom 0.56 m of the column was set at 10.0 ± 0.2 °C, the temperature between 1.10 and 1.45 m was set at 20.0 ± 0.2 °C. Masses of 0.30 ± 0.01 kg Carbokalk (7.48 ± 0.01 mol TOC m^{-2}) and 0.66 ± 0.01 kg wheat straw (9.3 ± 0.01 kg m^{-2}) were added to the treated mesocosm in a hollow cylinder attached to the sediment surface. The laboratory-scale mesocosms were incubated in darkness for the duration of the 12 month study conducted between January, 1998 and January, 1999. Detailed information on experimental setup is shown in Figure 1.

Field-scale mesocosms were realized as in-lake enclosures comprised of large tubes separating a defined water column of the lake. The tube consisted of 5 mm thick dump covercoat (high density polyethylene) burdened with a heavy high-grade steel frame at the lower end to form a stable contact with the sediment. Untreated (control) and treated mesocosms were defined in such a way to represent a section of lake from the sediment surface to the water surface covering an area of 4.14 m^2. The water depth in the control and treated enclosures was 6.40 ± 0.05 m. Mass transport between the enclosure and the remainder of the lake can be excluded. The detailed construction of a field-scale mesocosm is outlined in Figure 1.

Field experiments started in June, 1998 with the application of 8.05 ± 0.05 kg Carbokalk (3.5 ± 0.05 mol TOC m^{-2}) and 35.0 ± 0.5 kg wheat straw (8.5 ± 0.5 kg m^{-2}) to the treated mesocosm. Carbokalk was suspended with enclosed lake water and distributed via a plastic tube on the sediment surface. Three bales of wheat straw with edge lengths of 0.50 m were then placed in contact with the lake sediment. The results from the first 17 month investigation carried out in December, 1999 are given here.

Water Sampling and Analysis. Water samples were collected with a syringe through the sampling ports of the laboratory-scale mesocosms, and with a modified 5L-Friedinger-sampler (Limnos, Germany) at the water surface, as well as at 4 and 6 m water depths, in the field-scale mesocosm.

pH, acidity and alkalinity were determined at the start and conclusion of the experiments. *in situ* pH measurements in the lab were completed with an industrial glass electrode (Consort, Belgium); in the field, a multiparameter profiling probe (Idronaut, Italy) was used. Alkalinity ($K_{S4.3}$) and acidity ($K_{B8.2}$) were determined via volumetric titration according to procedure DIN 38409/H7 of the German Institute of Standardization.

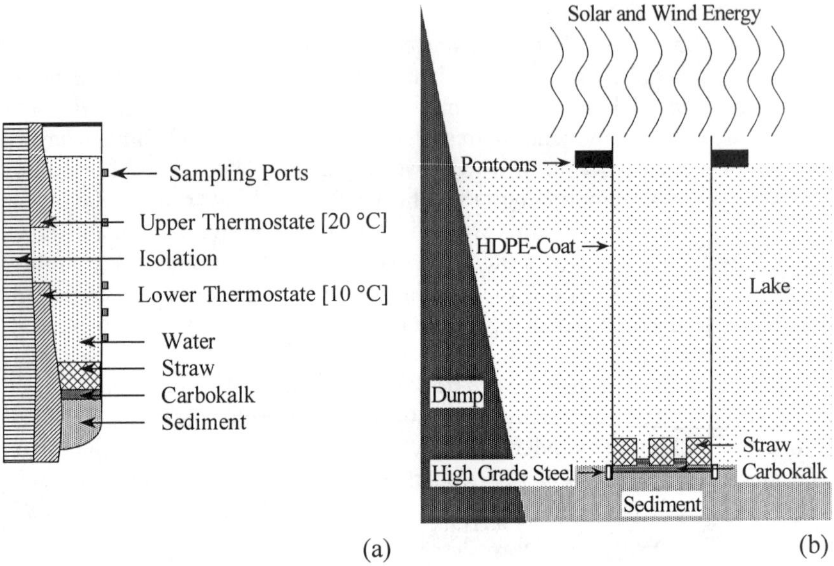

FIGURE 1. Cross sections of Mesocosm design.
(a) Laboratory-scale (b) Field-scale

Sediment Sampling and Analysis. Undisturbed sediment cores with a diameter of 0.09 m were collected from both the lab-scale and field-scale experiments. In the field, a modified Kajak-sediment corer (Uwitec, Austria) was used. Anoxic sediment cores were cut in 0.025 m layers to 0.10 m depth in the laboratory mesocosms, and in 0.01 m layers to 0.05 m depth and in 0.025 m layers between 0.05 and 0.10 m depth in field enclosures. The pH and oxidation-reduction potential (ORP) were measured in all sediment layers with a solid state glass and platinum electrode (Consort, Belgium). Prior to determination of the reactive ferrous iron ($Fe(II)_R$), acid volatile sulfide (AVS), chromium reducible sulfur (CRS) and elemental sulfur ($S^{(0)}$) all sediment samples were stored in liquid nitrogen.

AVS, CRS and $S^{(0)}$ were extracted sequentially in a three step method from fresh wet sediments. Extraction was modified according to Canfield (1989), Fossing and Jørgensen (1989) and Hsieh and Yang (1989). Depending on expected result (up to 20 mg g^{-1} S) for each reduced sulfur fraction, 2 to 5 g of fresh wet sediment was placed into a 250 ml flask. To the flask, 7.5 mL of degassed 1 M HCl was added, and the AVS extracted for 4 hours at room temperature. Next, a mixture of 7.5 mL degassed 12 M HCl and 7.5 mL degassed 2 M $CrCl_2$-solution in 1 M HCl was placed into the flask and the CRS extracted for 4 hours at room temperature. Finally, an additional 7.5 mL of degassed 2 M $CrCl_2$-solution in 1 M HCl and 7.5 mL degassed DMF (N,N-dimethylformamide) were added to the flask and the $S^{(0)}$ was extracted for 4 hours at a temperature of 60 °C.

In all three extraction steps the reduced sulfur fraction was driven from the extract with an argon gas stream as sulfide-S. For each extraction, the sulfide-S was absorbed in a separate gas trap. Fifty mL of alkaline buffer solution (sulfur antioxidant buffer, SAOB, consisting of 2 M NaOH, 0.18 M of ascorbic acid and 0.2 M EDTA) was placed into the gas trap (Cornwell and Morse, 1987). Sulfide analysis was then carried out in SAOB solution obtained from the gas trap. Differential pulse polarography (DPP) with the SDME (static mercury drop electrode) as the working electrode were employed to determine the sulfide concentration. The Ag/AgCl reference electrode was covered with a 1 M KNO_3 salt bridge to prevent AgS-precipitations at the diaphragm surface. A platinum rod served as the auxiliary electrode. The Radiometer Trace Lab® 32 system was used in conjunction with electrode stand MDE 150 (Tacussel, France).

Reactive ferrous iron ($Fe(II)_R$) was determined in the fresh sediment according to Lovley and Phillips (1987). An aliquot was displaced in 0.5 mol L^{-1} hydrochloric acid and the extraction proceeded for 2 hours in darkness at room temperature. After extraction, the sediment was centrifuged (10 min, 14000 rpm) and the ferrous iron in the supernatant determined photometrically with Ferrozin (Stookey, 1970).

Determination of the dry residues (T_R) as well as the sediment densities (Δ) for the individual sediment layers was performed by drying a defined balanced fresh wet sediment volume at 105 °C.

Calculation of Neutralization Rates. The calculation of neutralization rates (NR) was based on the presented accumulation period of non sulfidic ferrous iron, iron monosulfide sulfur (AVS), pyrite sulfur (CRS) and intermediate sulfur ($S^{(0)}$) in the lake sediment. Diurnal and seasonal variations of microbial activities were included.

For the treated lab- and field-scale mesocosms neutralization equivalents were calculated via production of reduced iron and sulfur compounds. For each mole of inorganic sulfur reduced, 2 moles of protons were consumed (Berner, 1970; Berner, 1984; Canfield and Raiswell, 1991) (equation 1). According to equation 2, the same proton consumption was assumed per mole non sulfidic reactive ferrous iron (Canfield and Raiswell, 1991; Vile and Wieder 1993). The aqueous iron of equation 2 must be fixed in iron phosphates or carbonates.

$$36\langle CH_2O\rangle + 12FeOOH + 18SO_4^{2-} + 36H^+ \rightarrow 36CO_2 + 12FeS + 6S^0 + 60H_2O \quad (1)$$

$$\langle CH_2O\rangle + 4FeOOH + 8H^+ \rightarrow CO_2 + 4Fe^{2+} + 7H_2O \quad (2)$$

By defining a reference area and time relation for neutralization equivalents compared to the control mesocosm, neutralization rates per square meter and year can be determined using equation 3.

$$NR = \frac{V \times 20}{t \times A} \sum_{i=1}^{n} \left([TRIS]_i + [Fe(II)_R]_i - [Fe(II)_{AVS}]_i \right) \Delta_i T_i \quad (3)$$

Where NR = neutralization rate [mol m^{-2} yr^{-1}]
[TRIS] = accumulation of total reduced inorganic sulfur [mol g^{-1}]
$[Fe(II)_R]$ = accumulation of reactive ferrous iron [mol g^{-1}]
$[Fe(II)_{AVS}]$ = accumulation of monosulfide iron [mol g^{-1}]
Δ_i = density of fresh sediment layer [kg L^{-1}]
T_{Ri} = dry residue of fresh sediment layer [%]
V = volume of fresh sediment layer [L]
A = area considered [m^2]
t = time of accumulation considered [yr]
i = sediment layer considered
n = number of sediment layers considered

RESULTS AND DISCUSSION

Acidity, Alkalinity and pH of the Water. At the termination of the treated lab-scale mesocosm experiment in January, 1999 the waterphase pH was measured between 5.9 and 6.4 as shown in Figure 2(a).

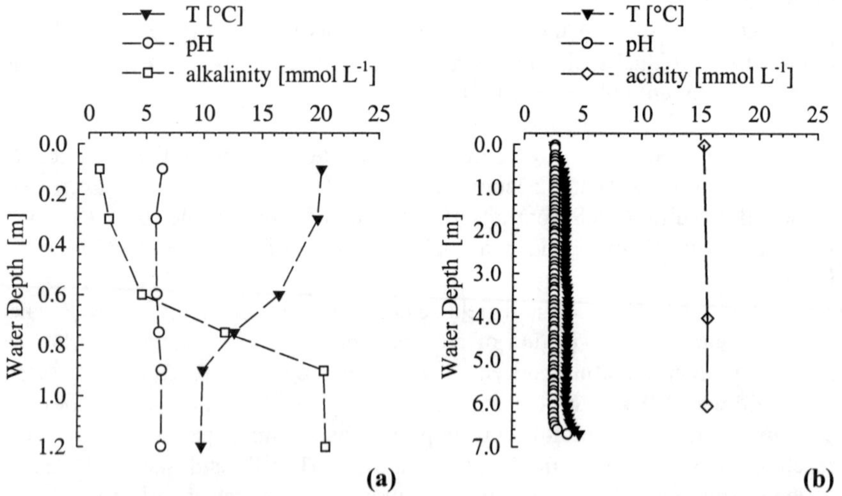

FIGURE 2. Plots of Temperature (T), pH, Acidity and Alkalinity *vs.* Water Depth of Treated Mesocosms. (a) Laboratory-Scale, (b) Field-Scale

This observed deacidification was the result of intense microbial activity in the lab-scale mesocosm. Following the first four weeks of the lab-scale experiment, anoxic conditions developed between water at a depth of 0.90 m and the sediment/water interface. In this region, total aqueous sulfide concentrations as high as 1.6 mmol L^{-1} were observed. Very low ORP (corrected for SHE and pH 7) with values less then - 0.3 V were also observed (Frömmichen, 2001). The increase in alkalinity was the result of supersaturation of the waterphase with microbially-produced inorganic carbon. Between 0.90 m and the sediment surface the content of inorganic carbon was 110 mmol L^{-1} and the acidity was approximately 20 mmol L^{-1}. In comparison, acidity at the top of the treated mesocosm was lower than the initial average value of 16.5 mmol L^{-1}. This was due to the lack of reduction of iron and aluminium concentrations (Frömmichen, 2001).

In the 17 month field scale experiments, pH and acidity in the free waterphase were not altered compared to the untreated mesocosm. Within 0.20 m of the sediment/water interface of the treated mesocosm the pH increased to a value of more than 3.0 (Figure 2(b)).

Reduced Iron and Sulfur Species of the Sediment. The average for the sum of the AVS, CRS and $S^{(0)}$ contents (TRIS content) taken over all sediment layers for the two non-treated controls in the lab-scale and the field-scale mesocosms were lower than 6 μmol g^{-1}. The reactive ferrous iron content in the untreated lab sediment was 63.0 ± 11.7 μmol g^{-1}, on average. In the untreated field-scale mesocosm the content of the same species was 23.9 ± 3.9 μmol g^{-1}.

The accumulation of reduced inorganic sulfur species and reactive ferrous iron in the sediments of treated mesocosms was in the same order (Figure 3 and 4).

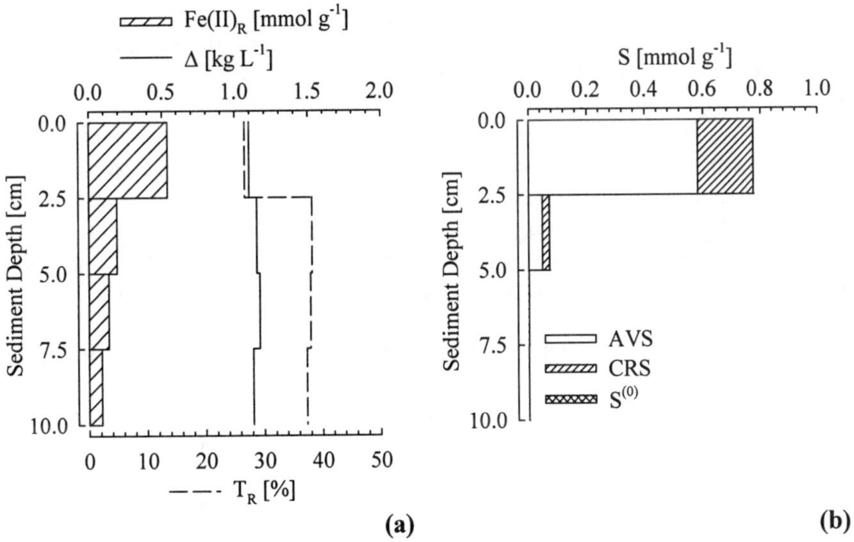

FIGURE 3. Plots of (a) Reactive Ferrous Iron ($Fe(II)_R$), Sediment Density (Δ), Dry Residue (T_R) and (b) Acid Volatile Sulfide (AVS), Chromium Reducible Sulfur (CRS), Elemental Sulfur ($S^{(0)}$) *vs.* Depth for Sediment of the Treated Laboratory-Scale Mesocosm.

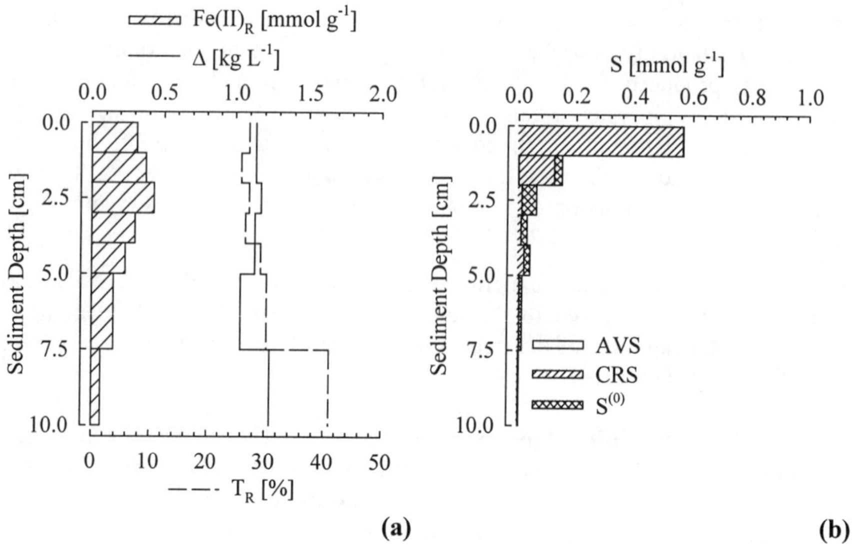

FIGURE 4. Plots of (a) Reactive Ferrous Iron ($Fe(II)_R$), Sediment Density (Δ), Dry Residue (T_R)and (b) Acid Volatile Sulfide (AVS), Chromium Reducible Sulfur (CRS), Elemental Sulfur ($S^{(0)}$) *vs.* Depth for Sediment of the Treated Field-Scale Mesocosm.

Differences in the determined species distribution between the lab- and field-scale mesocosms were observed. In the lab, AVS (mainly iron monosulfide sulfur) was identified as the main sulfur species (Figure 3(b)), while field results indicated high contents of CRS (pyrite sulfur) (Figure 4(b)). Therefore, the reactive iron in the lab sediment outlined in Figure 4(a) is primarily bound non-sulfidically. These differences between the lab and field observations are thought to be a result of variations in environmental conditions at the sediment water interface. In the spring and autumn circulation phases, oxygen can diffuse into the sediment and H_2S autooxidized to elemental sulfur reacting with iron monosulfide to pyrite. In the lab-scale mesocosm, anoxic conditions rendered this mechanism negligible.

Neutralization Rates. Calculated neutralization rates for the lab- and field-scale mesocosms were in the same order of magnitude (Table 2), but were higher by one to two orders of magnitudes than those calculated with corresponding methods for atmospherically acidified lakes (Giblin et al., 1990) or other open cast lignite mining lakes of different ages (Peine, 1998). At the evaluated neutralization rate, an noticeable increase of pH throughout the water column in the treated field-scale mesocosm will be observed over time. The predicted remediation period for the mining lake ML-111 is about 10 years.

Hydrological alterations in groundwater influx may transport ferrous iron and sulfate into the lake. The oxidation of ferrous iron and subsequent precipitation of ferric iron with or without sulfate as jarosite, schwertmannite or goethite may produce further acidity. The erosive supply of sulfate with dump material from the banks may lead to a similar situation in the lake. All of these factors have not been evaluated in this preliminary data. Since these factors have not been properly evaluated to date, conclusions about remediation and deacidification should be approached with caution.

TABLE 2. Acidity measured at start up *vs.* Neutralization Rates calculated at end of treated laboratory-scale and field-scale mesocosms

Acidity ($mol\ m^{-2}$)		Neutralization Rates ($mol\ m^{-2}\ yr^{-1}$)	
Laboratory-scale	**Field-scale**	**Laboratory-scale**	**Field-scale**
20.1	95.4	17.1	12.7

Furthermore, current neutralization rates can not be calculated for the littoral zone of lake ML-111. It is assumed that neutralization and acidity rates per square meter are proportional to water depth. For evaluation of realistic neutralization and acidity rates, hydrological research and information about microbial activity from the profundal to littoral zones stimulated with Carbokalk and straw are required.

Acknowledgements The authors thanks Mrs Corinna Scholz, Mrs Renate Görling and Mr Martin Wieprecht for the excellent technical assistance. This work was funded by the Ministry of Education and Research of the Federal Republic of Germany.

REFERENCES

Berner R.A. 1970. "Sedimentary pyrite formation." *American Journal of Science. 268*: 1-23.

Berner R.A. 1984. "Sedimentary pyrite formation: An update." *Geochimica et Cosmochimica Acta. 48*: 605-15.

Canfield D.E. 1989. "Reactive iron in marine sediments." *Geochimica et Cosmochimica Acta. 53*: 619-32.

Canfield D.E., and R. Raiswell. 1991. "Pyrite formation and fossil preservation." In P.A. Allison & D.E.G Briggs. *Topics in Geobiology. 9*: 337-387

Cornwell J.C., and J.W. Morse. 1987. "The characterization of iron sulfide minerals in anoxic marine sediments." *Marine Chemistry. 22*: 193-206.

Fossing H., and B.B. Jørgensen. 1989. "Measurement of bacterial sulfate reduction in sediments: Evaluation of a single-step chromium reduction method." *Biogeochemistry. 8*: 205-22.

Frömmichen R. 2001. "*In situ*-Sanierungsstrategie zur Förderung der mikrobiellen Entsäuerung von geogen schwefelsauren Bergbaurestseen – Mesokosmosstudien." Ph.D. Thesis, Technical University of Dresden. Dresden

Frömmichen R., K. Wendt-Potthoff, K. Friese, and H. Klapper. 2000a. "Verfahren zur mikrobiellen Sanierung von schwefelsauren Bergbaurcstseen." *German Pate and Application.* DE 199 07 002 A1

Frömmichen R., K. Wendt-Potthoff, K. Friese, and H. Klapper. 2000b. "Method for microbially cleaning up sulphuric-acid residual mining pools." *PCT Pate and Application.* WO 00/48949

Giblin A.E., G.E. Likens, D. White, and R.W. Howarth. 1990. "Sulfur storage and alkalinity generation in New England lake sediments." *Limnology and Oceanography. 35*(4): 852-69.

Herzsprung P., K. Friese, G. Packroff, M. Schimmele, K. Wendt-Potthoff, and M. Winkler. 1998. "Vertical and annual distribution of ferric and ferrous iron in acidic mining lakes." *Acta hydrochimica hydrobiologica. 26*: 253-262.

Hsieh Y.P., and C.H. Yang. 1989. "Diffusion methods for the determination of reduced inorganic sulfur species in sediments." *Limnology and Oceanography. 34*(6): 1126-1130.

Klapper H. (Ed.). 1992. *Eutrophierung und Gewässerschutz.* Gustav Fischer Verlag. Jena

Lovley D.R., and E.J.P. Phillips. 1987. "Rapid assay for microbially reducible ferric iron in aquatic sediments." *Applied and Environmental Microbiology. 53*(7): 1536-1540.

Packroff G. 1998. "Protozoen in Tagebaurestseen - Qualitative und Quantitative Aspekte." *Tagungsbericht zur DGL-Tagung, 1997.* 346-350, Frankfurt/M.

Peine A. 1998. "Saure Restseen des Braunkohletagebaus - Charakterisierung und Qualifizierung biogeochemischer Prozesse und Abschätzung ihrer Bedeutung für die seeinterne Neutralisierung" Ph.D. Thesis, University of Bayreuth. Bayreuth

Schultze M., K. Friese, R. Frömmichen, W. Geller, H. Klapper, and K. Wendt-Potthoff. 1999. "Tagebaurestsseen-schon bei der Entstehung ein Sanierungsfall." *GAIA. 8*(1): 32-43.

Stookey L.L. 1970. "Ferrozine-a new spectrophotometric reagent for iron." *Analytical Chemistry. 42*(7): 779-781.

Vile M.A., and R.K. Wieder. 1993. "Alkalinity generation by Fe(III) reduction versus sulfate reduction in wetlands constructed for acid mine drainage." *Water, Air and Soil Pollution. 69*: 425-441.

BIOREMEDIATION OF METALS FROM AN ACID MINE DRAINAGE AT CANE CREEK, COAL VALLEY SITE

Victor M. Ibeanusi, Errol Archibold, Latoya Hannon,
Erica Garry, April Hines and Adrianna Sola
Spelman College, Atlanta, Georgia

ABSTRACT: A demonstration research project on the treatment and removal of metals from an acid mine drainage (AMD), and the feasibility of increasing the pH of the AMD from acidic (< 3.5) to elevated pH conditions was conducted at the Cane Creek, Coal Valley Site, in the state of Alabama. The oxidation of pyrite from the AMD results in run off with considerable acidity and high concentrations of dissolved toxic metals. The current treatment method employed at this site, uses a limestone bed and a constructed wetland for the purpose of increasing the pH of the AMD. In this treatment process, the AMD was directed to flow through limestone beds, and a constructed wetland for the primary purpose of elevating the pH of the AMD before it flows into Cane Creek. However, this treatment process has not achieved the desired pH values that would cause the precipitation of the metals from the AMD. The purpose of this present demonstration project was to investigate the feasibility of using a patented in-situ/on-site bioremediation process for the purpose of detoxifying and removing the dissolved metals in the AMD. The results from the bioremediation process indicated that the pH of the AMD was elevated from <3.52 to 7.59 in a batch treatment. In the continuous flow treatment, the pH was increased to 6.27. Similarly, the redox of the AMD was changed from a highly oxidizing condition (190mV) to a reduced state of –26mV. Consequently, the reduced chemical environment provided a conducive setting for the precipitation of the dissolved metal ions in the AMD. As a result of these changes (elevation of pH, reducing redox condition, and the reduction of the metal concentrations) in the demonstration pond, we observed a return to a normal aquatic life in the treatment pond.

INTRODUCTION

As the cost of energy continues to soar, due to increasing limited oil reserves, the presence of abundance of coal in many parts of the world has re-kindled interest in over a century old coal technology. However, coal mining and the use of coal in general has an adverse effect on the environment (Ibeanusi, and Wilde, 1998). Wastewater from underground coal mining and coal piles are among the major sources of industrial pollution in the mid western states, such as Pennsylvania, West Virginia, Ohio and Kentucky (Hobbs, 1993; Kadnuck, 1995). Apart form the generation of greenhouse gases that contribute significantly to global warming and ozone depletion, the oxidation of pyrite in coal has adverse effects on the quality of surface and groundwater (Ackman and Jones, 1991; Aljoe, 1994; Bos et al., 1992; Hawkins, 1995; Kendorski, 1994; Rogowski and Pionke, 1984). The toxicity effects result from both biotic and abiotic oxidation of pyrite in the acid mine drainage (AMD) which results in a run off with high acidity and dissolved toxic metals. The major metals of concern include Al, Fe, and Mn, which may exceed drinking water standards by several thousand folds (Ibeanusi and Wilde, 1998). Other dissolved metals include Cr, Cd, Pb, Ni, Se, and Zn.

Furthermore, the impact of AMD on aquatic life and on the adjoining ecosystems has been well-documented (Haines, 1981; Kleinmann and Watzlaf, 1988; Nicholas and Bulow, 1973). Areas mostly affected include farmlands, corrosion and incrustation of pipes, and fish death. The degree and extent of toxicity has been attributed to the buffering capacity of the soil, pH, volume, frequency and rate of flow (Haines, 1981). In this present project, a patented bioremediation system, which employs a bacterial-mediated pH-redox reactions was used in a constructed pond to demonstrate the effectiveness of an in-situ/on-site bioremediation of AMD.

Objective. The specific objective of this project was to increase and maintain the pH of the AMD through a biological control of the redox condition of the wastewater. Consequently, the goal is to reduce the concentrations of the dissolved metal ions in the AMD. Because regions where soils are relatively acidic and low in carbonates offer the least natural attenuation, the use of a redox-mediated process offers a long-term guarantee for a sustained increased pH.

TABLE 1. Batch treatment of acid mine drainage.

Test Period (weeks)	Sample ID	pH	Redox (mV)	Temperature (^{0}C)
Week 0 (April 15)	AMD	3.68±0.76	186±0.78	15.5±0.02
	TP	3.82±0.73	183±0.58	15.7±0.01
Week 1 (April 22)	AMD	3.98±0.74	179±0.98	18.9±0.45
	TP	5.56±0.53	139±0.87	17.9±0.67
Week 2 (April 29)	AMD	3.52±0.63	176±0.63	22±0.05
	TP	7.59±0.76	-36±0.80	23±0.03
Week 3 (May 6)	AMD	3.54±0.52	182±0.86	24±0.36
	TP	7.59±0.85	-38±0.91	24±0.35

AMD = Control; AMD from the constructed wetland. No bacterial strains were added.

TP = Treatment pond; pond inoculated with bioremediation bacterial strains.

MATERIALS AND METHODS

The AMD from the constructed wetland was directed into a demonstration treatment pond with a dimension of 12 sq. ft x 2 ft. One-inch pipes connected to two 100-gallon feeder-tanks were used to inoculate the treatment pond with the bioremediation bacterial strains. Each feeder tank contained 100 liters of the bacterial mix at a standard bacterial inoculum of 10 mL/L. In the batch treatment demonstrations, one of the feeder tanks was used for the full-capacity treatment of the pond. While the two feeder tanks were used in the continuous-flow treatment process at the flow rate of 0.57 mL/sec. The AMD from the constructed wetland served as the control. Random triplicate samples from both the treatment pond and the constructed wetland were taken for a period of four weeks. Results in Tables 1, 2, and Figure 1, represent average results and the standard deviations.

Redox and pH Analysis. The oxidation-reduction potential (Eh) and pH of the inflow AMD, the treatment pond, and the effluent from the treatment pond were determined using a meter with a platinum-KCl electrode. Sampling was performed at different points and average and standard deviations were obtained.

Metal Analysis. Prior to metal analyses, water samples from the constructed wetland (control) and the treatment pond were digested with a Microwave Digestion System (CEM Incorporation). Digested samples were subsequently analyzed using Inductively Coupled Plasma (Plasma 400, Perkin Elmer).

RESULTS AND DISCUSSION

Batch Treatment. The batch treatment was conducted for four weeks, starting from April 15 to May 6, 2000. The data recorded on April 15 represented the initial results of both the constructed wetland (control) and the treatment pond. Subsequently, the bioremediation demonstration was initiated by inoculating the treatment pond with bacteria. After a minimum retention time of the AMD in the treatment pond for seven days, the effects of the bioremediation treatment became apparent on April 22. Stable pH and redox conditions were recorded on April 29 and May 6 (Table 1). Since daily readings were not taken, it was not known when the increased pH peaked at 7.59. However, previous laboratory and treatment of a coal pile run off at Savannah River Site indicated that the process generally takes place between 3-7 days (Ibeanusi and Wilde, 1998).

Continuous Flow Treatment. The continuous flow treatment was initiated on June 1 with the initial flow rates as indicated on Table 2. The flow rate for the AMD indicated the flow from the constructed wetland (control) into the treatment pond. The flow rate for the treatment pond (TP) represented the effluent from the treatment pond into Cane Creek. Similar to the batch treatment, the pH and redox in the treatment pond, steadily increased and changed, respectively. The peak pH in the treatment pond was 6.35. The redox condition continued to decrease considerable after 7 days of treatment, thus allowing for the precipitation and removal of the dissolve metals in the AMD.

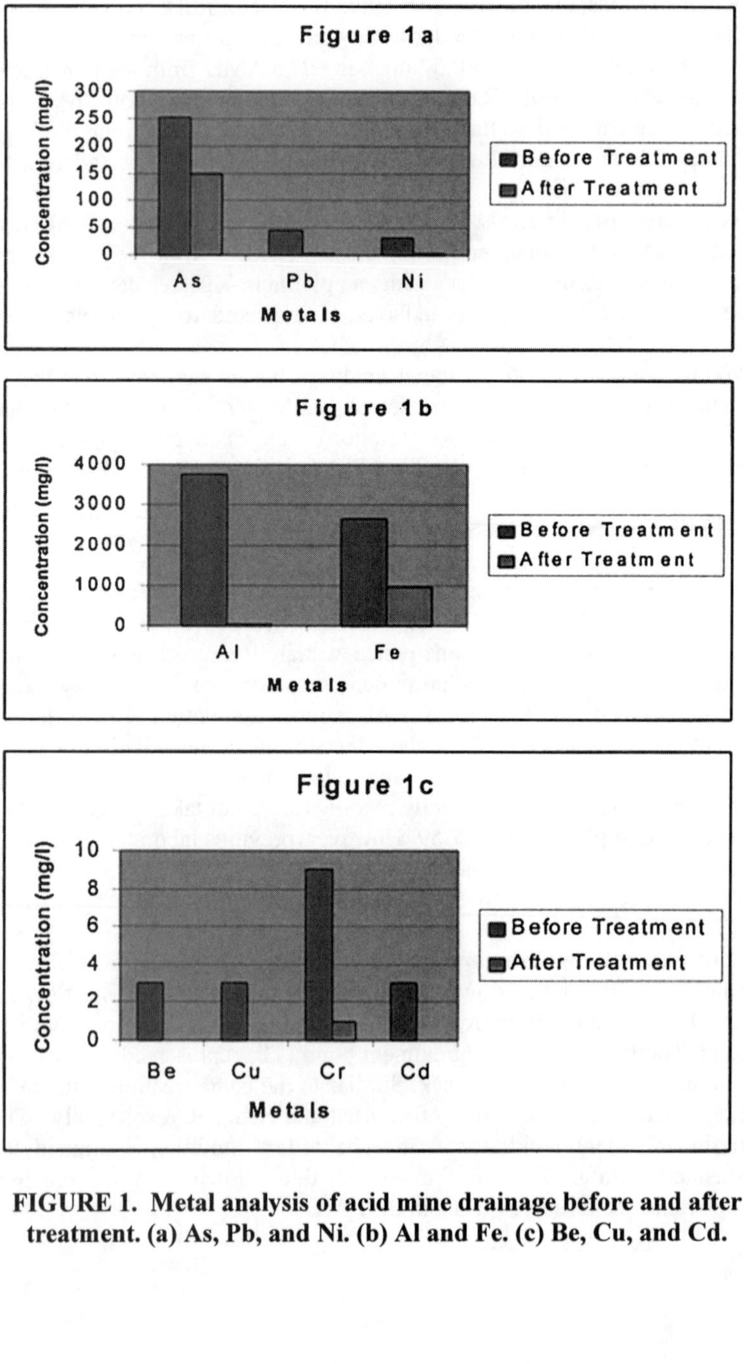

FIGURE 1. Metal analysis of acid mine drainage before and after treatment. (a) As, Pb, and Ni. (b) Al and Fe. (c) Be, Cu, and Cd.

Metal Analysis. The results of the metal analysis represented the concentrations of dissolved metals before and after treatment process. The results showed a significant reduction on the concentrations of the metals in the treated AMD.

TABLE 2. Continuous flow treatment of acid mine drainage

Test Period (Weeks)	Sample ID	Flow Rate (mL/sec)	pH	Redox (mV)	Temp. (^{0}C)
June 1	**AMD**	**17.7**	**4.08±0.20**	**138±0.86**	**25**
(week 0)	**TP**	**5.79**	**4.57±012**	**100±0.32**	**24**
June 8	**AMD**	**17.7**	**3.79±0.25**	**153±0.43**	**23**
(week 1)	**TP**	**5.79**	**5.31±0.38**	**94 ±0.54**	**25**
June 16	**AMD**	**25**	**3.76±0.28**	**189±0.53**	**27**
(week 2)	**TP**	**24**	**6.27±0.32**	**-27±0.28**	**25**
June 23	**AMD**	**25**	**3.76±0.28**	**189±0.53**	**23**
(week 3)	**TP**	**24**	**6.35+0.46**	**-17+0.43**	**24**

AMD = Control; AMD from the constructed wetland. No bacterial strains were added.
TP = Treatment pond; pond inoculated with bioremediation bacterial strains.

CONCLUSIONS

A bioremediation process for the treatment of acid mine drainage sites has been demonstrated. The results of both the batch and continuous treatment methods demonstrated the effectiveness of the prescribed bioremediation process for removing dissolved toxic metals from acid mine drainage. The use of bacterial-mediated reactions in controlling the redox states of the AMD was found to be effective in maintaining the elevated pH conditions observed in the treatment pond. As a result of the elevated pH and changes in the redox conditions, significant reduction of dissolved metal ions was obtained in the treated water (Figure 1).

The results obtained in this demonstration project were consistent with those obtained in an earlier demonstration at the Department of Energy's Savannah River Site, Aiken, SC, where the bioremediation process was used in treating a coal pile run off (Ibeanusi and Wilde, 1998).

In addition, it is important to note that the bioremediation system described here was driven by bacterial processes which evolved not as single cell culture but rather as a consortium of metal-resistant bacterial species functioning as a cooperative ecosystem unit. This kind of ecosystem approach offers opportunities for reclaiming waste sites to their original natural conditions.

From a practical standpoint, this bioremediation process offers substantial advantages over conventional methods such as lime precipitation or pump-and treat. Opportunities for larger scale demonstrations at waste sites are needed to fine-tune the engineered process of this technology.

ACKNOWLEDGEMENT

Our profound gratitude to Mr. Larry Barwick, Supervisor, Abandoned Mine Land Reclamation Division, State of Alabama, and his staff. Their support and personal involvement made it possible for us to use the Coal Valley site for this demonstration project. Special thanks to Fares El-Katri for his technical support services. Partial supports for this project was funded by the Office of Surface Mining, Bureau of Land Management, and the U.S. Environmental Protection Agency, Faculty Development Program.

REFERENCES

Ackman, T.E., and Jones, J.R. 1991. "Methods to Identify and Reduce Potential Surface Stream Water Losses into Abandoned Underground Mines". ***Environ. Geol. Water Science 17 (3).***

Aljoe, W.W., 1994. "Hydrologic and Water Quality Characteristics of a Partially-Flooded Abandoned Underground Coal Mine". ***U.S. Bureau of Mines Special Publication SP-06A-94.***

Bos, P., F.C. Boogerd'. J. G. Kuenen. 1992. " Microbial Desulfurization of Coal". ***Applied Environmental Microbiology. (15) 575-403***

Haines, T.A. 1981. "Acid precipitation and its consequences for aquatic ecosystems". ***A Review. Trans. Am. Entomol.*** Soc. 110: 669-707

Hawkins, J.W. 1995. "Impacts on Groundwater Hydrology from Surface Coal Mining in Northern Appalachia". ***In: Proceedings of the 1995 Annual Meetings of the 1995 Annual Meeting of the American Institute of Hydrology, Denver Co.***

Hobbs, W.A. 1993. "Effects of underground mining and mine collapse on the hydrology of selected basins in West Virginia". ***U.S. Geology Water Supply . Paper 2384***

Ibeanusi, V.M., and E.W. Wilde. 1998. " Bioremediation of Coal Pile Runoff Waters Using an Integrated Microbial Ecosystem" ***Biotechnology Letters*** 20(11) 1077-1079.

Kadnuck, L. M. 1995. " Hydrologic response to underground coal mining". ***Geoscience 1 (1) 101-107***

Kendorski, F.S. 1994. "Effects of High-extraction Coal Mining on Surface and Groundwater". ***In: Proceedings, 12th Conference on Ground Control in Mining. Charleston, W. VA.***

Kleinmann, R.L.P., and G.R. Watzlaf. 1988. "Should Effluent Limits for Mn be Modified?". ***In: Proceedings of the 1988 Mine Drainage and Surface Mine Reclamation Conference, American Society for Surface Mining and Reclamation and U.S. Department of the Interior (Bur. Mines and OSMRE).*** Pittsburgh, PA. pp. 305-310.

Nichols, L.E., and F.J. Bulow, 1973. "Effects of Acid Mine Drainage on the Stream Ecosystem of the east Fork of Obey River, Tennessee". ***J. Tenn. Acad. Sci. 48: 30-39.***

Rogowski, A.S., and H.B. Pionke, 1984. "Hydrology and Water Quality on Strip-mined Lands". ***U.S. Environmental Protection Agency, EPA-IAG-D5-E763, Washington, D.C. p. 183.***

REMOVAL OF HEAVY METALS AND ORGANIC COMPOUNDS FROM ANAEROBIC SEDIMENTS

J H P Watson and D C Ellwood
(University of Southampton, Southampton, SO17 1BJ, England)

Abstract: In earlier work studies were made of the adsorption of metal ions from solution by the anaerobic sulphate-reducing bacteria (SRB) and the adsorption by iron sulfide material produced by SRB. The SRB were grown from inocula obtained from semi-saline marine sediments. It was found that heavy metal ions are removed from solution to residual levels often less than 1 mg/L. Further each bacterium can take on a considerable fraction of its own weight. The iron sulfide precipitates produced by the SRB have a large specific surface area and will strongly adsorb heavy metals but also chloro- and fluoro- carbon compounds. The following metal ions were studied, U, Ru, Sr, Co, Pu, Am, Ce, Cs, Tc and Zr of interest to the nuclear industry, Pt, Pd, Rh, Ru, Ir, Au, Ag, Fe, Co, Ni, and Cu of interest to the precious metals industry and a number of pollutant metals such as Hg, As and Cd. It was possible to reduce many of the above metal ion concentrations from 10 mg/L to 1-2 μg/L and often to less than 1 μg/L. As most marine sediments are anaerobic the occurrence of SRB in these sediments is wide-spread, this means that powerful concentrating and immobilising mechanisms should exist within the anaerobic sediments. Hamilton has found that radioactive 'hot' spots are formed in anaerobic esturine muds and marshes near the Sellafield nuclear reprocessing plant on the north-west coast of England. The presence of these radioactive 'hot' spots is strong evidence that concentrating processes are present. Using high gradient magnetic separation (HGMS) it is possible to remove a substantial fraction (> 80%) of these 'hot' particles as most of the radioactive metals encountered are paramagnetic, consquently, when they are immobilised in a SRB cluster, the cluster becomes paramagnetic. After the sediment is dredged the application of HGMS will concentrate the heavy metals and chloro- and fluoro-carbon compounds into a small magnetic fraction of the throughput which greatly reduces the disposal and treatment problem of the non-magnetic part of the sediment.

INTRODUCTION

Hamilton has found that radioactive 'hot' spots(Hamilton, 1981; Hamilton and Clarke, 1984) are formed in anaerobic esturine muds and marshes near the Sellafield nuclear reprocessing plant situated on the north west coast of England on the Irish Sea. The samples were taken from a core in the tidal estuary of the River

Esk about 10 km south of Sellafield.

The presence of these 'hot' spots is strong evidence that concentrating processes are present within the sediment. For reasons outlined below, the suggestion is made here is that the action of sulfate-reducing bacteria (SRB), which are wide-spread in anaerobic sediments, provides an explanation for this concentrating mechanism. As iron is a common element in these sediments it will be precipitated as an iron sulfide by the action of SRB, many of the iron sulfides are ferrimagnetic(Watson et al., 2000), and in consequence may be removed from the sediments by magnetic separation. The attempt to remove the radioactive contamination from the esturine silts from the River Esk using high gradient magnetic separation (HGMS) is described in more detail below.

The action and metabolism of SRB in marine and salt marsh sediments have been studied extensively so here, because of space limitatins, we refer only two of studies concerned with the precipitation and immobilisation of heavy metals. Revis et al.(Revis et al., 1991) have immobilised Hg in soils, sediments, sludge and water using the SRB present in the system. They also suggested the addition of gypsum(calcium sulfate) to increase the population of SRB and thereby increase the immobilisation of Hg. Arakaki and Morse(Arakaki and Morse, 1993) have observed the coprecipitation and adsorption of Mn(II) with fine-grained mackinowite(FeS) produced by SRB in anoxic sediments. SRB have been observed to produce appreciable amounts of mackinowite within a bioreactor(Watson et al. 2000).

Immobilisation of metal ions and halogenated hydrocarbons from solution by the SRB has been studied extensively (Ellwood et al., 1992; Watson and Ellwood, 1994; Watson et al., 1995; Watson et al., 2000; Watson et al., 2001). The SRB used in this work came from a number of anaerobic semi-saline marine sediments from different places on the south coast of England. In the previous work it was found that heavy metal ions with insoluble sulfides are removed from solution to residual levels often less than 1 mg/L. Further each bacterium was observed in many cases to immobilise 3 to 4 times their own wet weight of the metal sulfides such as iron or uranium(Watson and Ellwood, 1994). The iron sulfide precipitates produced by the SRB have a large specific surface area(500 m^2/g) and are ferrimagnetic(Watson et al., 2000) and will strongly adsorb heavy metals but also halogenated hydrocarbon compounds. The following metal ions were studied, U, Ru, Sr, Co, Pu, Am, Ce, Cs, Tc and Zr of interest to the nuclear industry, Pt, Pd, Rh, Ru, Ir, Au, Ag, Fe, Co, Ni, and Cu of interest to the precious metals industry and a number of pollutant metals such as Pb, Hg, As and Cd. Most of these metals appear to be chemisorbed(Watson and Ellwood, 1994; Watson et al., 1995). It has been possible to reduce the above metal ion concentrations from 10 mg/L to 1-2 μg/L and in many cases less than 1 μg/L.

MATERIALS, METHODS AND EXPERIMENTAL RESULTS

In the early 1970s the so-called high gradient magnetic separation(HGMS) was developed (Kolm, 1971; Watson, 1973) which made it possible to extract weakly magnetic colloidal particles from a fluid, liquid or gas stream. HGMS is a process in which magnetisable particles are extracted onto the surface of a fine ferromagnetic wire matrix which is magnetised by an externally applied magnetic field. The process, which is used to improve the brightness of kaolin clay, was developed for and in conjunction with the kaolin industry in the U.S.A. This process allows weakly magnetic particles of colloidal size to be manipulated on a large scale at high processing rates so there are, in addition to the clay industry, a large number of potential applications in fields as diverse as the cleaning of human bone marrow, nuclear fuel reprocessing, sewage and waste water treatment, industrial effluent treatment, industrial and mineral processing, extractive metallurgy and biochemical processing(Watson, 1998).

Transuranic radionuclides (Pu, Am and Cm) which were present in effluents discharged into the north-east Irish Sea by British Nuclear Fuels plc(BNFL), Sellafield, Cumbria, England are found in the sediment and biota in the estuary of the River Esk, ~ 10 km to the south. The occurrence of these materials has been studied by Hamilton (Hamilton, 1981). Hamilton drilled core samples from the sediment in the estuary of the Esk from a position of in the esturary, which has been the subject of intensive study by Hamilton and Clarke(Hamilton and Clarke, 1984) who determined that the site is representative of a site of continuous accretion over the last 60 years which has taken place at a sedimentation rate of ~ 2.5 cm/year.

In preparing the samples of sediment for measurement, a measured volume of wet sediment was taken and the washed in distilled water before freeze-drying. The activity was then measured and the resulting U content was expressed as mg/L (dry weight) in Table 1. Measurements of the total concentration of uranium in the water above core sample, 31% sea water, is 2.8±0.3 μg uranium/L whereas the concentration on the average in the area investigated is 1.4 mg/cm^3 or approximately 500 times higher than in the seawater; however, as shown in Table 1, the values measured in any particular sample can be appreciably higher than the average value. This suggests the presence of a strong concentrating mechanism for uranium. The studies of the surface sediment from the Esk shows an association between uranium and the presence of anaerobic conditions and organic material. In the regions where organic material is absent, little uranium is retained.

The radioactive analysis used a dielectric α-track detector in association with conventional chemical separation techniques and α-particle spectrometry using surface barrier detectors. The method involves placing a sample in contact with a sheet of detector and, after a suitable period of exposure, the α-particles with a sheet of detector and, after a suitable period of exposure, the α-particles recorded as tracks are revealed by etching at 80°C in 6.0 M NaOH for 3 h. After etching the detector is viewed with the naked eye and the α-particle tracks associated with the "hot" particles can easily be observed against a background of randomly distributed tracks; analysis of the α-particles track distributions can be determined using a conventional light microscope. Quantitative analysis of track densities can be done using a Quantimet image analyser. The ^{241}Am content of the samples was determined by γ- spectrometry.

TABLE 1. Silt sample from the River Esk at 2 cm deep into the silt some of the measured radioactive elements were as follows:

mg/L(dry wt.)	pCurie/L (dry wt.)				
U	^{241}Am	^{144}Ce	^{137}Cs	^{134}Cs	^{106}Ru
3.4	127.1	128.6	255.8	13.9	39

The HGMS equipment followed the conventional lay-out. It consisted of a feed vessel containing the slurry stirred with a high-speed mixer keeping the particles in suspension. From the feed vessel the slurry passed, under gravity, into a canister containing a ferromagnetic matrix situated in an applied external magnetic field of 5 Tesla supplied by a superconducting solenoid. The canister was situated within the room temperature bore of the solenoid with dimensions 7.62 cm in diameter and 46 cm long. The matrix was a grade 430 ferromagnetic stainless steel mesh of expanded metal having a saturation magnetisation, M_S, of 1.7 Tesla. The hole in the mesh was diamond shaped, with maximum and minimum lengths of 5.84 and 1.59 mm, respectively. The strands of the mesh were 150 μm in diameter. The matrix occupied 5% of the canister volume. The plane of the grid was oriented perpendicular to the applied field which was parallel to the flow velocity and to the axis of the solenoid.

The slurries were prepared by mixing approximately 320 g of sediment into 9 L of sea water. The slurry was filtered to remove coarse extraneous material from the slurry such as sticks and grass. This extraneous material was not radioactive. Prior to the start of each experiment, air is removed from the canister by filling the canister with sea water. The valve at the bottom of the canister was opened allowing the sea water to flow out at a controlled rate followed smoothly by the slurry. When slurry vessel was almost empty, sea water was poured into the canister so that after all the slurry had passed through, the canister remained full of sea water. This procedure prevented water-air interfaces passing through the canister which are extremely effective at stripping the magnetics from the matrix even in the presence of the applied magnetic field. After the slurry had been displaced from the canister by the additional sea water and collected for analysis, the magnetic field was switched off. The canister was then flushed with sea water in order to remove the magnetics extracted from the slurry and retained by the matrix. After the magnetics had been collected, both the slurry passing through and the magnetics were flocculated, filtered, dried, weighed and sent for chemical analysis. The results are shown in Table 2(a) and (b).

TABLE 2(a). The results of high gradient magnetic separation applied to radioactive sediments from the estuary of the River Esk, Cumbria, England. Velocity through the matrix 4 cm/s and magnetic field 4 Tesla.

FEED		MAGNETICS		EFFLUENT	
MASS %	ACTIVITY %	MASS %	ACTIVITY %	MASS %	ACTIVITY %
100	100	4	90	96	10
100	100	5	84	95	16

TABLE 2(b). The results of high gradient magnetic separation applied to radioactive sediments from the estuary of the River Esk, Cumbria, England. Velocity through the matrix 8 cm/s and magnetic field 4 Tesla

FEED		MAGNETICS		EFFLUENT	
MASS %	ACTIVITY %	MASS %	ACTIVITY %	MASS %	ACTIVITY %
100	100	2	50	98	50

As shown in Table 2(a) and (b), HGMS gives a good recovery of the radioactive material at a high concentration, for the slurry velocities of 4 cm/s, in which the concentration of the radioactivity in the magnetics is between 216 and 100 greater than in the non-magnetic fraction. When the slurry velocity is increased to 8 cm/s, the ratio of the concentrations falls to 49. At the low velocity, the recovery of radioactivity into the magnetics is greater than 84%

DISCUSSION OF THE RESULTS

Considering the results of the application of magnetic separation to the sediments from the estuary of the Esk, it can be said that magnetisable particles exist in these sediments and, further, that the radioactive heavy metals present are predominantly associated with these particles. This is amply illustrated by the fact that 90% of the radioactivity can be removed with the magnetic fraction. An application of the theory of magnetic separation to these results is of interest.

A theory of capture of paramagnetic particles was developed previously based on the interaction between a paramagnetic particle carried by a fluid past a ferromagnetic wire magnetised by a uniform applied magnetic field H_o (Watson, 1973). In the analysis it was found that the overall extraction efficiency of a filter composed of these wires, depended on the ratio V_m/V_o, where V_o is the background velocity of the fluid. V_m, the magnetic velocity, is given by equation (1) below. When $V_m \geq V_o$ the capture of the particles is very strong and the separator behaves like a filter.

$$V_m = (2/9)\{\mu_o \chi b^2 M_s H_o/\eta a\} \qquad (1)$$

where M_s and H_o in A/m, all lengths are in meters and η is in units of Pascal-second. (Water at 20°C has $\eta = 10^{-3}$ Pascal-second), $\mu_o = 4\pi\,(10)^{-7}$ Henry/m, the magnetic constant and the particles have an radius b and magnetic susceptibility χ

In the remainder of the paper χ will be used to denote the effective magnetic susceptibility of a composite particles with an effective radius b expressed in μm. If $V_m/V_o = 1$ then capture is strong. As as much as 84-90% of the radioactivity is captured at V_o=4 cm/s so V_m must be greater than V_o. Setting V_m = 4 cm/s, equation (1) gives a lower estimate for χb^2. With $\mu_o M_s = 1.7$ Tesla, $\mu_o H_o = 4$ Tesla, $\eta = 10^{-3}$ Pascal-s and the wire radius a = 75 μm then for 84-90% of the particles $\chi b^2 > 2.5(10^{-3})$. Similarly, from the high velocity results 50% of the particles $\chi b^2 > 5(10^{-3})$

In Figure 1 using equation (1), χb^2 is plotted versus the ratio of the mass of US_2 (Grindler, 1962) and the mass of the microorganism. The magnetic susceptibility of US_2 is $1.2(10^{-3})$(Weast and Astle, 1978). Figure 1 clearly indicates the values of χb^2 are not large enough to allow significant capture of individual microorganisms at these velocities and illustrates convincingly the unlikelihood of individual microorganism existing in sediments and showing they occur in clusters. Nevertheless using Figure 1 and the observed values of χb^2 the minimum size of the clusters which are easily captured can be estimated. From the flow velocity of 8 cm/s, $\chi b^2 > 5(10^{-3})$ and in addition assume that uptake mass to bacterial mass is 2 and using a density for US_2 of 10^4 kg/m^3, then $\chi b^2 = 1.7(10)^{-3}$ and, assuming the susceptibility of the microorganism can be neglected, the effective susceptibility χ is approximately $2.4(10)^{-4}$ meaning that for 50% of the clusters b >4.6 μm and for the case with the velocity 4 cm/s case, 84-90% of the clusters must have b > 3.2 μm.

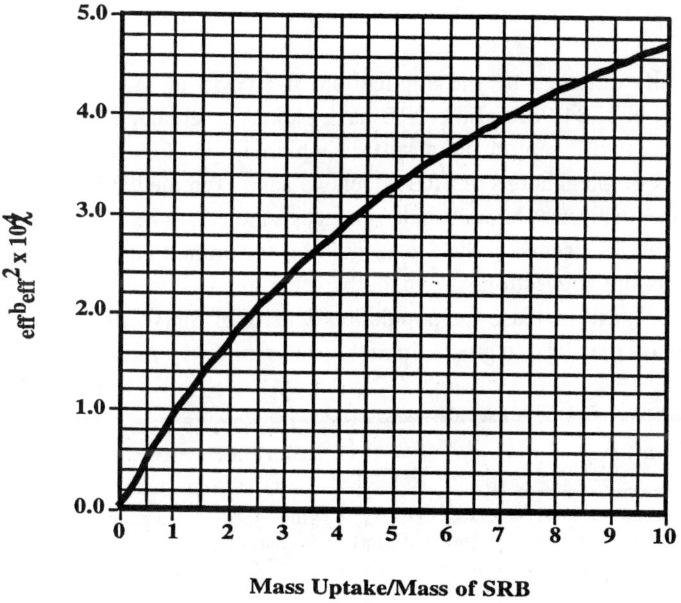

FIGURE 1. The product of the effective magnetic susceptibility and the square of effective particle radius $\chi_{eff}b_{eff}^2$ for the SRB loaded with US_2 versus the weight of US_2/ wet weight of SRB.

It has been found that in the sediments of the Esk the uranium concentration is 1.4 $\mu g/cm^3$. If we assume that each SRB captures 2 times its own mass and if 84-90% of the activity is removed, then the number density of the SRB in the sediment can be estimated. If the mass of SRB is approximately. $1.25.(10^{-15})$kg then each microorganism carries $2.5(10^{-15})$kg of US_2 therefore the number of SRB $/cm^3$ of sediment is approximately $5.6(10^5)$. This is rather a small number compared with the densities of microorganisms within a chemostat which are of the order of 10^9 SRB/cm^3. It is important therefore, in any particular sediment which must be treated to discover the limiting nutrient of the sediment and try to add it to the sediment, together with ferrous sulfate sufficiently long before dredging to allow the population density of the SRB to increase and more heavy metal immobilisation to take place.

CONCLUSIONS

(1) Simple adsorption to the cell walls of organic material will not occur for concentrations of uranium in sea water much below mg/L levels.

(2) Precipitation of uranium as a sulfide by SRB will occur down to less than 1 mg/L level (Watson et al., 2001) and can therefore be considered as a mechanism for the precipitation of uranium from sea water in the anaerobic regions of the sludge.

(3) The results obtained by the application of HGMS at low velocity can be explained can be if 84-90% of the SRB have picked up greater than 2 times their own wet weight of uranium as a sulfide which is very reasonable from our previous work on populations of SRB which we have examined individually.

(4) Work done on river water and sludges suggests there is to be expected an ample number of naturally occurring SRB to account for the uranium concentration in the Esk sediments but suggestions are made here for the addition of reagents to the sediment which will increase the population density of SRB within the sediment and this increase may lead to greater immobilisation of heavy metals.

Text Citations	References
Arakaki and Morse (1993)	Arakaki, T. and J. W. Morse 1993. "Coprecipitation and adsorption of Mn(II) with mackinawite (FeS) under conditions found in anoxic sediments." *Geochimica et Cosmochimica Acta 57*: 9-14.
Ellwood et al. (1992)	Ellwood, D.C., M.J. Hill and J.H.P. Watson 1992. "Pollution Control using Microorganisms and Magnetic Separation" In J.C. Fry, G.M. Gadd, R.A. Herbert, C.W. Jones and I.A. Watson-Craik(Eds.), *Microbial Control of Pollution*, Cambridge University Press, Cambridge, UK.
Grindler (1962)	Grindler, J.E. 1962. *Radiochemistry of Uranium,* National Academy of Sciences, Nuclear Science, Report NAS-NS 3050
Hamilton (1981)	Hamilton, E. I. 1981. "Alpha-particle radioactivity of hot particles from the Esk estuary." *Nature 290*(April 23): 690-693.
Hamilton and Clarke (1984)	Hamilton, E. I. and K. R. Clarke 1984. "The recent sedimentation history of the Esk estuary, Cumbria, UK. : The application of radiochronology." *The Science of the Total Environment 35*: 325-386.
Kolm (1971)	Kolm, H. H. 1971. U.S. Patent: No. 3 567 026.
Revis et al. (1991)	Revis, N.W., J. Elmore, H.Edenborn, T.Osborne, G.Holdsworth, C.Hadden and A.King 1991. "Immobilization of Mercury and other heavy metals in soil, sediment, sludge and water by sulfate-reducing bacteria" In H.M. Freeman and P.R. Sferra (Eds.), *Biological Processes*, Freeman. *3*: 97 - 105.
Weast and Astle (1978)	Weast, R.C. and Astle, M.J. (Eds.). 1978. *Handbook of Physics and Chemistry,* pp. E-126.CRC Press, Inc.: West Palm Beach, Florida

Watson (1973)	Watson, J.H.P. 1973. "Magnetic Filtration." *J.Appl.Phys.* *44*(9): 4209 - 4213.
Watson (1998)	Watson, J.H.P. 1998. "Superconducting magnetic separation" In B. Seeber (Ed.), *Handbook of Applied Superconductivity, 2*: 1371-1406, Institute of Physics Publishing, Bristol, UK.
Watson and Ellwood (1994)	Watson, J.H.P. and D.C. Ellwood 1994. "Biomagnetic separation and extraction process for heavy metals from solution." *Minerals Engineering* *7*(8): 1017-1028.
Watson et al. (1995)	Watson, J.H.P., D.C. Ellwood, Qixi Deng, S. Mikhalovsky, C.E.Hayter and J.Evans 1995. "Heavy metal adsorption on bacterially produced FeS." *Minerals Engineering* *8*(10): 1097-1108.
Watson et al. (2000)	Watson, J.H.P., B.A. Cressey, A.P. Roberts, D.C. Ellwood, J.M. Charnock and A.P. Soper 2000. "Structural and magnetic studies on heavy-metal-adsorbing iron sulfide nanoparticles produced by sulfate-reducing bacteria." *J. of Magnetism and Magnetic Materials* *214*(1-2): 13 -30.
Watson et al. (2001)	Watson, J.H.P., I.W. Croudace, P.E. Warwick, P.A.B. James, J.M. Charnock and D.C. Ellwood 2001. "Adsorption of radioactive metals by bacterially-produced strongly magnetic iron sulfide nanoparticles." *Separation Science and Technology, 36*, (12) In Press.

IN SITU TREATMENT OF METALS-CONTAMINATED GROUNDWATER USING PERMEABLE REACTIVE BARRIERS

David J.A. Smyth (University of Waterloo, Waterloo, Ontario, Canada)
D.W. Blowes (University of Waterloo, Waterloo, Ontario, Canada)
S.G. Benner (Stanford University, Stanford, California, USA)
C. J. Ptacek (University of Waterloo, Waterloo, Ontario, Canada)

ABSTRACT: Laboratory-scale tests and field demonstrations conducted during the past decade have shown that many metals and other inorganic contaminants can be removed from groundwater using permeable reactive barrier (PRB) systems. PRBs provide passive interception of contaminated groundwater and promote the *in situ* treatment of dissolved contaminants by inducing changes to geochemical conditions in the subsurface. A PRB system installed in 1995, to remediate groundwater impacted by acid-mine drainage near Sudbury, Ontario is treating groundwater containing several thousands of milligrams per liter of sulfate and as much as 2,000 mg/L of ferrous iron. This PRB, which contains plant-based compost, limestone and gravel, has enhances microbially mediated sulfate reduction removes iron and other metals through the precipitation metal-sulfides. The groundwater has been modified from net acid producing to net acid consuming, and the potential impacts on the quality of receiving surface water have been reduced. A field trial of a similar PRB has indicated effective removal of dissolved metals, including Cu, Zn, Cd, Co, Ni, and Pb, in sulfate-rich groundwater at a metal-processing facility in Vancouver, BC. PRBs containing zero-valent iron have demonstrated excellent removal of electroactive metals from groundwater. At Elizabeth City, NC, up-gradient chromium (VI) concentrations of as much as 5 mg/L have been decreased to <0.05 mg/L within a zero-valent iron PRB. The PRB was installed in 1996, and continues to meet objectives for treatment and hydraulic performance. Contaminant removal is achieved as a reduced-mineral precipitate or co-precipitate. Laboratory and field tests using zero valent iron have also shown excellent of other metals such as arsenic, selenium, technetium and uranium at influent concentrations of several milligrams per liter and flow velocities of the order of a foot per day. Laboratory batch and column tests with iron-based media have also indicated excellent potential for the removal of mercury at concentrations of as much as 1 mg/L or higher from groundwater.

INTRODUCTION

Permeable reactive barriers (PRBs) are being considered as an alternative for plume control and remediation at many contaminated sites. Permeable reactive barriers have two essential functions. The barrier must facilitate the interception or capture of a contaminant plume at some distance down-gradient of the source, and provide treatment or removal of contaminants to acceptable levels. Treatment is achieved within or down-gradient of the barrier by physical, chemical or biological processes. Potential treatment technologies for application in reactive barriers have been identified for a wide range of organic and inorganic contaminants (eg. Blowes et al., 2000). This paper provides background information about the use of reactive materials to remove metal and other inorganic contaminants from groundwater at both the field and laboratory scale.

PRBS FOR GROUNDWATER IMPACTED BY ACID-MINE DRAINAGE

Acid-mine drainage (AMD) is caused by the oxidation of residual sulfide minerals in mine waste rock and tailings. The oxidation of sulfide minerals including pyrite in the presence of oxygen and water in the vadose zone of tailings or waste-rock piles can be represented by the reaction:

$$2FeS_{2(s)} + 7O_2 + 2H_2O \rightarrow 4SO_4^{2-} + 2Fe^{2+} + 4H^+ \qquad (1)$$

AMD effluent is acidic and contains elevated concentrations of sulfate, ferrous iron (Fe(II)) and dissolved trace metals. The oxidation of Fe(II) to Fe(III) occurs upon discharge of AMD to receiving surface water. This generates additional acidity and results in the precipitation of ferric oxyhydroxides consistent with the reaction:

$$4Fe^{2+} + O_2 + 10\,H_2O \rightarrow 4Fe(OH)_{3(s)} + 8H^+ \qquad (2)$$

AMD discharge to surface water can have adverse impacts on aquatic ecosystems by lowering the pH and enhancing the mobility of trace metals in surface water.

Conventional control for AMD has involved the collection of the effluent followed by treatment using lime neutralization and settling. Within the past decade, the interception and treatment of AMD–impacted groundwater using PRBs has been suggested as a viable component of environmental mine-waste management programs (Benner et al., 1997; Blowes et al., 2000; Blowes and Ptacek, 1994; Smyth et al., 2001; Waybrant et al., 1998).

The PRB provides a source of carbon to enhance bacterially mediated sulfate reduction and promotes the precipitation of low solubility metal sulfides. The consumption of simple organic carbon in the sulfate reduction process produces hydrogen sulfide and alkalinity, and can be represented by:

$$SO_4^{2-} + 2CH_2O \rightarrow H_2S + 2HCO_3^- \qquad (3)$$

The presence of dissolved metals and hydrogen sulfide can result in the precipitation of low solubility metal sulfides:

$$Me^{2+} + H_2S \rightarrow MeS + 2H^+ \qquad (4)$$

Me^{2+} denotes a metal such as Fe, Cd, Ni, Co, Cu, Zn, As or Zn. If the PRB remains below the water table, the sparingly soluble metal sulfide precipitates are expected to be stable for the very long term.

Nickel Rim Mine. At the Nickel Rim Mine near Sudbury, Ontario, AMD-impacted groundwater has migrated from inactive pyrrhotite-rich tailings through an aquifer and discharged to a lake approximately 160 m from the tailings impoundment. The aquifer is fine-grained alluvial sand in a bedrock valley. The horizontal component of groundwater velocity in the vicinity of the PRB is of the order of 15 meters per year. The pH of the groundwater ranges between 4 and 6, and contained 1000 to 5000 mg/L sulfate and 200 to 2500 mg/L iron.

In 1995 Benner et al. (1997, 1999) installed a PRB system that was approximately 20 m long, 3.6 m deep and 4 m thick in the direction of groundwater flow across the shallow aquifer. The reactive materials consisted of a mixture of municipal compost (20% by volume), leaf mulch (20%), wood chips (9%), limestone (1%) and gravel (50%). Zones of sand approximately 1 m thick were placed on both the up- and down-gradient faces of the PRB. The PRB system was installed in a backhoe-excavated trench and, was capped by clay to minimize direct infiltration of surface water into the reactive materials. Benner et al. (1999) suggested that the PRB theoretically contained sufficient organic carbon to sustain sulfate-reduction and removal of contaminants for more than 100 years. Only a portion of the organic carbon, however, is likely available for sulfate-reducing bacteria.

Water quality has been improved significantly within and down-gradient of the PRB. The concentrations of sulfate, iron and other metals have been decreased by the PRB, and alkalinity has been increased. A zone of higher flow in which the residence time of groundwater is approximately 60 days was identified within the central portion of the PRB. Throughout the remainder of the PRB, the residence time of groundwater has been as much as 165 days (Benner et al., 1997: 1999). The variation in flow is a consequence of variability in groundwater flow in the aquifer and heterogeneity of hydraulic conductivity within the PRB. In the first year, alkalinity increased by as much as 2,500 mg/L (expressed as mg/L $CaCO_3$). During the initial year, the concentrations of sulfate decreased by as much as 3,000 mg/L, iron decreased by as much as 1,300 mg/L and nickel decreased by as much as 30 mg/L within the PRB (Benner et al., 1997). Using data for the first three years of performance in proximity to the central zone of higher flow, vertically averaged concentrations of sulfate and iron entering the PRB were approximately 2,600 mg/L (27 mmol L^{-1}) and 560 mg/L (10 mmol L^{-1}) respectively. Immediately down-gradient of the PRB, the vertically averaged concentrations increased from less than 1,630 mg/L (17 mmol L^{-1}) to more than 2,200 mg/L (23 mmol L^{-1}) for sulfate between the first and third years, and from less than 50 mg/L (1 mmol L^{-1}) to approximately 335 mg/L (6 mmol L^{-1}) for iron over the same period (Benner et al., 1999). The alkalinity increase was initiated in the up-gradient sand zone as a consequence of calcite dissolution, and was accompanied by an increase in pH to approximately 6. Most of the observed increases of alkalinity, however, occurred in the reactive organic carbon material as a consequence of sulfate reduction, and resulted in the generation of capacity for the groundwater to consume acidity in the down-gradient portion of the aquifer and in the receiving lake.

Evidence for sulfate reduction was provided by the presence of sulfate-reducing bacteria, the presence of dissolved sulfide (17 mg/L) and isotopic enrichment of ^{34}S in remnant sulfate (Benner et al., 1999). Iron mono-sulfide solids including mackinawite were identified in cores from within the PRB using scanning electron microscopy with energy-dispersive x-ray analysis (SEM-EDX) (Herbert et al., 2000). The total mass of solid-phase sulfide precipitates accumulating in the barrier are consistent with the calculated mass of S and Fe removal (Benner et al., 2000). Geochemical-speciation modelling also confirmed that the groundwater was saturated with respect to mackinawite and amorphous iron sulfide (Benner et al., 1999).

Spatial and temporal variations in the extent of sulfate and metals removal were caused by heterogeneity of hydraulic conductivity, loss of reactivity with time, and fluctuations of temperature within the PRB. There is a zone of less removal of sulfate and iron and less production of alkalinity in the middle portion of the PRB parallel to the direction of groundwater flow. Benner et al. (1997; 1999) attributed this pattern to the presence of higher hydraulic conductivity materials and higher rates of groundwater flow through this central zone. With lower residence time, less complete sulfate and iron removal occurs. The preferential flow of water through the central zone may also flush labile organic carbon from this part of the PRB, and decrease the reactivity of this portion of the barrier. Benner et al. (2000) also inferred higher levels of sulfate and iron removal in the late summer months relative to those in the post-winter months over four years of monitoring, and attributed this to warmer groundwater temperatures during the late summer and fall months.

Industrial Site, Vancouver. Sulfide-rich minerals were introduced to shallow soils from the historical storage of ore concentrate at a former industrial property in Vancouver, British Columbia. Most of the iron sulfide minerals had been removed prior to storage in the ore-concentration process, but the oxidation of the other sulfide minerals such as chalcopyrite and sphalerite has caused elevated concentrations of sulfate and metals including copper, zinc, cadmium, nickel and lead in the shallow groundwater. The site is adjacent to a tidally influenced inlet, and the surficial aquifer is composed of heterogeneous fill and alluvial sand and gravel, containing some cobbles and boulders (McGregor et al., 2000).

In 1997 a demonstration-scale PRB was installed. The test wall was constructed using guar-gum slurry-trenching techniques. The reactive materials consisted of a mixture of leaf compost (approximately 15% by volume), limestone (approximately 1%) and pea gravel (approximately 84%), and were placed using a clamshell bucket. The PRB was approximately 10 m in length, 6 m in depth and 2.5 m in thickness in the direction of groundwater flow. Using hydraulic head levels at low tide and data from single-well hydraulic-conductivity tests, McGregor et al. (2000) estimated the residence time of water within the PRB to be three days.

Monitoring over a three-year period indicated excellent attenuation of the metals of concern within the PRB. Locally, the concentration of sulfate entering the PRB exceeded 1,500 mg/L, and as much as several hundred of milligrams per liter was removed by sulfate reduction. Evidence for sulfate reduction included the development of low Eh conditions, the generation of dissolved hydrogen sulfide concentrations of as much as 1.7 mg/L and the production of alkalinity within the PRB. Geochemical modelling suggested that conditions within the PRB favored the precipitation of iron and copper sulfide minerals. After two years, average contaminant concentrations in groundwater decreased from 0.0159 to 0.0001 mg/L for cadmium, from 4.510 to 0.0077 mg/L for copper, from 0.118 to 0.0065 mg/L for nickel, from 2.396 to 0.082 mg/L for zinc and from 0.0038 to 0.0019 mg/L for lead in multi-level wells up- and down-gradient of the test PRB (McGregor et al., 2000).

In advance of full-scale implementation, column tests were performed in the laboratory using similar reactive materials and site groundwater. The limestone composition was increased from 1 to 5%, and the volume of gravel decreased by the same amount. The full-scale PRB was installed at the site in late 2000 and early 2001, and it consisted of approximately 400 linear meters of reactive wall consisting of compost, gravel and limestone. In zones of higher contamination, a second reactive wall that included zero-valent iron in the reactive mixture was also installed. The initial performance data will be collected within several months of the full-scale PRB installation.

ZERO-VALENT IRON AND OTHER REACTIVE MATERIALS FOR PRB TREATMENT OF METALS

Blowes et al. (1999ab) provide detailed documentation of the installation and performance of the zero-valent iron PRB that was installed at Elizabeth City, NC, to remediate chromium (VI) and several chlorinated solvents in groundwater. Installed in 1996 in co-operation with the United States Environmental Protection Agency and the United States Coast Guard, the PRB is a permeable wall approximately 46 m long, 7 m deep and 0.6 m in thickness in the direction of groundwater flow. Influent concentrations of Cr(VI) have been as high as 5 mg/L, but removal of chromium to non-detectable levels (<0.0025 mg/L) has been achieved since installation of the PRB. Cr(VI) was not detected in multi-level monitors approximately 0.01 m into the zero-valent iron zone. Eh conditions are strongly reducing and pH increases from approximately 6.5 up-gradient to between 9 and 11 within the PRB. Laboratory investigations indicated the removal mechanism for Cr(VI) is abiotic and is achieved by reduction to Cr(III) and the precipitation as Cr(III) oxyhydroxide or co-precipitation with iron oxyhydroxide minerals on the iron-grain surfaces (Blowes et al., 1997). The monitoring program identified the decrease in concentrations of alkalinity, calcium, magnesium and manganese as groundwater migrated through the wall. Geochemical and reactive transport modelling suggests that the precipitation of secondary carbonate minerals was responsible for these observed decreases in concentrations, and that the secondary precipitates may contribute to declining hydraulic conductivity and reactivity of the zero-valent iron with time. Reactive-transport modelling indicated that the porosity of the entry zone of the PRB may decrease from the initial 0.5 to 0.36 after 20 years (Mayer, 1999).

In addition to Cr(VI), pilot and full-scale PRB systems have been installed to treat technetium (Tc(VII)), uranium (U(VI)) and molybdenum (Mo(VI)) in groundwater (Blowes et al., 2000; Naftz et al., 2000). Although other media have been evaluated, zero-valent iron has provided excellent conditions for reductive precipitation of these electroactive metals. At the laboratory scale, the removal of arsenic, selenium and mercury has also been demonstrated. McRae et al. (1999) observed rapid removal of arsenic and selenium from water in static batch tests using zero-valent iron. For arsenic, As(V) concentrations decreased from 1 to less than 0.003 mg/L in a two-hour period. Similar removal was observed for As(III) and As(V)/As(III) solutions. For selenium, the concentration of Se(VI) decreased from 1.5 to less than 0.1 mg/L in several hours. In a column containing 10 wt. % iron in silica sand, McRae et al. (1999) observed the removal of approximately 0.5 mg/L of both As(III) and As(V) in simulated groundwater. Effluent concentration of total As was less than 0.003 mg/L for more than 320 pore volumes. On the basis of energy dispersive x-ray (EDX) analysis and x-ray photoelectron spectroscopy of iron surfaces, McRae et al. (1999) suggest that the removal mechanism for arsenic is reductive precipitation on the grain surfaces and in secondary precipitate coatings of the grains.

Using contaminated groundwater from industrial sites containing arsenic at concentrations of as much as 15 mg/L and simulated groundwater velocities of tens of centimeters per day, zero-valent iron mixtures in columns have yielded total arsenic concentrations of less than 0.002 mg/L in the effluent. Similar tests with selenium-contaminated groundwater have indicated that removal of selenium from input concentrations of 2 mg/L to less than 0.001 mg/L in the effluent can be achieved using zero-valent iron. These column tests have shown excellent removal potential for the electroactive metals, but have also indicated that the formation of secondary precipitates such as carbonates may have significant impact on the performance and potential longevity of iron-based PRBs.

Preliminary laboratory batch testing has indicated that the removal of Hg from contaminated groundwater also occurs in the presence of zero-valent iron. The testing was conducted using groundwater from a mine site in the western United States that water was amended with an inorganic mercury salt to increase concentrations prior to testing. Initial concentrations of mercury were several tens of mg/L in one series of tests, and of the order of 0.2 mg/L in the other. The solid materials in the reaction vessels consisted of a 50:50 mixture of zero-valent iron and aquifer sand, and the mass of iron was approximately 0.1 of the mass or volume of water. In the high-concentration tests, Hg concentrations decreased from several tens of mg/L to approximately 0.1 mg/L within 3 to 6 days and to less than 0.03 mg/L within tens of days. In the second series of tests, the concentration of Hg decreased from approximately 0.2 mg/L to approximately 0.02 mg/L within 4 to 6 hours. Geochemical modeling suggested that mercury was removed from solution in conjunction with the precipitation of other mineral phases, most likely ferrihydrite or related iron oxyhydroxides. Further laboratory testing has also confirmed removal of mercury to concentrations of less than 0.000025 mg/L can be achieved in by zero-valent iron in a dynamic column test with approximately 0.100 mg/L Hg in sulfate-rich laboratory-synthesized input solution. The test operated at a simulated groundwater flow velocity of approximately 0.02 m per day. A profile of mercury concentration with distance in the column is consistent with those for arsenic, selenium and other metals in previous testing, and confirms removal of mercury within a thin zone at the up-gradient interface of the zero-valent iron.

REFERENCES

Benner, S.G., D.W. Blowes and C.J. Ptacek. 1997. "A Full-Scale Porous Reactive Wall for Prevention of Acid Mine Drainage". *Ground Water Monitoring and Remediation, 17*(4): 99-107.

Benner, S.G., D.W. Blowes, W.D. Gould, R.B. Herbert Jr., and C.J. Ptacek. 1999. "Geochemistry of a Reactive Barrier for Metals and Acid Mine Drainage". *Environmental Science and Technology. 33*(16): 2793-2799.

Benner, S.G., D.W. Blowes, C.J. Ptacek and U. Mayer. 2000. "A Permeable Reactive Wall for Treatment of Mine Drainage: Long-Term Results". In*: ICARD 2000, Proceedings from the Fifth International Conference on Acid Rock Drainage, (Society of Economic Geologists), Denver, Colorado, May 21-24.* pp.1221-1225. Society for Mining Metallurgy and Exploration: Littleton, CO.

Blowes, D.W. and C.J. Ptacek. 1994. *System for Treating Contaminated Groundwater.* U.S. Patent 5,362,394 filed March 3, 1992, issued Nov. 8, 1994, Canada patent 2,062,204, filed March 3, 1992, issued July 7, 1998, Europe Patent 92103559.8, filed March 2, 1992, allowed, January 2000.

Blowes, D.W., C.J. Ptacek and J.L. Jambor. 1997. "In-Situ Remediation of Cr(VI) Contaminated Groundwater Using Permeable Reactive Walls: Laboratory Studies". *Environmental Science and Technology. 31*(12): 3348-3357.

Blowes, D.W., Puls, R.W., Gillham, R.W., Ptacek, C.J., Bennett, T.A., O'Hannesin, S.F., Hanton-Fong, C.J., Paul, C.J. and Bain, J.G. 1999a. *An In-Situ Permeable Reactive Barrier for the Treatment of Hexavalent Chromium and Trichloroethylene in Ground Water: Volume 1 Design and Installation.* United States Environmental Protection Agency, Cincinnati, OH, Report EPA/600/R-99/095a.

Blowes, D.W., Puls, R.W., Gillham, R.W., Ptacek, C.J., Bennett, T., Bain, J.G., Hanton-Fong, C.J., and Paul, C.J. 1999b. *An In-Situ Permeable Reactive Barrier for the Treatment of Hexavalent Chromium and Trichloroethylene in Ground Water: Volume 2 Performance Monitoring.* United States Environmental Protection Agency, Cincinnati, OH, Report EPA/600/R-99/095b.

Blowes, D.W., C.J. Ptacek, S.G. Benner, C.W.T. McRae and R.W. Puls. 2000. "Treatment of Dissolved Detals and Nutrients Using Permeable Reactive

Barriers". *Journal of Contaminant Hydrology. 45* (1): 123-137.

Herbert Jr., R.B., Benner, S.G., and Blowes, D.W. 2000. "Solid Phase Iron-Sulfur Geochemistry of a Reactive Barrier for Treatment of Acid Mine Drainage". *Applied Geochemistry*: *15*: 1331-1343.

McRae, C.W.T., Blowes, D.W., and Ptacek, C.J. 1999. "*In Situ* Removal of Arsenic from Groundwater Using Permeable Reactive Barriers: A Laboratory Study". *Sudbury '99- Mining and the Environment II Conference, September 13-17, 1999, Sudbury, Ontario*: pp. 601-609.

Mayer, K.U. 1999. *A Numerical Model for Multi-Component Reactive Transport in Variably Saturated Porous Media*. Ph. D. Thesis, Department of Earth Sciences, University of Waterloo, Waterloo, Ontario. 286 p.

McGregor, R.G., Blowes, D.W., Ludwig, R. and Choi, M. 2000. "The Use of an *In Situ* Porous Reactive Wall to Remediate a Heavy Metal Plume". In: *ICARD 2000, Proceedings from the Fifth International Conference on Acid Rock Drainage, (Society of Economic Geologists), Denver, Colorado, May 21-24*: pp. 1227-1232. Society for Mining Metallurgy and Exploration, Littleton, CO.

Naftz, D.L., C.C. Fuller, J.A. Davis, M.J. Piana, S.J. Morrison, G.W. Freethey and R.C. Rowland. 2000. "Field Demonstration of Permeable Reactive Barriers to Control Uranium Contamination in Groundwater". In: *Chemical Oxidation of and Reactive Barriers C2-6, Proceedings of the Second International Conference on Remediation of Chlorinated and Recalcitrant Compounds, Monterey, CA. May 22-25*: pp. 281-290. Battelle Press, Columbus, OH.

Smyth, D.J.A, D.W. Blowes S.G. Benner and A.M.H. Hulshof. 2001. "*In Situ* Treatment of Groundwater Impacted by Acid Mine Drainage Using Permeable Reactive Materials". In: *Tailings and Mine Waste '01 (Proceedings of the Eighth International Conference, Fort Collins, Colorado, January 16-19*: pp. 313-322. A.A. Balkema Publishers, Rotterdam.

Waybrant, K.R., Blowes, D.W., and Ptacek, C.J.. 1998. "Prevention of Acid Mine Drainage Using *In Situ* Porous Reactive Walls: Selection of Reactive Mixtures". *Environmental Science and Technology 32*(13): 1972-1979.

REMOVAL OF HEAVY METALS USING A ROTATING BIOLOGICAL CONTACTOR

Shauna C. Costley (University of Natal, Pietermaritzburg, South Africa)
F. M. Wallis (University of Natal, Pietermaritzburg, South Africa)
M. D. Laing (University of Natal, Pietermaritzburg, South Africa)

ABSTRACT: A laboratory-scale investigation of the use of a rotating biological contactor (RBC) to treat heavy metal polluted wastes was conducted. The effect of disc rotation speed, flow rate on metal removal as well as the sorption capacity, long term use and rejuvenation capacity of the biofilm was investigated. The system was found to operate successfully at 10 rpm with a 24h hydraulic retention time (HRT). A metal ion affinity series: of Cu >> Zn > Cd occurred in all experiments conducted. The system was shown to operate efficiently over 84d and metals were successfully recovered using a dilute acid. The desorption procedure employed did not adversely affect the sorption capacity of the biofilm, evident in the continued uptake of all three metals after metal desorption. The results obtained suggest that the RBC can be used successfully to treat metal contaminated wastes.

INTRODUCTION

Heavy metal releases to the environment have been increasing continuously as a result of industrial activities and technological development posing a significant threat to the environment and public health because of their toxicity, accumulation in the food chain and persistence in nature (Braune, 1994). This together with their wide use has highlighted the importance of the efficient removal of metal ions from wastewater resulting in a widely studied research area where a vast number of technologies have been developed over the years, including ion exchange, precipitation, filtration and membrane systems (Belhateche, 1995). All these approaches have their inherent advantages and limitations including high cost particularly when dealing with large volumes, as well as the difficulties encountered in the solid waste generated (Kratochril et al., 1997). Biosorption, on the other hand, employs biological material to remove heavy metals from solution by a variety of physical, chemical and biological mechanisms (Gadd, 1988), and does not entail high operational costs and many potential sources of suitable biological material are cheaply and readily available (Hobson and Poole, 1988).

Although metal removal from industrial effluents by means of biosorption has been studied extensively (Gadd, 1990; Volesky and Holan, 1995), very few studies have been carried out examining metal removal using a RBC. The objective of this study was to determine the feasibility of using a RBC to treat a synthetic wastewater containing cadmium(Cd), copper (Cu) and zinc (Zn).

MATERIALS AND METHODS

A laboratory scale RBC as described by Costley and Wallis (2000) was built and operated as a single-stage bioreactor (Figure 1).

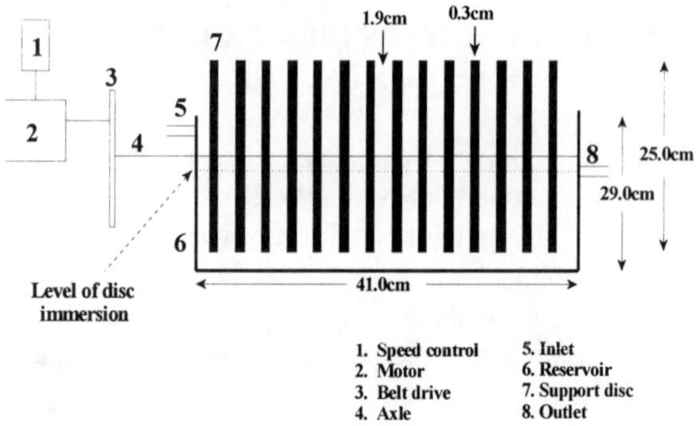

FIGURE 1. Diagram of the Rotating Biological Contactor (RBC).
(Source:Costley and Wallis, 1999).

Rotation Speed. After development of a suitable biofilm 3 different rotation speeds (3, 15 and 25 rpm) were implemented and their affect on metal removal from a synthetic effluent (10% (v/v) nutrient broth spiked with 100mg/l Cd, Cu and Zn) determined as described by Costley and Wallis (1999).

Hydraulic Retention Time (HRT). An experiment was conducted as outlined in Costley and Wallis (2000) to investigate the influence of 4 HRTs (3, 6, 12 and 24h) on metal removal. All subsequent experiments were conducted using a HRT of 24h.

Sorption Cycle 1 (SC1). The RBC was drained, washed and set up to run continuously for a period of 12 weeks (equivalent to 84 cycles). Samples of both the influent and effluent were taken daily for AAS analysis using aVarian Spectr AA-200 Series Atomic Absorption Spectrophotometer.

Desorption Cycle 1 (DC1). After completion of SC1 the RBC was drained, washed, and filled with 0.1M hydrochloric acid (HCl) solution and run, batch mode, for a period of 3h. These steps were repeated once more where after the RBC was filled with fresh 0.1M HCl and run for 18h. The RBC was then filled with 0.5M HCl for 12h and then rinsed with distilled water for a period of 12h. Samples were taken after each water or acid wash for AAS analysis.

Sorption Cycle 2 (SC2). After completion of DC1 the experiment was repeated as for SC1.

Desorption Cycle 2 (DC2). After completion of SC2 the RBC was drained, washed, filled with 0.5M HCl and run, batch mode, for 3h. The acid was then drained and replaced with fresh acid solution and the RBC run for 12h where after it was rinsed with distilled water for 3h. Samples were taken as for DC1.

Sorption Cycle 3 (SC3). The experiment was repeated as for SC1 and 2.

Desorption Cycle 3 (DC3). The RBC was drained, washed and filled with 0.5M HCL and run for a period of 3h. These steps were repeated twice more and then the RBC was washed with distilled water for 3h. Samples were taken as for DC1.

RESULTS AND DISCUSSION

Rotation Speed. Metal removal rates for each of the 3 metals exhibited similar patterns at all 3 rotation speeds investigated (Costley and Wallis, 1999). Although sorption of all 3 metals occurred at each speed, the biofilm showed a definite selectivity for Cu, removal efficiencies averaging 92% in each run. A decrease in Cd concentration occurred immediately after contact of the biofilm with the synthetic effluent followed by an increase to levels above that initially added. The initial decline was most probably due to the initial rapid phase of sorption which is metabolically independent involving physical adsorption (Gadd, 1988). The subsequent increase in Cd concentration was most probably due to desorption and resolubilisation of previously sorbed Cd ions as well as the affect of competition for sorption sites introduced by the presence of Cu and Zn (Collins and Stotzky, 1989). Zn was accumulated by the biofilm but to a lesser extent than Cu. This may have been due to the presence of competing ions since Harris and Ramelow (1990) showed Zn binding to be affected by the presence of Cd and Cu, either alone or, more strongly, when in combination. Although removal rates didn't differ greatly between rotation speeds, higher rotation speeds were found to be associated with biofilm detachment and hence for all further experiments a rotation speed of 10rpm was implemented. (Motor limitations required rotation speeds no < 10rpm).

HRT. The contact time between metal and biomass is very important in determining metal removal efficiencies, insufficient residence times markedly decrease metal removal levels (Zhou and Kiff, 1991). From the results (Costley and Wallis, 2000) it was clear that the highest metal removal occurred when a 24h HRT was used. At shorter HRTs negative removal rates were recorded for both Cd and Zn possibly due to competition from other metal ions present (Collins and Stotzky, 1989). The low pH recorded (pH < 4) at HRTs < 12h may also have affected binding of these ions. At a low pH the presence of H^+ ions reduces surface binding and intracellular influx. The H^+ ions are able to out-compete other cations for binding sites and hence occupy many potential metal binding sites resulting in poor metal biosorption results (Gadd, 1988). At a 24h HRT the pH increased to between 5 and 5.5, a pH range more conducive to removal of each of the three metals, optimal removal of Cd (Sag and Kutsal, 1989), Cu (Panchanadikar and Das, 1993) and Zn (Harris and Ramelow, 1990) occurring at pH values > 5.

Sorption Cycle 1(SC1). The long-term sorption capacity of the biofilm is important if the system is to become a viable means of treating metal polluted wastes and compete successfully with physico-chemical means of treatment. The continued removal of all three metals over a 84d period confirmed the potential of a RBC for treating metal wastes (Figure 2).

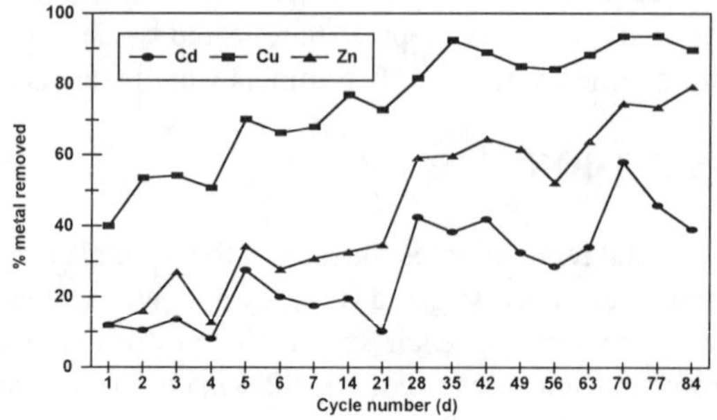

FIGURE 2. Percentage metals removed by the biofilm over 84d.

Although Cd and Zn removal was relatively low over the first 21d (≈15% and ≈25% respectively), a sizeable increase in the removal of both occurred after 28d contact, at which stage removal efficiencies of ≈42% and ≈59% respectively were recorded. Cu removal efficiencies remained high throughout the experimental period, increasing from ≈58% during the initial 7d contact to above ≈84% thereafter.

The relatively high removal levels recorded for Cd and Zn could be explained on the basis that both have well known metabolic functions and hence may be sorbed with greater efficiency than Cd, an ion with no known metabolic functions known to affect a series of cellular functions (Trevors et al., 1986). Consequently cells may exhibit resistance mechanisms to enable them to withstand high concentrations of such metals and hence result in low sorption capacities.

During SC1 there were two occasions (cycles 21 and 52) when the RBC stopped due either to power or motor failure. Neither interruption was longer than 12h and the supply of synthetic wastewater to the RBC was not interrupted. The biofilm, which became dehydrated during the stoppage, was quick to recover, removal rates of all 3 metals returning to within the range recorded prior to the stoppage within 24h of reactivation.

Both Cu and Zn removal was relatively stable over the last 28d of the experimental period suggesting that the biofilm may have reached an equilibrium as far as sorption of these metals was concerned. Cd removal however fluctuated over the last 28d possibly due to a number of reasons including: weak binding of Cd to the biofilm; saturation of binding sites of the biofilm with either Cu or Zn ions; repeated sorption-desorption of Cd ions; or alternatively the maximum sorption capacity of the biofilm for Cd had not yet been reached.

Desorption Cycle 1 (DC1). The ability to and ease of recovery of sorbed ions from the biofilm has a large influence on the potential viability of the treatment system. Dilute acids (< 1M) have the ability to remove sorbed ions from a biomass without causing much damage to the biomass (Cotoras et al., 1992). The protons made available by the HCl may dislodge metal ions from active sites, by making the bond between the metal and the biosorbent labile (Mathiekal et al., 1991).A comparison of the total metal sorbed and that consequently desorbed after contact with dilute HCl illustrated the effectiveness of dilute HCl as a desorbing agent (Table 1). More than ≈90% of each of the three metal ions sorbed was recovered suggesting accumulation by means of passive binding to the cell walls and associated extracellular material of the biofilm.

TABLE 1. Concentration of the total metals sorbed by the biomass (SC1) and subsequently desorbed (DC1).

Total Concentration (mg)	Cd	Cu	Zn
Metal sorbed	1918.8	4423.3	3467.3
Metal desorbed (%)	1723.2 (89.8)	4125.8 (93.3)	3260.8 (94.0)

Sorption Cycle 2 (SC2). The reusability of a biomass is a vital component of a potentially successful treatment mechanism (McHale and McHale, 1994). Removal efficiencies were erratic over the initial 28d but appeared to stabilize after approximately 35d (Figure 3).

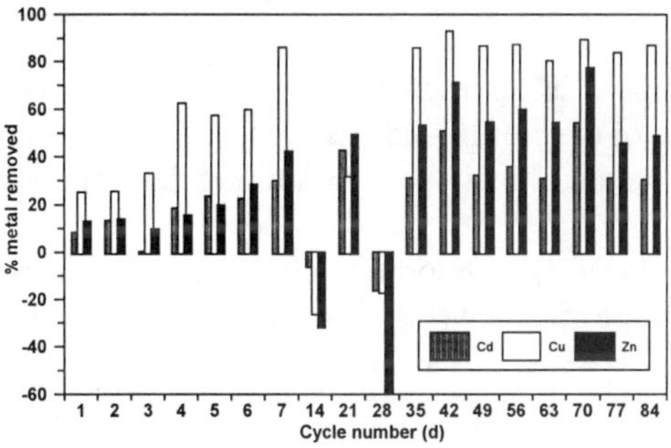

FIGURE 3. Percentage metals removed by the regenerated biofilm (SC2).

Average removals during the 84d period were 34%, 74% and 51% for Cd, Cu and Zn respectively compared with average values of 34% for Cd, 85% for Cu and 57% for Zn recorded during SC1 suggesting no long term adverse affect on the biofilm. Negative removal rates recorded on days 14 and 21 corresponded with interruptions to the operation of the RBC and were most probably due to the desorption of previously sorbed ions resulting in an increase in metal concentration in the outflow.

Desorption Cycle 2 (DC2). In order to decrease the time required for desorption of metals and hence to increase the economic viability of this system, a stronger form of HCl (0.5M) was used as the eluting agent and the entire DC was reduced from 48h to 18h. Desorption efficiencies did not differ greatly from those recorded during DC1. For example DC1 resulted in 93.3% recovery of Cu whereas DC2 resulted in 93.7% recovery. Similarly with Zn, DC1 recovered 94.0% whereas DC2 recovered 96.4%. Cd recovery appeared to have diminished in DC2, however closer inspection showed that the desorption efficiencies did not differ greatly, DC1 recovered 98.8% whilst DC2 recovered 83.2%. It must be noted that although the recovery efficiencies did not differ to any great degree on employment of the stronger eluting agent, recovery times were 50% shorter with the stronger acid.

Sorption Cycle 3 (SC3). Comparison of average metal removal rates recorded after the 84d period with those obtained during SC1 and SC2 showed little substantial differences (Table 2) suggesting that the stronger eluting agent (0.5M HCl) had not adversely affected the sorption capacity of the biofilm.

TABLE 2. Comparison of average percentage metal removal recorded after completion of each sorption cycle (SC). (Source: Costley and Wallis, 2001).

Sorption Cycle	Cd	Cu	Zn
SC1	34.0	84.6	57.3
SC2	33.7	73.7	50.8
SC3	30.1	81.8	49.7

Any differences in removal efficiencies may have been due to structural changes in the cells of the biofilm induced by HCl (Wong et al., 1993), the changes affecting surface binding of metal ions by altering the sorption sites available.

Desorption Cycle 3 (DC3). Desorption efficiencies similar to those recorded during DC1 and 2 were recorded (Figure 4), although the total desorption cycle was reduced to 12h (DC1 - 48; DC2 - 18h).

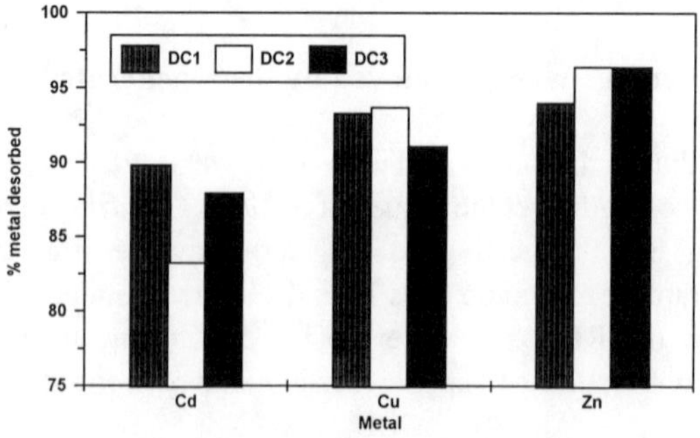

FIGURE 4. Percentage metal desorbed by each desorption cycle (DC).

Closer inspection of metal removal after each acid wash revealed that ≈50% of the desorption occurred during the first acid wash suggesting that the length of the desorption procedure may be reduced further.

The high metal sorption removal efficiencies recorded suggest that the RBC could be used as a means to treat heavy metal contaminated wastes. The rejuvenation capacity of the system as well as the ability to recover previously sorbed metal ions clearly accentuates the potential employment of this system in industrial clean-up operations. Further studies are however required on scale-up of the system to pilot size as well as an investigation into the installation and running costs of such a system.

REFERENCES

Belhateche, D.H. 1995. "Choose Appropriate Wastewater Treatment Technologies." *Chem. Eng. Prog. 91(8)*: 32-51.

Braune, E. 1994.*Approaches for Groundwater Quality Protection in South Africa.* Department of Water Affairs and Forestry.

Collins, Y.E. and G. Stotzky. 1989. "Factors Affecting the Toxicity of Heavy Metals to Microbes". In "*Metal Ions and Bacteria.*" Beveridge, T.J. and R.J. Doyle. (Eds). John Wiley, New York. pp 31-90.

Costley, S.C. and F.M. Wallis. 1999. "Effect of Disc Rotational Speed on heavy Metal Accumulation by Rotating Biological Contactor (RBC) Biofilms." *Lett. Appl. Microbiol. 29*: 401-405.

Costley, S.C. and F.M. Wallis. 2000. "Effect of Flow Rate on Heavy Metal Accumulation by Rotating Biological Contactor (RBC) Biofilms." *J. Ind. Microbiol. Biotechnol. 24*: 244-250.

Costley, S.C. and F.M. Wallis. 2001. "Bioremediation of Heavy Metals in a Synthetic Wastewater Using a Rotating Biological Contactor." *Water Res.* In Press.

Cotoras, D., M. Millar, P. Viedma, J. Pimentel and A. Mestre. 1992. "Biosorption of Metal Ions by *Azotobacter vinelandii.*" *World J. Microbiol. Biotechnol. 8*: 319-323.

Gadd, G.M. 1988. "Accumulation of Metals by Microorganims and Algae." In "*Biotechnology - A Comprehensive Treatise. Vole 6b.*" Rehm, H.J. and G. Reed (eds). VCH Publishing, New York. pp 401-433.

Gadd, G.M. 1990. "Heavy Metal Accumulation By Bacteria and Other Microorganisms." *Exper. 46*: 834-840.

Harris P.O. and G.J. Ramelow. 1990. "Binding of Metal Ions by Particulate Biomass Derived from *Chlorella vulgaris* and *Scenedesmus quadricauda*." *Environ. Sci Technol. 24(2)*: 220-228

Hobson, P.N. and N.J. Poole. 1988. "Water Pollution and its Prevention." In "*Microorganisms in Action: Concepts and Applications in Microbial Ecology*." Lynch, J.M. and J.E. Hobbie (eds). Blackwell Scientific Publications, Oxford. pp 302-321.

Kratochril, D., B. Volesky and G. Demopoulos. 1997. "Optimizing Cu Removal/Recovery in a Biosorption Column." *Water Res. 31(9)*: 2327-2339.

Matheikal, J.T., L. Iyengar and C. Venkobachar. 1991. "Sorption and Desorption of Cu(II) by *Ganodeerma lucidium*." *Water Pollut. Res. J. Can. 26*: 187-200.

McHale, A.P. and S. McHale. 1994. "Microbial Biosorption of Metals: Potential in the Treatment of Metal Pollution." *Biotechnol. Adv. 12*: 647-652.

Panchanadikar, V.V. and R.P. Das.1993. "Biorecovery of Zinc from Industrial Effluent Using Native Microbes." *Int. J Environ. Stud. 44*: 251-257.

Sag, Y. and T. Kutsal. 1989. "The Use of *Zoogloea ramigera* in Wastewater Treatment Containing Cr(VI) and Cd(II) Ions." *Biotechnol. 11*: 145-148.

Trevors, J.T., G.W. Stratton and G.M. Gadd. 1986. "Cadmium Transport, Resistance, and Toxicity in Bacteria, Algae and Fungi." *Can. J. Microbiol. 32*: 447-464.

Volesky, B. and Z.R. Holan. 1995. "Biosorption of Heavy Metals." *Biotech. Prog. 11*: 235-250.

Wong, P.K., K.C. Lam and C.M. So. 1993. "Removal and Recovery of CU(II) from Industrial Effluent by Immobilized Cells of *Pseudomonas putida* II-11." *Appl. Microbiol. Biotechnol. 39*: 127-131.

Zhou, J.L. and Kiff, R.J. 1991. "The Uptake of Copper from Aqueous Solution by Immobilized Fungal Biomass." *J Chem. Technol. Biotechnol. 52*: 317-330.

HEAVY METAL IN-SITU BIOPRECIPITATION & ADSORPTION ON A MANUFACTURING SITE (BELGIUM)

Dirk Nuyens (Environmental Resources Management – ERM, Brussels, Belgium), Leen Bastiaens, Johan Vos, Johan Gemoets and Ludo Diels (Flemish Institute for Technological Research - VITO, Mol, Belgium)

ABSTRACT: Bench-scale feasibility tests were carried out to evaluate if in-situ bioprecipitation and adsorption should be considered as remediation treatment for heavy metal impacted groundwater. The remedial goals are 1) controlling further off-site spreading, 2) mitigating source areas, and 3) reducing risks. In-situ metal bioprecipitation batch tests, based on the sulphate reduction and metalsulphide precipitation process using sulphate-reducing bacteria, were carried out. High metals removal rates were obtained using an appropriate electron donor or bio-augmentation (*Desulfovibrio desulfotomaculum* Dd8301 and the Postgate medium). The use of a sorption barrier (secondary polishing step) to prevent further spreading of the heavy metal groundwater contamination has been validated using batch tests with various aluminosilicates (zeolites and Metasorb) and aerobic/anaerobic compost (with/without crushed limestone for pH adjustment). The compost mixtures demonstrated significant heavy metal removal, the best results were obtained with the anaerobic compost. Feasibility tests were done in batch and at column level.

INTRODUCTION

Heavy metal contamination related to ore extraction, steel and non-ferrous production, and metal-related manufacturing processes are widespread (U.S. EPA, 1996). Certain heavy metals (lead, chromium, cadmium, …) are known to be toxic, mutagenic, and/or carcinogenic. To reduce human and environmental exposure to these contaminants, efforts are under way to develop new effective and cost-efficient remediation technologies as alterative to pump & treat (Evanko and Dzombak, 1997).

The feasibility of heavy metal remediation by in-situ bioprecipitation and adsorption is being studied. In-situ bioprecipitation involves sulphate reduction and metalsulphide precipitation process using sulphate-reducing bacteria (SRB). It is important to obtain and maintain optimal growth conditions (pH, redox, sulphate content, organic substrates, nutrients) for the SRB. The need for such amendments must be determined on a case-by-case basis.

Interception barriers using chemical oxidation/reduction or adsorption are currently installed to reduce off-site migration and exposure risks related to heavy metal contaminated groundwater (U.S. EPA, 1998; U.S.EPA and NATO/CCMS, 1998; U.S. EPA, 1999). Positive results were obtained with the compost (municipal, leaf, …); wood chips, limestone, aluminosilicates, molasses, and zerovalent iron. Bench-scale tests are required for the optimal barrier material feasibility evaluation and barrier remedial design.

Bench-scale studies examined the feasibility of in-situ bioprecipitation and adsorption to remediate heavy metal impacted groundwater. Batch and column results are presented.

Objective. The objective of this study is to determine whether in-situ bioprecipitation and adsorption should be considered as remediation treatment for heavy metal impacted groundwater. The goal is to achieve heavy metal concentrations in the impacted phreatic aquifer below the risk based remediation target concentrations in a technically sustainable and cost-efficient way.

Site Description. Distinct historical releases of heavy metals (Cd, Pb, Zn, Cr and Cu; observed concentrations in mg/L level), acids and ironsulphates on an active production site have resulted in groundwater contamination. The site geology consists of silty/clayey sand material followed by a clay aquitard at 4-5 m bgs (meters below ground surface). The groundwater at the site varies between 1.5 and 2 m bgs. Off-site migration has been observed.

MATERIALS AND METHODS

Groundwater Sampling and Analysis. Samples were collected for the bench-scale testing based on the results of the site characterization and delineation studies. Additional chemical analysis (pH, Oxidation-Reduction Potential or ORP, conductivity, dissolved oxygen, temperature, heavy metals (Zn, Cr, Pb, As, Cd, Ni, Fe and Cu), calcium, sulphate, phosphate, total phosphor, nitrate, and sulphide) and microbial characterization test were carried out.

Soil Sampling and Analysis. Aquifer material (soil) was sampled and analyzed for pH, water-content, clay-fraction, heavy metals (Zn, Cr, Pb, As, Cd, Ni, Fe and Cu), and total N/P/S.

ORP-Lowering Tests. Pretests were carried out to determine the required amendments to lower the ORP (acetate and oxygen-removal using N_2-gas) in order to achieve optimal SRB-conditions. Under aerobic conditions acetate (100/500 µL of K-acetate 25%) was added to 10 g aquifer material and 20 ml groundwater. The ORP was measured and subsequently all oxygen was removed by flushing with N_2-gas. The change in ORP was measured during the anaerobic incubation period. Contaminated groundwater from another industrial site was used as control sample.

SRB Batch Tests. Batch tests were performed on the SRB-containing groundwater showing promising in-situ bioprecipitation parameters. Aquifer material and groundwater were anaerobically incubated after addition of acetate, the SRB *Desulfovibrio desulfotomaculum* Dd8301, Postgate C medium, and/or optional an ORP-Lowering Compound (OLC). A poisoned control was also set-up. The pH, ORP en heavy metal concentration in the water phase were measured at the start (T0), after 4 weeks (T4), 8 weeks (T8), 12 weeks (T12) and after 19 weeks (T19).

Batch and Column Adsorption Tests. Adsorption tests were carried on heavy metal impacted groundwater using zeolites (clino, mordenite, Na-mordenite/liant, and zeolite X), compost (aerobic/anaerobic with limestone), limestone, and Metasorb (altered aluminosilicate). Small vials (25 mL) of groundwater were used and the adsorbent was added. Heavy metal concentrations (filtrated over 0.45 μm) were measured 24 hours after sample preparation, pH and ORP were also measured. The compost samples were also measured after 1 and 2 months testing. Column tests were performed in glass columns of 100 mm filled with18 g of anaerobic compost, impacted groundwater was flowing through the column at 12.5 mL/day or 18 cm/day (observed groundwater flow rate). ORP, pH and heavy metal concentrations were measured at the column inlet/outlet.

RESULTS AND DISCUSSION

Groundwater Analysis. Strong variations in pH, conductivity, ORP, dissolved oxygen, sulphate and temperature were observed in the groundwater samples used for the in-situ bioprecipitation and adsorption testing (Table 1).

TABLE 1. Groundwater results for SRB/adsorption studies.

Analysis	GW sample 01/06/01/07	GW sample 01/06/01/23
pH	2.66	4.76
Conductivity (μS/cm)	11,010	1,976
ORP (mV)	246	141
Dissolved oxygen (mg/L)	0.95	3.35
Temperature (°C)	16.1	22.5
Zn (μg/L)	12,800	17,800
Cr (μg/L)	783	<5
Cr(IV) (μg/L)	<5	<5
Pb (μg/L)	<10	31
As (μg/L)	<10	<10
Cd (μg/L)	14,500	10
Ni (mg/L)	3,430	122
Fe (μg/L)	3,810,000	<50
Cu (μg/L)	2,430	57
Calcium (mg/L)	424	152
SO_4^{2-} (mg/L)	14,090	315
PO_4^{3-} (mg/L)	ND	<0.5
Total P (mg/L)	<0.05	<0.05
NO_3-N (mg/L)	ND	33.7
S^{2-} (mg/L	<0.07	<0.07

ND = non detect

Increased Zn, Cr, Cd, Ni, Fe and Cu concentrations were measured, the sulphate content shows a strong variation. The measured pH and ORP for the low-sulphate groundwater sample 01/06/01/23 are favorable for the SRB batch scale testing. Microbial test (>4 10^4 cfu/g aquifer) confirmed the presence of SRB's. The high-sulphate groundwater sample 01/06/01/07 has a higher ORP and acid pH, no significant microbial activity (<10^2 cfu/g aquifer) was observed.

Aquifer Material Analysis. The heavy metal concentrations in the soil aquifer material are low. The soil sample from 01/06/01/07 showed an acid pH, high sulphur content and high buffering capacity (Table 2).

TABLE 2. Aquifer soil sample results for SRB/adsorption studies.

Analysis	Aquifer soil sample 01/06/01/07	Aquifer soil sample 01/06/01/23
pH	2.5	4.8
Water content %	17.2	16.1
Clay fraction %	5.87	11.2
Zn (mg/kg)	7.6	124
Cr (mg/kg)	26	52
Pb (mg/kg)	18	44
As (mg/kg)	<2	<2
Cd (mg/kg)	<0.5	<0.5
Ni (mg/kg)	7	21
Fe (mg/kg)	11,200	<16,500
Cu (mg/kg)	3.3	17
TOC (% C)	0.05	0.45
TIC (% C)	<0.01	<0.01
TC (% C)	0.05	0.45
Total N (mg/kg)	437	495
Total P (mg/kg)	55	204
Total S (mg/kg)	7,450	481
Buffering capacity		
pH 4.3 mmol/L	7.22	0.04
pH 8.3 mmol/L	17.2	0.1

ORP-Lowering Tests. Acetate addition and nitrogen flushing results in anaerobic conditions, the ORP decreased from 270 mV to 183 mV. Higher acetate dosage (500 μL) does not result in a steeper ORP-decrease (Table 3). The control sample demonstrates oxygen removal, ORP dropped from 404 mV to –212 mV, acetate addition induced a stronger ORP decrease.

TABLE 3. Influence of acetate addition on ORP during anaerobic incubation.

Sample	ORP Aerobic (mV) Bottle			ORP Anaerobic (mV)	
		Start	5 days	3 weeks	5.5 weeks
Control	404	332	194	188	-212
Control Ac100μL		187	115	147	-68
Control Ac 500 μL		216	94	-31	-245
23	270	321	95	5	-146
23 Ac 100 μL		96	9	-50	-54
23 Ac 500 μL		115	88	-14	-183

SRB Batch Tests. High metals removal rates (below risk-based remedial values) were obtained using acetate and bio-augmentation with the sulphate reducing bacterium *Desulfovibrio desulfotomaculum* Dd8301 and Postgate medium (Table 4), i.e. R3 and R5)

TABLE 4: Results batch test (01/06/01/23) - heavy metals

	Pb (μg/L)		Ni (μg/L)		Zn (μg/L)		Cd (μg/L)	
	T0	T19	T0	T19	T0	T19	T0	T19
R1: aquifer + GW	54	70	195	239	25,100	28,200	21	24
R2: aquifer + GW + $HgCl_2$	45	67	190	226	23,900	26,900	20	23
R3: aquifer + GW + Kacetate	52	4.0	238	51	26,700	2.680	23	<0.5
R4: aquifer + GW + Kacetate + Dd8301	34	69	222	206	27,900	24,500	0.63	17
R5: aquifer + GW + postgate C + Kacetate + Dd8301	82	<0.5	202	6.5	17,700	6.0	0.66	<0.5
R6: aquifer + GW + OLC	51	0.63	203	2.1	24,700	42	20	<0.5
R7: aquifer + GW + $HgCl_2$ + OLC	51	2.4	202	28	25,300	38	19	<0.5
R8: aquifer + GW + Kacetate + OLC	33	<0.5	211	<2	26,400	5.2	20	<0.5
R9: aquifer + GW + Kacetate + OLC + Dd8301	24	<0.5	201	<2	22,400	<5	<0.5	<0.5

5 mL Kacetate (25%);.
T0 = start and T19 = 19 weeks

The addition of OLC also results in a fast and efficient removal of the heavy metal concentrations. The risk-based remediation goals were already obtained within 12 weeks after incubation. The heavy metals removal observed with OLC addition is a combined biotic/abiotic process. The fact that metal removal is observed in the poisoned run (R7) indicates an abiotic sorption process or abiotic reduction of metals/sulphates. The intermediary results demonstrate initial lower metal removal compared to the SRB-stimulated test samples (R8 and R9), indicating SRB-activity in these test samples.

The sulphate concentration at T0 is 200 mg/L, with the exception of R5 (Postgate medium C contains extra sulphate) (Table 5). The sulphate-content in the first four test runs (R1-R4) did not change after 19 weeks of incubation (T19), indicating that no SRB activity occurred. Nearly all sulphate was consumed in the bio-augmented R5-series, similar sulphate consumption was observed for the OLC runs, including the abiotic control run R7. The OLC series (R6-R9) indicate that sulphate was abiotically reduced to sulphide, with possible metal sulphide precipitation.

TABLE 5: Results batch test 2 (01/06/01/23) – sulphate

	Sulphate (mg/L)	
	T0	T19
R1: aquifer + GW	200	190
R2: aquifer + GW + $HgCl_2$	192	196
R3: aquifer + GW + Kacetate	201	237
R4: aquifer + GW + Kacetate + Dd8301	202	216
R5: aquifer + GW + postgate C + Kacetate + Dd8301	2,880	1.4
R6: aquifer + GW + OLC	196	<1
R7: aquifer + GW + $HgCl_2$ + OLC	189	1.1
R8: aquifer + GW + Kacetate + OLC	231	<1
R9: aquifer + GW + Kacetate + OLC + Dd8301	202	1.0

The batch results show that high metals removal rates (below risk-based remedial values) were obtained by 1) creating ideal SRB growing conditions using acetate and Postgate medium C and by 2) adding OLC. The reaction mechanism with OLC is probably a combined abiotic/biotic process. The effect of bio-augmentation with the sulphate reducing bacterium *Desulfovibrio desulfotomaculum* Dd8301 in R4 is still unclear. Stimulating the endogenous SRB bacteria with acetate activation also results in metal removal, the process is much slower however than with Postgate medium C and OLC addition. In other experiments it could be shown that endogenous bacteria had a better metal bioprecipitation potential than added *Desulfovibrio* strains.

Batch and Column Adsorption Tests. The batch adsorption tests carried out on well 01/06/01/07 groundwater only showed significant heavy metal removal using compost (aerobic/anaerobic with limestone) (Table 6), the best result was obtained for the anaerobic compost mixture. The observed Pb-increase is related to the limestone addition to the compost mixture. The Zn-, Ni-, Cd- and Fe-results are more positive after 1 month incubation. It is unclear whether the metal removal in the compost mixture is related to biological or chemisorption-precipitation processes.

Table 6: Batch adsorption results (01/06/01/07) – metal concentrations (μg/L)

	pH	E_h	Zn	Pb	As	Cd	Ni	Fe (mg/L)	Cu
0A. Prior to incubation	2.87	460	13,500	<10	<10	13100	2,930	3,920	1,870
1A. Control	3.29	375	12,900	<10	<10	13600	2,790	4,040	1,830
2A. Clino	3.13	375	13,000	<10	<10	13000	2,780	3,860	1,820
3A. Mordenite	3.20	370	12,700	<10	<10	12,200	2,780	3,700	1,750
4A. Mord-enite-liant	3.27	367	13,400	<10	<10	12,900	2,910	3,860	1,870
5A. Zeolite X	3.45	353	12,400	<10	<10	11,700	2,740	3,550	1,690
6A. Metasorb	3.92	282	12,900	<10	<10	12,500	2,810	3,610	1,240
7A. Compost aerobic-limestone	4.50	207	13,800	19	<10	4,810	1,350	2,470	5.6
8A. Compost anaerobic-limestone (24 hr)	5.44	68	7,580	56	<10	1,160	1,200	2,310	<5
9A. Compost-anaerobic - limestone (1 month)	5.90	-121	488	62	17	48	226	2,100	307
10A. Compost-anaerobic - limestone (2 months)	6.16	-178	777	59	< 10	5,2	47	1,180	< 5
11A. Limestone (24 hr)	5.12	242	12,300	<10	<10	12,800	2,630	3,660	1,620
12A. Limestone (1 month)	4.28	222	14,600	198	<10	13,300	3,250	3,510	2,590
13A. Limestone (2 months)	4.09	214	15,400	232	< 10	12,400	3,200	3,210	2,520

The column test (anaerobic compost), on the same groundwater sample and using the site-specific groundwater flow rate, shows relatively fast breakthrough of heavy metals between 3.4 and 5 times pore volume (Figure 1).

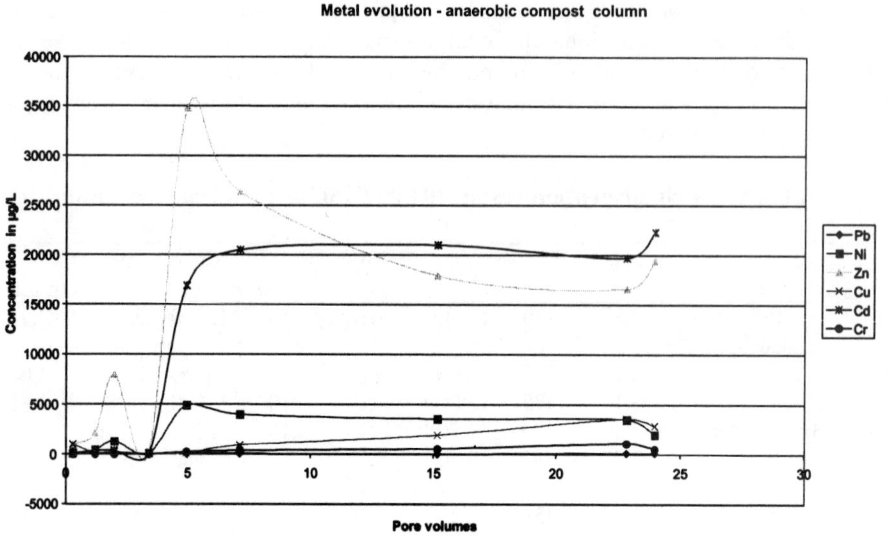

FIGURE 1: Breakthrough curves for Pb, Ni, Zn, Cu, Cd, and Cr – sample 01/06/01/07 and anaerobic compost and limestone

The following remedial feasibility and design considerations can be made based on the currently available bench scale results. Sulphate reducing bacteria (SRB's) show high metal removal rates (below the anticipated risk-based remedial values), it is however necessary to create ideal SRB growing conditions (using acetate with optional OLC addition). Anaerobic compost shows an important adsorption effect when tested in bench-scale, the breakthrough curves however clearly show the challenge for up-scaling it into a field application as compost reactive barrier.

An interesting approach for metal removal from the groundwater can be by the combination of in situ bioprecipitation by SRBs (to remove the bulk of the metals content) followed by a metals sorption barrier (to remove the lower rest concentrations of the metals. The removal of the bulk of the metals before the entrance in the sorption barrier will prevent the barrier from fast saturation and increase the sustainability of the barrier.

REFERENCES

Evanko, C. R., and D. A. Dzombak. 1997. Remediation of metals-contaminated soils and groundwater. Groundwater Remediation Technologies Analysis C, GWRTAC Technology Evaluation Report, TE-97-01, Pittsburgh, PA.

U.S. EPA & NATO/CCMS, 1998. Treatment walls and permeable reactive barriers, Nato/CCMS Pilot study. NATO/CCMS and U.S. Environmental Protection Agency Report, EPA 542-R-98-003, Washington DC.

U.S. EPA, 1996.Recent developments for in-situ treatment of metal contaminated soils. U.S. Environmental Protection Agency Report, EPA contract number 68-W5-0055, Washington DC.

U.S. EPA, 1998. Permeable reactive barrier technologies for contaminant remediation. U.S. Environmental Protection Agency Report, EPA 600-R-98-125, Washington DC.

U.S. EPA, 1999 Field applications of in-situ remediation technologies : permeable reactive barriers. U.S. Environmental Protection Agency Report, EPA 542-R-99-002, Washington DC.

IN SITU BIOREMEDIATION OF A SOIL CONTAMINATED WITH HEAVY METALS AND ARSENIC

S.N. Groudev (University of Mining and Geology, Sofia 1700, Bulgaria)
P.S. Georgiev (University of Mining and Geology, Sofia 1700, Bulgaria)
K. Komnitsas (National Technical University of Athens, 15710 Zografos, Greece)
I.I. Spasova (University of Mining and Geology, Sofia 1700, Bulgaria)
I. Paspaliaris (National Technical University of Athens, 15710 Zografos, Greece)

ABSTRACT: An experimental plot of about 800 m^2 consisting of a soil contaminated with heavy metals (copper, zinc, cadmium, lead) and arsenic was treated by a flushing system using acidified water leach solutions. The initial soil pH was in the range of about 5.4 – 6.0. The soil contained a rich indigenous microflora. The solubilization of contaminants was mainly a result of the activity of this microflora. This activity was enchanced by suitable changes in the levels of some essential environmental factors, such as water, oxygen and nutrient content of the soil. The removal of contaminants was efficient and within 18 months their residual concentrations in the soil, with the exception of that of the lead, were decreased below the relevant permissible levels for soils of such type. The pregnant soil effluents were treated by a natural wetland located near the experimental plot.

INTRODUCTION

Some agricultural land located near the copper deposit Elshitza in Central Bulgaria has been contaminated with heavy metals (copper, zinc, cadmium, lead) and arsenic as a result of mining and mineral processing activities carried out in the deposit for a long period of time. Laboratory experiments carried out with soil samples from this land revealed that an efficient remediation of the soils was achieved by means of a flushing system connected with the solubilization of the contaminants as a result of the activity of the indigenous soil microflora. This activity was enchanced by suitable changes in the levels of some essential environmental factors such as water, oxygen and nutrient content of the soil. Acidified leach solutions were used to facilitate the solubilization and to remove the dissolved contaminants from the soil profile. The pregnant soil effluents were then treated by a passive system of the type of constructed wetlands and the depleted solutions were recycled to the soil.

In 1996-97 the above-mentioned method was applied under real field conditions in an experimental plot located near the copper deposit. Some data about this study are shown in this paper.

MATERIALS AND METHODS

A detailed sampling procedure was carried out to characterize the soil and the subsurface geologic and hydrogeologic conditions of the site. Surface and bulk soil samples up to a depth of 2 m were collected by an excavator. Drill hole samples were collected up to a depth of 10 m. Elemental analysis in the samples was performed by digestion and measurement of the ion concentration in solution by atomic absorption spectrometry and induced coupled plasma spectrometry. Mineralogical analysis was carried out by X-ray diffraction techniques.

The main geotechnical characteristics of the site such as permeability and wet bulk density were measured in situ using the sand-core method (U.S. Environmental Protection Agency, 1991). True density measurements were carried out in the laboratory using undisturbed core samples. Such samples were also used for determination of their acid generation and net neutralization potentials using static acid-base accounting tests. The bioavailable fractions of the pollutants were determined by leaching the samples with DTPA and EDTA. The mobility of the pollutants was determined by the sequential extraction procedureл. The toxicity of soil samples was determined by the EPA Toxicity Characteristics Leaching Procedure (U.S. Environmental Protection Agency, 1990).

The isolation, identification and enumeration of soil microorganisms were carried out by methods described previously (Karavaiko et al., 1988; Groudeva et al., 1993)

The experimental plot had a rectangular shape and was 800 m^2 in size (40 m x 20 m). Water acidified with sulfuric acid to pH of 3.5 – 4.0 was used as leach solution. The upper soil layers were ploughed up periodically to enhance the natural aeration.

The flowsheet included also a system to collect the soil drainage solutions and to avoid their seepage and the distribution of contaminants into the environment. The system consisted of several ditches and wells located in suitable sites in the experimental plot. The soil effluents collected by this system were then treated by a natural wetland located near the experimental plot to remove the dissolved contaminants. This wetland was characterized by an abundant water and emergent vegetation and a divers microflora. *Typha latifolia, Typha angustifolia, Juncus spp.* and different algae were the main plant species in the wetland.

A plot with the sole size and shape was used as a control. This plot also consisted of polluted soil. This soil, however, was not treated during the whole experimental period.

RESULTS AND DISCUSSION

The soil profile was approximately 90 cm deep (horizon A, 30 cm; horizon B, 45 cm; horizon C, 15 cm). The soil profile was underlined by intrusive rocks consisting of dense diorite and gabbro. The filtration properties of these intrusives were very low. The filtration coefficient was approximately 8 x 10^{-8} m/s, and the deeply located rocks were even more impermeable.

The ground water level was located generally 7 to 12 m below the surface. The surface and ground waters in the site were well separated by the impermeable rocks.

It was assumed that both the geologic and hydrogeologic conditions in the site were suitable for the application of an in situ method for soil remediation.

Data about the chemical composition and some essential geotechnical parameters of the soil are shown in Table 1. The concentrations of contaminants were higher in the upper soil layers (mainly in the horizon A) (Table 2). The contaminants were present mainly in forms susceptible to biological and/or chemical leaching but considerable portions were refractory to solubilization. The primary sulphide minerals were essential components of these inert fractions. The soil exhibited negative net neutralization potential, indicating a relatively high potential of acid generation.

The permeability of the soil was high and rainwater infiltrated into and created conditions favorable for the dissolution of elements. The pore water quality was poor with high concentrations of contaminants such as toxic heavy metals (copper, zinc, cadmium, lead), arsenic, free sulphuric acid and sulphates. Prior laboratory experiments with soil samples from the experimental plot treated in this study had shown that the dissolution of contaminants was connected with the activity of the indigenous soil microflora, mainly with the activity of acidophilic chemolithotrophic bacteria. These bacteria were able to oxidize sulphide minerals and to solubilize their metal components.

Copper, zinc, cadmium and arsenic were removed mainly in this way as the relevant sulphates. However, portions of these contaminants were removed as complexes with some organic compounds, including different microbial metabolites. Lead was removed mainly as such complexes.

TABLE 1. Characteristics of the soil used in this study

Parameters	Horizon A + Horizon B	Parameters	Horizon A + Horizon B
Chemical composition (in %)			
- SiO_2	64.8	Bulk density, g/cm^3	1.52
- Al_2O_3	10.6	Specific density, g/cm^3	2.93
- Fe_2O_3	9.7	Porosity, %	49
- CaO	0.32	Moisture capacity, %	47
- MgO	0.37	Permeability, cm/s	8×10^{-2}
- K_2O	1.49	pH (H_2O)	5.5
- Na_2O	0.17	Net neutralization	
- S total	0.99	potential, kg $CaCO_3$/t	-150
- S sulphidic	0.82		
- humus	2.6		

The analysis of the soil microflora revealed that it included a rich variety of microorganisms (Table 3). The mesophilic, acidophilic bacteria related to the species *Acidithiobacillus ferrooxidans,* *Thiobacillus thiooxidans* and *Leptospirillum ferrooxidans* were the prevalent microorganisms in the top soil layers, but some basophilic chemolithotrophic species (mainly *T. thioparus* and *T. denitrificans*) and some heterotrophs were also present in high numbers. In the deeply located soil layers (subhorizon B_2), the total number of microorganisms was much lower but various anaerobes were well present.

TABLE 2.Toxic elements in the horizon A of the soil before and after the treatment

Parameters	Cu	Zn	Cd	Pb	As
Content of toxic elements, ppm:					
- before treatment	380	161	3.2	150	62
Permissible levels for soils with pH 5.5, ppm	60	95	2.0	70	25
- after treatment	37	53	1.0	48	17
Permissiblre levels for soils with pH 4.4, ppm	40	60	1.5	40	25
Bioavailable fraction, ppm:					
a. by DTPA leaching:					
- before treatment	44	32	0.6	4.4	10
- after treatment	5.1	3.7	0.03	0.7	1.2
b. by EDTA leaching:					
- before treatment	27	12	0.3	7.1	5.3
- after treatment	1.7	1.5	0.01	0.9	0.3
Easily leachable fractions (exchangeable + carbonate), ppm:					
- before treatment	77	59	1.2	37	21
- after treatment	6.0	5.1	0.03	3.5	2.4
Inert fraction, ppm:					
- before treatment	107	73	1.9	9.5	28
- after treatment	32	46	0.9	41	12
Toxic elements solubilized during the toxicity test, ppm:					
- before treatment	7.90	7.12	0.12	0.21	3.25
- after treatment	0.30	0.28	0.01	0.07	0.19

TABLE 3. Concentration of microorganisms related to different physiological groups in the soil (horizon A) before and after the treatment

Microorganisms	Before treatment	After treatment
	Cells/g dry soil	
Aerobic heterotrophic bacteria	$10^4 – 10^7$	10^3 - 10^6
Oligocarbophiles	$10^3 – 10^6$	$10^2 – 10^4$
Cellulose-degrading microorganisms	$10^2 – 10^6$	$10^2 – 10^5$
Nitrogen-fixing bacteria	$10^2 – 10^4$	$10^2 – 10^4$
Nitrifying bacteria	$10^2 – 10^4$	$10^1 – 10^3$
$S_2O_3^{2-}$ oxidising chemolithotrophs (at pH 7)	$10^3 – 10^6$	$10^3 – 10^7$
S^o – oxidising chemolithotrophs (at pH 2)	$10^3 – 10^5$	$10^3 – 10^7$
Fe^{2+} - oxidising chemolithotrophs	$10^1 – 10^2$	$10^1 – 10^5$
Anaerobic heterotrophic bacteria	$10^2 – 10^5$	$10^2 – 10^4$
Bacteria fermenting carbohydrates with gas production	$10^1 – 10^3$	$10^1 – 10^3$
Denitrifying bacteria	$10^2 – 10^4$	$10^1 – 10^3$
Sulphate- reducing bacteria	$10^2 – 10^4$	$10^1 – 10^4$
Fe^{3+} - reducing bacteria	$10^1 – 10^4$	$10^1 – 10^3$
Mn^{4+} - reducing bacteria	$10^1 – 10^3$	$10^1 – 10^3$
Methanogenic bacteria	$0 – 10^2$	$0 – 10^2$
Streptomycetes	$10^2 – 10^5$	$10^1 – 10^4$
Fungi	$10^2 – 10^5$	$10^2 – 10^6$
Total cell numbers	$1 \times 10^7 – 2 \times 10^8$	$6 \times 10^5 – 3 \times 10^8$

Note: The quantitative determination of the different physiological groups of microorganisms was carried out by the spread plate technique on solid nutrient media or by the most probable number method using end-point dilutions. The total number of microbial cells was determined by epifluorescence microscopy.

It was found that under neutral conditions the bacterial oxidation of sulphide minerals in the soil proceeded continuously but at relatively low rates. The number and activity of the acidophilic chemolithotrophs in the soil were limited by some essential environmental factors such as the relatively high soil pH, shortage of oxygen inside the soil horizons, insufficient soil moisture during relatively long periods of time, absence of some important nutrients such as nitrogen and phosphorus sources.

The treatment of the contaminated soil was connected with increasing the number and activity of the indigenous microorganisms by suitable changes in the levels of the above-mentioned environmental factors. This was achieved by regular ploughing up and irrigation of the soil and by addition of some essential nutrients. The optimum soil humidity was about 50% of the moisture capacity of the soil, but periodic flushing with slightly acidified water (pH about 3.5 – 4.0) was needed to remove the soil contaminants from the soil profile. Zeolite saturated with ammonium phosphate was added to the soil (in amounts of 3 – 5 kg/t dry soil) to provide the microorganisms with ammonium and phosphate ions and to improve the physico-mechanical properties of the soil.

The treatment of the soil was started in the middle of April 1996 and within 18 months (in the middle of October 1997) the residual concentrations of contaminants, with the exception of that of the lead, were decreased below the relevant permissible levels for soils of such type. The removal of contaminants strongly depended on the temperature of the soil. The highest rates of contaminant solubilization were achieved during the warmer summer months (June-August) when the soil temperatures exceeded 20°C. However, these processes were efficient even at temperatures as low as 5-10°C (in April and November 1996 and in March and April 1997). During the cold winter months (December 1996 – February 1997) the rates of solubilization were negligible and the operation was practically ceased for a period of about 2 months. The soil pH after the treatment was decreased to about 4.4. There were even zones with pH less than 4.0. The chemical composition, structure and physico-mechanical and water properties of the soil were altered only to a small extent.

In 1998 the experimental plot was subjected to some conventional remediation procedures such as grassing of the treatment soil, addition of suitable fertilizers and animal manure as well as with periodical ploughing up, liming and irrigation, As a result of this, the quality of the soil was completely restored. No soluble forms of the above-mentioned contaminants in concentrations higher than the relevant permissible levels were detected so far (October 1999) in the soil pore and drainage waters after rainfall.

The pregnant soil effluents were efficiently treated by the natural wetland located near the experimental plot. The contaminants were deposited in this wetland mainly as a result of processes such as the microbial dissimilatory sulphate reduction and the sorption by the plant and microbial biomass and by the clay minerals.

The removal of contaminants in the control plot was negligible. The number of the different microorganisms in this plot was much lower than the respective values in the experimental plot.

The results obtained during this study showed that the removal of heavy metals and arsenic from polluted soils using an appropriate flushing treatment connected with the enhanced activity of the indigenous microflora is an attractive method for remediation *in situ* of such soils.

ACKNOWLEDGEMENTS

A part of this study was funded by the National Science Fund (Research Contracts Nos. TH-809/98 and MY-X-09/96).

REFERENCES

Groudeva, V.I., I.A. Ivanova, S.N. Groudev and G.C. Uzunov. 1993. "Enhanced Oil Recovery by Stimulating the Activity of the Indigenous Microflora of Soil Reservoirs". In A.E. Torma, M.L. Appel and C.L. Brierley (Eds.), *Biohydrometallurgical Technologies*, vol. II, pp. 349-356. TMS Minerals. Metals & Materials Society, Warrendale, PA.

Karavaiko, G.I., G. Rossi, A.D. Agate, S.N. Groudev and Z.A. Avakyan, (Eds.) 1988, *Biogeotechnology of Metals. Manual,* Center for International Projects GKNT, Moscow.

Sobek, A.A., W.A. Schuller, J.R. Freeman and R.M. Smith, 1978. *Field and Laboratory Methods Applicable to Overburden and Mine Soils*, US Environmental Protection Agency, Report 600/2-78-054.

Tessier, A., P.G.C. Campbell and M. Bisson, 1979, "Sequential Extraction Procedure for Speciation of Particulate Trace Metals", *Analytical Chemistry*, 51(7): 844–851.

U.S. Environmental Protection Agency, 1990, *Characteristics of EP Toxicity*, Paragraph 261.24, Federal Register 45 (98).

U.S. Environmental Protection Agency, 1991. *Description and Sampling of Contaminated Soils – A Field Pocket Guide.* EPA/625/12-91/002 Technology Transfer, Centre for Environmental Research Information, U.S. Environmental Protection Agency, Cincinnati, OH.

IN SITU BIOREMEDIATION OF METALS-CONTAMINATED GROUNDWATER USING SULFATE-REDUCING BACTERIA: A CASE HISTORY

James A. Saunders (Auburn University, Auburn, Alabama)
Ming-Kuo Lee (Auburn University, Auburn, Alabama)
Jill M. Whitmer (Geosyntec Consultants, Boca Raton, Florida)
Robert C. Thomas (University of Georgia, Athens, Georgia)

ABSTRACT: Contamination of groundwater by toxic metals from both natural and anthropogenic sources is a growing problem worldwide. One promising new way to remediate such groundwater is to precipitate the contaminants *in situ* as solid (mineral) phases. A field-scale experiment was used to demonstrate that indigenous bacteria can be stimulated to remove contaminants by injection of water-soluble nutrients into a shallow contaminated water-table aquifer. Groundwater at the site is under aerobic conditions and contained high levels of lead, cadmium, zinc, copper, and sulfuric acid (pH = 3.1) derived from a car-battery recycling plant. Injected nutrients initiated bacterial sulfate reduction by one principal species (*Desulfosporosinus orientis)* under these acidic conditions. Lead, cadmium, zinc, and copper were almost completely removed by precipitation of solid sulfide phases as pH increased to 4.1. Geochemical reaction path modeling was used to trace the geochemical evolution of sulfate reduction and sorption processes. The results show that sulfate reduction produced desired geochemical conditions (i.e., low Eh and high pH) for the precipitation of metal sulfides and sorption processes. Results have implications for *in situ* remediation strategies for other sites where anthropogenic or natural sources of metals contaminate groundwater.

INTRODUCTION

Anaerobic sulfate-reducing bacteria (SRB) long have been known to have an important control on the geochemistry of iron and sulfur in nature (Berner, 1970). SRB metabolism causes iron sulfide minerals to precipitate in recent marine sediments, wetlands, lacustrine sediments, or wherever there is a source of sulfate and an easily-degradable source of organic carbon (Jakobsen and Postma, 1994; Morse et al., 1987). In addition, zones of active biogenic sulfate reduction (and iron-sulfide precipitation) in groundwater have been documented around leaking landfills (Bottrell et al., 1995; Lyngkilde and Christensen, 1992). The precipitation of sulfide minerals by SRB metabolism could be a potential metal scavenger in natural attenuation. For example, a number of trace elements (As, Co, Ni, Hg, etc.) are contained in the reactive iron-sulfide phases (or pyrite, FeS_2) of anoxic marine and lake sediments (e.g., Cooper and Morse, 1998). Similarly, SRB metabolism caused the co-precipitation of a number of trace elements in iron sulfide minerals (pyrite, marcasite, and pyrrhotite containing As and Tl), sphalerite (ZnS with coprecipitated Cd, Ag), and galena (PbS) in Gulf Coast salt dome cap rocks (Saunders and Swann, 1994). Further, we have documented that SRB have removed As, Co, and Ni from shallow groundwater in an alluvial aquifer in central Alabama, USA, by coprecipitating them in biogenic pyrite (Saunders et al., 1997). SRB are ubiquitous in natural aqueous environments even under *predominantly aerobic conditions* (Canfield and Des Marais, 1991), perhaps much more so than generally recognized. We propose that naturally occurring SRB might be useful in remediating metal-contaminated groundwater *in situ* because of their well-known effects on metals in natural settings. This paper presents the results of a field-scale experiment involving the stimulation of naturally occurring bacteria to effect remediation under induced anaerobic conditions.

Objective. The objective of this study is to inject required nutrients to stimulate the growth and metabolic activity of SRB that we hypothesized existed in groundwater under aerobic and low-pH conditions. We test the hypothesis that sulfate reduction can be stimulated to the point that amorphous solid sulfide will precipitate, which in turn lead to metal removals by coprecipitation in solids. The study documents the principle biogeochemical processes that occur at the test site after injection. This includes: (1) tracking changes in water chemistry (major and trace elements, sulfur isotope compositions, pH, Eh); (2) identifying changes in bacterial populations over time using genetic sequencing (16S rDNA) of bacteria filtered from groundwater; (3) identifying solid "mineral" phases produced by biochemical processes.

FIELD SITE

Groundwater at the site near Troy, Alabama (Figure 1) was initially contaminated by lead, cadmium, zinc, and sulfuric acid released in a large car-battery recycling operation in southeastern Alabama, USA, which accounts for ~15% of the current US lead production. Contaminated groundwater at the site occurs in a shallow water-table aquifer set in a sandy formation containing abundant iron oxides. Depth to the water table is typically about 5 m, and an underlying clay-rich confining layer occurs at a depth of about 12 m. Contaminated groundwater discharges southeastward toward a natural wetland and has caused extensive killing of natural vegetation. The plumes of Pb and Cd observed in 1998 show the "chromatographic" (i.e., separation of the components of plume) effect of retardation (Figure 1). The cadmium, which is not sorbed at low pH environments, can be seen to have moved significantly downgradient beyond the strongly sorbed lead. A large scale pump-and-treat remediation operation has been ongoing for a decade but with little success in improving water quality. In December 1999, we began an *in situ* bioremediation experiment at the site to remediate the metals-contaminated groundwater.

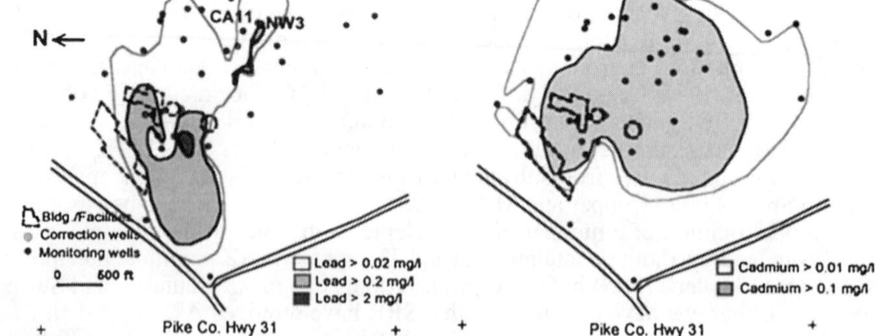

FIGURE 1. Observed lead (a) and cadmium (b) plumes at Troy site.

METHODS

Approximately 825 L of a solution containing 45 kg of sucrose and 2.5 kg of diammonium phosphate (herein called the CNP solution for the elements added) was injected by gravity feed into a well screened in a shallow sand aquifer. The well chosen (NW3, Figure 1) for the injection experiment was an existing well at the site known for having the highest metal concentrations and lowest pH. Groundwater geochemical changes (major ions, trace elements, pH, Eh) were monitored weekly in the injection well for 10 weeks. Genetic sequencing of bacteria was conducted on 4 samples using 16S rDNA techniques before and after the injection. A reaction path model was developed to describe the geochemical evolution of redox conditions, mineral precipitation, and sorption processes during the injection experiment.

RESULTS

Groundwater Geochemistry Changes. Within a month, major changes in the aquifer geochemical conditions occurred including the development of a smell characteristic of hydrogen sulfide produced by bacterial sulfate reduction. Groundwater pH had increased about 1 log unit and Eh had dropped dramatically (e.g., from aerobic to highly reducing; Figure 2a). Water samples collected from the well no longer had their typical orange color from the suspended hydrous ferric oxides (HFOs) normal for aerobic, iron-rich water samples. The principal contaminants Pb and Cd showed dramatic drops in their concentrations, with lowest dissolved values of 1.5 ppb and 0.5 ppb, respectively (Figure 2b). Similarly, other chalcophile ("sulfur-loving") elements (Cu, Zn) showed similar decreases, consistent with all of these elements forming or coprecipitating in insoluble sulfide phases.

Concentration of iron increased significantly after injection and did not drop to pre-injection levels during the course of the experiment (Figure 2c). The iron increase indicated that the CNP solution initially stimulated indigenous dissimilatory iron-reducing bacteria (DIRB) to reductively dissolve HFOs in the aquifer and release ferrous iron to solution. The lack of a significant drop in iron concentration after bacterial sulfate reduction began indicates that iron was not precipitated as an iron-sulfide phase. Similarly, siderophile ("iron-loving") elements Co, Ni, As, and Cr showed no significant drop in concentration during the experiment (Figure 2d). The abundance of lithophile elements (affinity for silicate melts or minerals) Cr and V also were little changed during the course of the experiment (Figure 2d). In contrast, selenium concentration dropped significantly (Figure 2e). "Incompatible" lithophile elements uranium and thorium concentrations decreased significantly as the Eh values dropped during the course of the experiment (Figure 2e). Similarly, Al concentration also dropped during the course of the experiment, apparently due to pH increase (Figure 2f) which probably caused precipitation of an aluminum hydroxide phase.

Sulfur Isotope Composition. Because SRB have a well-known kinetic isotope fractionation effect on dissolved sulfate and resulting biogenic hydrogen sulfide (Ehrlich, 1997), we analyzed groundwaters both before and during the course of the experiment for the $\delta^{34}S$ ratio of dissolved sulfate to track the progress of bacterial sulfate reduction. Bacterial reduction of sulfate leads to enrichment of ^{34}S in dissolved sulfate relative to H_2S because SRB preferentially use the lighter ^{32}S in generating H_2S gas. Although observed fractionation was not great (of ~0.5 ‰), we observed a linear trend where dissolved sulfate became isotopically heavier as the sulfate content decreased. This trend is consistent with sulfur-isotope fractionation attending bacterial sulfate reduction.

Bacteria Populations. Genetic sequencing of bacteria from groundwater was conducted on 4 samples: one baseline sample taken prior to injection of CNP solution and three additional samples during the course of the field experiment. Three samples taken after the injection contained sequences resembling *Desulfosporosinus* *orientis*, which has 96% homology to type strain DSM 765 (Y11570), and *Alicyclobacillus hesperidensis* (90% homology). *D. orientis* can reduce sulfate to H_2S and is a heterotrophic anaerobe with glucose as its preferred organic carbon source. Sulfate reducing bacteria were not detected before injection using the most probable number (MPN) method. However, the number of SRB was increased to 512 cells/mL 42 days after the injection.

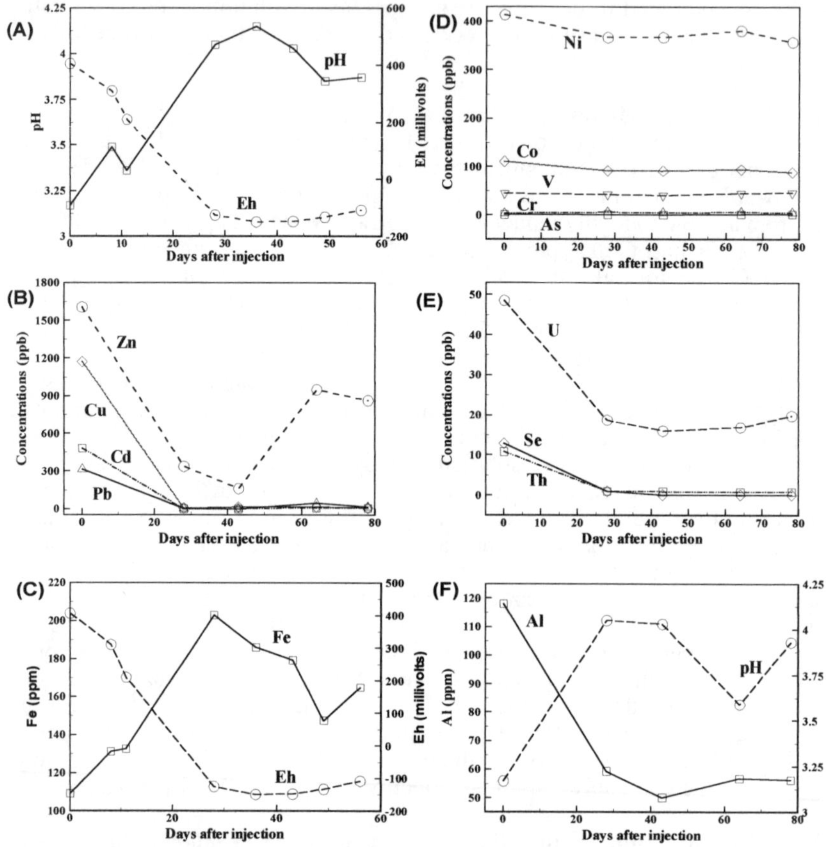

FIGURE 2. Changes in geochemical parameters with time after the injection of sucrose and diammonium phosphate solution.

Solid Metal-Sulfide Precipitation. Filters used on groundwater samples turned black suggesting that metal-sulfide colloids were present in the water. Filters were first analyzed by X-ray diffraction and they showed only a diffuse and broad peak at Cu-Kα d-spacing of 3.12 A, which is the principal ("100") peak for sphalerite (ZnS). Both extremely small grain size and not-completely crystalline phases can both cause the broad and diffuse X-ray diffraction peaks and both conditions may have prevailed with these recent precipitates. Imaging of the filter surfaces with an SEM failed to reveal any obvious crystals, but energy-dispersive X-ray analysis (EDAX) of filtered material consistently resulted in peaks for Cu, Zn, Fe, and S, along with Si (Figure. 3). Relative peak heights for metals and sulfur suggest that perhaps a single Zn-Fe-Cu-S phase could be present, or more likely, mixed Cu-Fe-S and Zn-Fe-(Cd?)-S phases given the results of X-ray diffraction data.

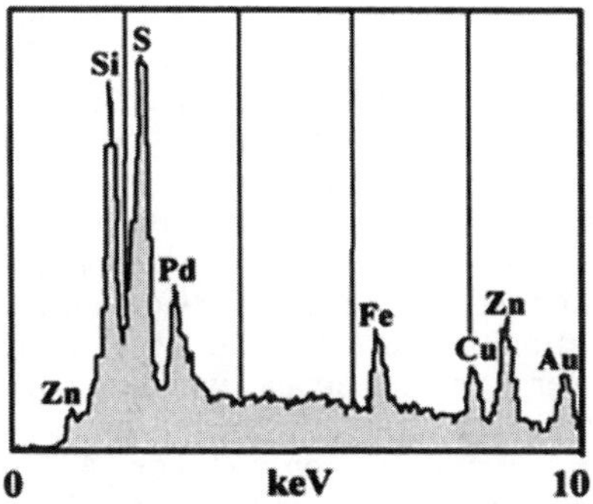

FIGURE 3. Energy dispersive X-ray (EDAX) spectra of filtered materials from groundwater after bacterial sulfate reduction began.

REACTION PATH MODELING

Geochemist's Workbench (Bethke, 1996) was used to investigate how bacterial sulfate reduction induces the precipitation of metal sulfides from the contaminated groundwater. The calculations also predict the sorption of metals onto the surface of HFOs in the aquifer. The surface complexation reaction modeling was based on the double layer model presented by Dzombak and Morel (1990). To calculate the effect of SRB metabolism, the fluid redox potential Eh increases linearly from its initial values of 400 mV to -150 mV at the end of the reaction path, as observed during bioremediation. The initial system contain 1 kg of contaminated fluid (Table 1, pH = 3.17) and a small amount of hematite (Fe_2O_3), a proxy for hydrous ferric oxides. HFOs used in the simulation have specific surface areas of 600 m^2/g for sorption.

TABLE 1. Chemical composition of contaminated groundwater used in geochemical modeling at 25°C.

Ion	Concentrations (ppm)	Ion	Concentrations (ppb)
Cl^-	590.0	Pb^{+2}	314.2
SO_4^{-2}	800.0	Cd^{+2}	480.4
Na^+	214.3	Zn^{+2}	1605.0
Ca^{+2}	163.6	Cu^+	1172.0
Mg^{+2}	13.4	Co^{+2}	110.5
K^+	3.2	Ni^{+2}	412.8
Al^{+3}	81.6	Cr^{+3}	3.8
Fe^{+2}	88.5	As^{+3}	2.0
Mn^{+2}	4.0	Se^{+4}	12.9
		Ba^{+2}	16.6
		Cs^+	3.2

The modeling results show that bacteria sulfate reduction produces desired geochemical changes including a drop in Eh and an increase in pH, as commonly observed in the field (Whitmore and Saunders, 2000). Sulfide produced by sulfate reduction react with metals to form minerals including pyrite (FeS_2), galena (PbS), sphalerite (ZnS) and others at Eh below -50 mV (Figure 4a). Although precipitation of metal sulfide minerals in the natural system is not expected over a short period of time, the formation of metal-sulfide colloids observed (Figure 3) agrees with the geochemical evolution predicted by the modeling. The calculation also shows that various metals are sorbed over a wide range of pH during sulfate reduction (Figure 4b). Metal Pb is strongly sorbed at relatively low pH (< 4), which is consistent with its removal observed during the field remediation (Figure 2b). Up to 20% of aqueous Cu can be also sorbed on HFOs in the similar pH range (Figure 4b). The modeling result explains why the Pb plume is retarded in migration with respect to the Cd plume (Figure 1). The sorption of Zn, Cd, Co, and Ni takes place only at relatively neutral pH conditions (Figure 4b) and has thus little effects in metal attenuation. The effects of sorption on chemical evolution of metals can be seen by comparing the results of a model that combines precipitation and sorption (Figure 4c) and a model that considers precipitation only (Figure 4d). Sorption effectively removes Pb in low pH conditions. Concentration curves of most metals in both models do not depart largely from each other. It is apparent that Zn and Cd are removed mainly by direct precipitation of sulfide solids via sulfate reduction.

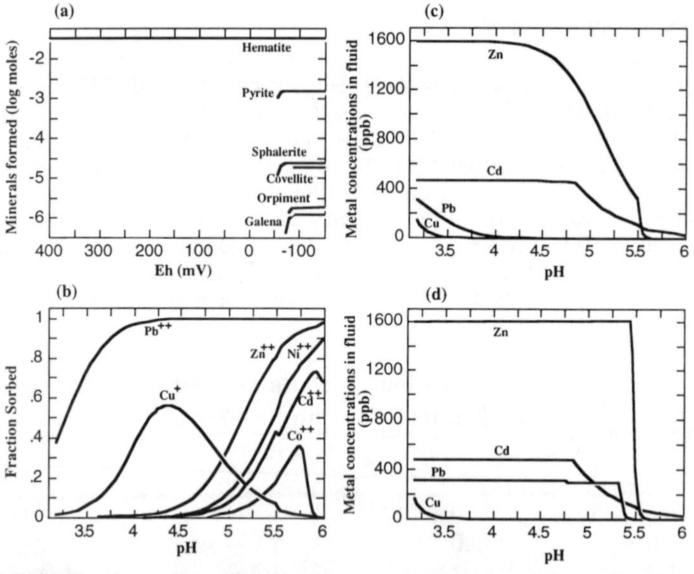

FIGURE 4. (a) Predictive cumulative mineral assemblage precipitated as Eh decreases as a result of bacterial sulfate reduction. (b) Computed curves for fraction sorption of metals on HFOs. The effect of sorption can be seen by comparing the concentration curves of metals which are sorbed (c) with the curves of the same metals with no sorption (d).

DISCUSSIONS

Microbiological data and sulfur isotope analyses confirmed that SRB metabolism led to removal of chalcophile elements by precipitating solid metal-sulfide phases including ZnS, PbS, and CuS. However, we could discern only one crystalline phase (sphalerite, ZnS) in the X-ray diffraction analysis, consistent with the fact that from a mass standpoint, more Zn was removed than any other element (Figure 2b). Given the similar geochemical behavior of Zn and Cd, it is likely that

Cd would have coprecipitated in sphalerite. The relatively abundant siderophile elements Fe, Ni, Co showed no significant decrease in their concentrations during the course of the experiment, nor did As. Iron concentration *increased* initially as the aquifer first became anaerobic and apparently iron-reducing bacteria led to the dissolution of HFOs in the aquifer. We had hypothesized that both Cr and V would decrease in concentration as they tend to precipitate under reducing conditions. It is possible that given the short duration of this experiment in the field that Cr and V reduction and mineral precipitation were too sluggish to have been observed. However, there seemed to be little kinetic inhibition for reducing and precipitating U and Th (Figure 2e), which also form oxides under reducing condition. Dissolved Al also decreased during the course of the experiment, which we interpret to be more of an influence of increasing pH (Figure 2f). Aluminum typically is much more soluble at low and high pH than at circum-neutral pH. In the experiment, Al concentration is inversely related to pH, suggesting Al-hydroxide was precipitating as pH increased.

From the data gathered, it appears that the trace elements analyzed exhibited three different behaviors during bacterial sulfate reduction: 1) chalcophile elements (Zn, Cd, Cu, Pb, Se) with a strong affinity for sulfur precipitated as low-solubility metal-sulfide phases; 2) siderophile elements (Co, Ni, Fe, As) remained in solution apparently because the iron-sulfide ion activity product ($aFe^{2+} \bullet aS^{2-}$) did not exceed the K_{sp} value for precipitating FeS(am), which generally controls dissolved iron and sulfide solubility in aqueous systems; and 3) redox-sensitive uranium and thorium apparently were removed from groundwater as an indirect consequence of lowering the ambient Eh. The induced geochemical changes also lead to the sorption of some metals on HFOs. However, the modeling and field data suggest that sorption does not play a significant role as a metal scavenger as direct precipitation during our bioremediation process (Figures 4c and d).

Given the iron-rich nature of the aquifer and groundwaters in this study, the site appears to behave from the biogeochemical standpoint like the "Fe-dominated" system as defined by Cooper and Morse (1998). Under Fe-dominated conditions, the *solid* HFO phases behave as a "sink" for sulfide, perhaps by a reaction like this:

$$FeO(OH) + H_2S + 1/2H_2 \Rightarrow FeS(am) + 2H_2O \qquad (1).$$

At the site of the experiment, reactive iron phases in the aquifer apparently have the effect of lowering the dissolved sulfide concentration below that necessary for precipitating FeS(am), but keeps it great enough to precipitate sulfides of the chalcophile elements. So it would appear that in order to precipitate siderophile elements in an Fe-dominated freshwater aquifer system, SRB have to be stimulated to produce enough H_2S to first consume reactive solid iron phases prior to precipitation of FeS(am), the kinetically favored precursor to the very common mineral pyrite. The fact that aquifer conditions changed from aerobic to highly anaerobic for SRB metabolism suggests that SRB might exist in micro-environments where they can out-compete DIRB's and then diffusion of H_2S outward lowers the redox state of the aquifer. This in turn causes dissolved chalcophile elements and uranium to precipitate, DIRB metabolism to be inhibited, and conversion of some of the solid reactive iron phases to iron sulfide.

Our results show that indigenous SRB have been stimulated in higher pH micro-environments created by artificial injection. Further, results of the field experiment show that bacteria can be selectively stimulated to effect groundwater remediation. In this study, chalcophile metals lead, cadmium, copper, zinc, and selenium were removed from contaminated groundwater by precipitating solid metal-sulfide phases. Our results, along with observations from natural environments, indicate that siderophile elements such as cobalt, nickel, arsenic (here defined as siderophile) and others can also be removed from groundwater by SRB. However, their precipitation as solid sulfide phases apparently requires producing enough H_2S to first convert solid reactive iron phases (HFO) to iron sulfides before the *aqueous* ion activity product for ferrous iron and sulfide can exceed the K_{sp} required for precipitating FeS (am) from solution. The removal of cobalt, nickel, and cadmium by sorption on HFOs requires near neutral pH conditions (Figure 4b) induced by SRB metabolism. Results have implications for a possible strategy of using SRB to cheaply remove various heavy metals from contaminated groundwater.

REFERENCES

Berner, R. A. 1970. "Sedimentary Pyrite Formation." *Am. J. Sci. 268*(1): 1-23.

Bethke, C. M. 1996. *Geochemical Reaction Modeling*. Oxford Press, New York.

Bottrell, S. H., P. J. Hayes, M. Nannon, and G. M. Williams. 1995. "Bacterial Sulfate Reduction and Pyrite Formation in a Polluted Sand Aquifer." *Geomicrobiol. J. 13*(2): 75-90.

Canfield, D. E., and D. J. Des Marais. 1991. "Aerobic Sulfate Reduction in Algal Mats." *Science*. *251*(5): 1471-1473.

Cooper, D. C., and J. W. Morse. 1998. "Biogeochemical Controls on Trace Metal Cycling in Anoxic Marine Sediments." *Environ. Sci. & Tech. 32*(3): 327-330.

Dzombak, D. A., and F. M. M. Morel. 1990. *Surface Complexation Modeling: Hydrous Ferric Oxide*. John Wiley & Sons, New York.

Ehrlich, H. L. 1997. *Geomicrobiology*. Marcel Dekker, New York.

Jakobsen, R., and D. Postma. 1994. "In Situ Rates of Sulfate Reduction in an Aquifer (Romo, Denmark) and Implications for the Reactivity of Organic Matter." *Geology*. *22*(12):1103-1106.

Lyngkilde, J., and T. H. Christensen. 1992. "Redox Zones in a Landfill Leachate Pollution Plume." *J. Contam. Hydrol. 10*(3): 273-289.

Morse, J. W., F. J. Millero, J. C. Cornwell, and D. T. Rickard. 1987. "The Chemistry of Hydrogen Sulfide and Iron Sulfide Systems in Natural Waters." *Ear. Sci. Rev. 24*(1):1-42.

Saunders, J. A., and C. T. Swann. 1994. "Mineralogy and Geochemistry of Cap Rock Zn-Pb-Sr-Ba Mineralization at Hazlehurst Salt Dome, Mississippi."*Economic Geology*. *89*(2):381-390.

Saunders, J. A., M. A. Pritchett, and R. B. Cook. 1997. "Geochemistry of Biogenic Pyrite and Ferromanganese Stream Coatings: A Bacterial Connection?" *Geomicrobiology Journal*. *14*(2):203-217.

Whitmore, J. M., and J. A. Saunders. 2000. "Bioremediation of Groundwater Contaminated by Phosphate Mining and Extraction: A New Approach Using Indigneous Bacteria." *Proceeding of 17th American Society For Surface Mining and Reclamation Annual Meeting*, pp. 107-116. Tampa, Florida.

SOIL HEAVY METAL CONTAMINATION AND PRACTICAL APPROACHES TO REMEDIATION IN SOME PARTS OF CHINA

Qingren Wang, Yiting Dong, and Yanshan Cui (Research Center for Eco-Environmental Sciences, Chinese Academy of Sciences, Beijing, China)
Xiumei Liu (Shandong Agricultural University, Tai'an, Shandong, China)

ABSTRACT: This paper mainly deals with the current status of soil contamination with heavy metals, sources of pollutants, and some practical approaches to reduce metal contents in the food chain and soil remediation accepted by Chinese farmers. For the whole country, soil contamination is not an important issue but in some individual sites, *e.g.*, mining, industrial sites, or areas with wastewater irrigation, heavily contaminated soils have been found. The metals are usually from untreated wastewater irrigation, sewage sludge, industrial or domestic waste disposal, or metal-containing agrochemical application in agriculture, *e.g.* Cd-abundant phosphates, besides atmospheric deposition of metal dusts in the vicinity of industrial areas. In regards to soil remediation for such a country with a large population as in China, keeping high output or production of enough food from agriculture is still a bottleneck-like problem. To prevent food contamination from polluted soil, a series of practical approaches has been developed. Among them, non-edible crop growth, exploiting as breeding sites, choosing tolerant crops, etc. have showed a wide acceptability and benefit to farmers before a more hopeful approach, *e.g.*, phytoremediation is put into practice.

INTRODUCTION

Soil heavy metal contamination in some industrial sites or wastewater irrigation areas has become a serious problem around China and much more attention has been paid to the pollution status and control strategies than ever before for human health concern (McGrath, 2000; Zhao, 2000). For example, in some locations, up to 145 mg kg^{-1} of Cd in soil has been found, which could cause a serious consequence if used for growth of edible crops, such as paddy rice. As a matter of fact, in some contaminated soils near smelters or waste disposal areas, patches of edible crops are still cultivated by individual farmers. However, these soils are not allowed to grow food crops according to governmental guidelines for limiting metal contents in soils, foods, irrigation water, and sewage sludge application, etc., set up by the CSEPA (China State Environmental Protection Administration).

For soil remediation, the current approaches, such as leaching with water alone or with some chelants, or replacing contaminated soil with clean, were costly and less practicable in those areas. In addition, the soil has to be excavated and a risk of secondary contamination should be considered when the leaching process is carried out. Phytoremediation is a favored measure with low costs and without the need to excavate the soil. The only step is to deal with the plants properly after harvest, *e.g.,* buried deeply or recovering the corresponding metals from the ash after burning if the content is high enough. Unfortunately, the hyperaccumulators are rare and their biomass is usually low with a tediously long time required to get a site restored. Based on this point, some alternative approaches have to be applied to resolve the emergent problems. Therefore, this paper presents a general assessment on soil pollution in cooperation with major sources of the contaminants and some practical approaches adopted in China.

SOIL HEAVY METALS IN CHINA

Compared with some other countries, soil heavy metal concentrations in a national scope of China are fortunately not so severe, more or less higher than those in USA, but quite similar to those in Japan and Britain (Table 1). However, a lot more attention has to be paid to the contaminated sites, especially suburbs of large cities and heavy industries since some heavily contaminated sites exist there. For example, up to 145 mg kg^{-1} of soil Cd has been found in those areas through an extensive investigation (Wang, 1997; Wang et al., 2001).

TABLE 1. Soil metal contents in China and some other countries (mg kg^{-1})

Element	China			U. S. A.			Japan	Britain
	Range	A.M.*	G.M.*	Range	AM	GM	AM	AM
As	0.01-62.6	11.2	9.2	< 0.01-97	7.2	5.2	9.0	11.3
Cd	0.001-13.4	0.097	0.074	--	--	--	0.413	0.62
Co	0.01-93.9	12.7	11.2	< 0.3-70	9.1	6.7	10.0	12
Cr	2.20-1209	61.0	53.9	1-2000	54	37	41.3	84
Cu	0.33-272	22.6	20.0	<1-700	25	17	37	25.8
Hg	0.001-45.9	0.065	0.040	< 0.01-4.6	0.089	0.058	0.28	0.098
Mn	1-5888	583	482	<2-7000	550	330	583	761
Ni	0.06-627	26.9	23.4	<5-700	19	13	28.5	33.7
Pb	0.68-1143	26.0	23.6	<10-700	19	16	20.4	29.2
Zn	2.6-593	74.2	67.7	<5-2500	60	48	63.8	59.8

* AM and GM represent arithmetic and geometric means, respectively.

SOURCES OF HEAVY METALS

Mining. Wastewater, air, and tailings discharge from mining play an important role in contributing to soil contamination by metals. For example, Baiyin, Gansu province is one of the northwestern country's important bases of metal mining and smelting, which has disposed of a large amount of wastewater, gas, and sludge causing serious soil contamination in this area. More than 40 to 60% of samples have the metal content over background values in this area. The maximum concentration for various metals is Pb >700, Cu >300, As >120 and Cd >15 mg kg^{-1}, respectively (Xu and Yang, 1995). In Shuikoushan area of Hunan province, a large Pb/Zn ore is located where the mining history can be traced to the Ming dynasty (1368-1644). After more than 350 years of mining, up to 115.1, 360.4, 15.6, and 3,888 mg kg^{-1} of Cu, Pb, Cd, and Zn, respectively as maximum concentrations have been found in the area (Xu and Yang, 1995).

Untreated Wastewater Irrigation. In arable land near an ore mining area, the content of heavy metals increased dramatically through irrigation with wastewater from mining (Table 2). In Zhangshi area of Shenyang suburb, 20-year municipal sewage water irrigation caused Cd contamination of 2,500 ha of arable land. In this area, more than 300 ha with Cd contents of 5 to 7 mg kg^{-1} in the soil, 0.4 to 1.0 mg kg^{-1} as average and some even reached as high as 3.4 mg kg^{-1} in the rice (Zhou, 1987). In Jing-Jin-Tang area, southeastern Beijing, sewage water irrigation has caused about 60% of soil samples and 36% of rice samples from the area to be contaminated (Cai et al., 1985).

TABLE 2. Soil contamination results from industrial wastewater nearby a smelter.

	Cd	Zn	Pb	Cu
Irrigated with clean water (mg kg^{-1})	0.37	121	44	31
Irrigated with wastewater (mg kg^{-1})	12.1	3700	1600	133
Concentration in the wastewater (mg l^{-1})	0.023	3.8	0.47	0.04

Sewage Sludge or Solid Waste Disposal. Improper application or disposal of sewage sludge from polluted streams or water treatment plants can cause soil heavy metal contamination as well; however, farmers preferred to apply that to their farmland because the contents of nitrogen (N) and phosphorus (P) account for 3.8% and 2.2%, respectively, in sludge dry solids. It is estimated that approximately 20,000 tons of N and 11,000 tons of P have been recycled to farmland each year in the UK through the agricultural use of sludge (Smith, 1996). In addition to the beneficial effects of plant nutrients, agricultural land application of sewage sludge could also improve the physical condition of barren soils (Wong and Wong, 1997), but some heavy metals deposited in the sludge are transferred to the soil simultaneously. Long-term accumulation could probably cause soil contamination, although the contents from different countries or areas are below the "clean sludge" standard (Table 3).

TABLE 3. Concentration of heavy metals in sewage sludge from some countries as compared to the maximum in proposed "clean sludge" by USEPA (mg kg^{-1} DW) (Wong and Wong, 1997).

		China						
	Australia	Mainland	Hong Kong	New Zealand	UK	USA		Max in proposed "Clean sludge"
Metal	Mean	Mean	Median	Mean	Median	Range	Median	
As	5.5	-	-	21	3.4	0.3-315.6	6	100
Cd	3.9	4.6	2.99	9.2	3.2	0.7-8220	7	18
Cr	81	95	70.5	1,209	86	2.0-3750	40	2000
Cu	1427	122.8	479	571	479	6.8-3120	463	1200
Hg	5.1	-	-	1.8	3.0	0.2-47.0	4	15
Ni	25	51.4	64.2	179	37	2.0-976	29	500
Pb	140	134.4	54.7	579	204	9.4-1670	106	300
Se	10.6	-	-	-	1.2	0.5-70.0	5	32
Zn	839	1756.3	2123	1,826	843	37.8-68000	723	2700

Among the potentially toxic elements, Zn, Cu, and Ni are essential to plants but become phytotoxic in excess. Cadmium, Hg, Pb, and Cr are considered as non-essential for plant growth, but they are of primary concern because of their potential hazard to animals and humans through plant accumulation. Boron, Mo, and As are of concern for both plant and animal health and Se is of concern for animal health (McGrath, 2000; Wong and Wong, 1997).

Besides sewage sludge, domestic or industrial wastes are the main contribution of soil heavy metals as well (Wang and Li, 1999; Wang, 2000). Heavy metals in compost are a major concern, thereby limiting their use (Table 4).

TABLE 4. Composition of domestic wastes from major cities of China.

	Beijing	Tianjin	Shanghai	Guangzhou	Means from 22 compost samples through the country
Hg	3.57	□	□	0.43	9.23
Cd	0.27	0.9	0.37	0.39	2.14
Cr	20.37	20.8	□	46.75	26.50
Pb	54.30	41.0	45.80	87.80	39.60
As	5.53	□	□	22.42	93.00

Other Sources. Atmospheric deposition in the vicinity of heavy industry, mining areas, or near smelters is an important source for soil metal contribution (Table 5) since metal dusts form through smelting or manufacturing processes. Another example is that soil Pb, Zn, and Cd contents exponentially decreased as distance from a smelter increased (unpublished data).

Metal-containing chemicals, such as Cd-containing phosphate fertilizers and Pb, As, Hg, or Zn-containing pesticides with long-term application in agriculture, could also result in soil contamination (McGrath, 2000; Wang, 1998; 2000; Wang and Li, 1999). For instance, approximately 80% of soil Cd in the soil surface layer is estimated to come from P-fertilization since soil Cd content is well associated with soil phosphate (Mulla et al., 1980).

TABLE 5. Contents of some heavy metals in different parts of paddy rice at the vicinity of a steel manufacture (mg kg^{-1}).

	Leaf			Stem			Grain		
	Cu	**Pb**	**As**	**Cu**	**Pb**	**As**	**Cu**	**Pb**	**As**
Control	38.4	0.8	0.9	41.1	1.2	0.7	14.2	0.6	Trace
Contaminated area	176.0	9.7	15.3	48.0	3.5	11.9	24.0	2.7	0.7
Times increased	3.6	11.1	16.0	0.2	1.9	16.0	0.7	3.5	-

PRACTICAL APPROACHES

To prevent food contamination from polluted soil and for soil remediation, some practical approaches have been introduced into the Chinese agriculture. Among them, phytoremediation is one of the excellent candidates whenever a hyperaccumulator has been found and adapted to the bioclimatic condition with a reasonable quantity of biomass production. However, in such a country with large population and arable land at only 0.08 ha per capita, one third of the world average (Zhao, 2000), food production and agricultural practice are still a major concern. Furthermore, a strategy acceptable and beneficial to the farmers would be easily carried out by all means.

Growing Non-edible Plants. In heavily contaminated sites, food crops or vegetables should be forbidden to grow but fiber plants (*e.g.*, cotton, flax etc.), cash crops, such as garden trees, flowers, or lawn grass are encouraged to grow. After certain years, edible crops can be grown again if the soil metal content drops below a certain level. However, to prevent the occurrence of secondary pollution, residues with high concentrations of metals in some specific plants should be dealt with properly; it is encouraged for deep burial after collected and burnt.

Exploiting as Breeding Sites. Exploiting the contaminated land as breeding sites for quality or high yield breeding sites is one of the effective approaches to prevent contaminants getting into the food chain. For instance, when wheat seeds bred from Cd-contaminated soil are grown in clean soil, the Cd content in the grains decreased dramatically below the guideline for "hygienic food" (< 0.2 mg kg^{-1}) through dilution and redistribution (Table 6).

Choosing Correct Crops or Varieties. There are some differences among crops in accumulating soil metals, especially in the edible parts of food crops (Table 7). This result implies that in Cd contaminated soils, growing maize is much safer than growing wheat since a larger amount of Cd accumulated in wheat grains rather than in maize at same rates of Cd application.

TABLE 6. Grain Cd decrease in wheat bred from contaminated sites (mg kg^{-1}) (Xu and Yang, 1995).

Samples	Seed bred from contaminated sites ①	Grain grown in uncontaminated soil using contaminated seeds from ①
Site A	0.5114	0.0044~0.0154
Site B	0.6473	0.0121~0.1759
Site C	0.6897	0.0004~0.0748

TABLE 7. Contents of Cd in different parts of wheat and maize (mg kg^{-1}).

Rates	Grain		Shoots		Roots	
	Maize	Spring wheat	Maize	Spring wheat	Maize	Spring wheat
CK	0.023	0.038	0.19	0.17	0.24	0.40
1	0.035	0.133	1.10	0.99	0.80	1.59
3	0.044	0.176	1.40	0.99	1.69	2.98
10	0.096	0.664	3.40	3.77	6.80	15.95
50	0.549	2.531	21.60	9.90	59.30	64.88
100	0.643	3.090	55.34	13.95	82.31	89.58
LSD($p<0.05$)	0.970		15.10		3.45	

Soil Amendments Applied. Liming or applying soil amendments, such as slag, to increase soil pH is an effective approach to reduce metal content in plants through decreased bioavailability. Among them, blast furnace slag from steel manufacture is more effective. For example, over 90% of Cd was reduced through application of 0.2-0.6% of the slag in a pot experiment (Table 8). The slag consists of Fe+FeO+Fe_2O_3 17.4%, CaO 53.1%, SiO_2 0.7%, and MgO 2.6%; the blast furnace slag consists of CaO 52% and SiO_2 32%. The higher amount of either total or extractable SiO_2 (23%) in blast furnace slag compared with 8% in the common slag possibly made it more effective. The pH value in the pot soil was increased from 5.5 to 7.2 (Chen, 1996).

CONCLUSION

From the above results we can conclude that some soils at mining sites, heavy industrial, or wastewater irrigation areas have been contaminated by heavy metals, although the contents based on the whole country are not very high compared with some other countries. Mining, wastewater irrigation, sewage sludge, or domestic waste disposal associated with atmospheric deposition and metal containing agrochemical application are the major contribution to soils.

TABLE 8. Cadmium content in the grain and straw of paddy rice with different amendments applied (mg kg^{-1}DW).

Treatment and rates (%)		Unpolished rice		Straw	
		mg kg^{-1}	% decreased	mg kg^{-1}	% decreased
CK		4.13	-	41.1	-
Liming	0.2	3.18	23.0	20.7	49.6
($CaCO_3$)	0.4	2.56	38.0	13.1	68.1
Common slag	0.2	2.91	29.5	30.9	24.8
	0.4	1.77	57.1	10.1	75.4
	0.6	0.43	89.6	2.57	93.7
Blast furnace	0.2	0.32	92.3	2.57	93.7
slag	0.4	0.34	91.8	2.33	94.3
	0.6	0.35	91.5	2.02	95.1
L.S.D.($p<0.05$)		0.79		3.43	
L.S.D.($p<0.01$)		1.08		4.71	

For safety, food production and contaminated land restoration, some practical approaches, such as growing non-edible plants, breeding site exploitation, choosing some tolerate varieties or crops with less amount of metals accumulated in the edible parts or applying some amendments are acceptable and beneficial to farmers. However, phytoremediation is a most hopeful and low-cost approach to remove the metals from soil as long as a hyperaccumulator has been put into practice, growing well with a reasonable quantity of biomass produced.

ACKNOWLEDGEMENT

The program was financially supported by a foundation from Chinese Academy of Sciences as a key project in Knowledge Innovation Strategy (KZCX2-401-1-1).

REFERENCES

Cai, S., S. Li, and X. Zhao. 1985. "Effect of Sewage Irrigation on Eco-environments and Strategies in Jing-Jin-Tang Area" (in Chinese). *Agric. Eco-Environment. 4*: 1-5.

Chen, H. 1996. *Heavy Metal Contamination in Soil--Plant Systems* (in Chinese). Scientific Press of China, Beijing.

McGrath, S.P. 2000. "Risk Assessment of Metals." In Y. Luo and S. P. McGrath (Eds.), *Proceedings of SoilRem 2000, Int. Conf. of Soil Remediation.* pp. 1-7 (Oct. 15-19, Hangzhou, China).

Mulla, D.J., A.L. Page, and T.J. Ganje. 1980. "Cadmium Accumulation and Bioavailability in Soils from Long Term Phosphorus Fertilization". *J. Environ. Qual. 9*: 408-412.

Smith, S.R. 1996. *Agricultural Recycling of Sewage Sludge and the Environment.* CAB international, Wallingford, UK.

Wang, K. 1997. "Current status of Cd contamination of farmland in China and strategies for remediation" (in Chinese). *Agro-Env. Protec.* 16 (6): 274-278.

Wang, Q. 1998. "Integrated Amendment and Ecological Restoration of Polluted Soil by Heavy Metals" (in Chinese). *Proceedings for Strategy of Soil Environmental Protection in the New Century of China.* (Oct. 26-28. Beijing, China).

Wang, Q. 2000. "Phytoremediation-A Unique Approach to Restoration of Contaminated Soil with Heavy Metals in China." In Y. Luo and S. P. McGrath (Eds.), *Proceedings of SoilRem 2000, Int. Conf. of Soil Remediation.* pp. 197-202 (Oct. 15-19, Hangzhou, China).

Wang, Q., and J. Li. 1999. "Fertilizer Proper Use and Sustainable Development of Soil Environment in China" (in Chinese). *Advances of Env. Sci. 7*(2): 116-124.

Wang, Q., Y. Dong, Y. Cui, and X. Liu. Instances of Soil & Crop Heavy Metal Contamination in China. Soil and Sediment Contamination, 2001 (in press).

Wong, M. H., and J. W. C. Wong. 1997. "Environmental Health of Sewage Sludge Recycling with Emphasis on Land Application." In Z. Cao (Ed.). *International*

Symposium on Soil, Human and Environment Interactions. pp. 223-236. China Sci. and Tech. Press, Beijing.

Xu, J., and J. Yang (Eds.). 1995. *Heavy Metals in Land Ecosystems* (in Chinese). China Environmental Science Press, Beijing.

Zhao, Q. 2000. "The Environmental Problems of China". In Y. Luo and S.P. McGrath (Eds.), *Proceedings of SoilRem 2000, Int. Conf. of Soil Remediation*. pp. 14-16 (Oct. 15-19, Hangzhou, China).

Zhou, X. 1987. "Polluted Status of Cd in Zhangshi Irrigation Area and Solutions" (in Chinese). *Agro-Env. Protec. 6*(2):17-19.

TREATMENT OF METAL-CONTAMINATED SOIL AND WATER BY SULPHATE REDUCING BACTERIA

Torbjörn Håkansson and Bo Mattiasson (Lund University, Lund, Sweden)

ABSTRACT: To mobilise heavy metals for recovery under organised conditions is an interesting challenge to soil remediation. Use of a chelating substance to transfer the metal ion into solution and a subsequent sulphide precipitation step has been studied. To precipitate the heavy metal ion from a chelated form is somewhat more complicated than from free water solution. The choice of the chelating substance is in this sense essential. A range of chelating substances has been used in treatment of contaminated soil samples and the precipitation efficiency by microbially produced sulphide to precipitate these ions has been evaluated.

INTRODUCTION

Heavy metals are causing severe environmental problems. Contrary to organic pollutants that eventually may get degraded, there is no such remedy for heavy metals. They have to be taken out of circulation in the ecosystem. Heavy metals are polluting soil and they also constitute a threat to ground water. A range of studies focusing on enrichment/precipitation of heavy metals, by biotechnological means, have been presented recently (Bayoumy et al., 1999; Kolmert et al., 1997; Steed et al., 2000). When selecting the method of choice to carry out precipitation, it is important that it is efficient and that it can be operated at reasonable costs. Precipitation of di- and tri-valent ions as sulphides is a well-known strategy in inorganic chemistry. In order to reduce costs and to make the process robust, a bacterial step has been used to generate the sulphide. This is done by using sulphate-reducing bacteria (SRB). (Hallberg, 1980; Kolmert et al., 1997)

Heavy metals in soil are often relatively immobile, but leakage takes place and that constitutes a threat to the surroundings. Two strategies have been applied: either to increase the immobility, or by facilitating mobility in order to harvest them downstream (Hong et al., 1995). This latter technique is the topic of this report.

The mobilised heavy metal ions present in a leachate are mixed with the effluent from a SRB-reactor. In the reactor sulphate has been reduced to sulphide. After the mixing, metal sulphides are formed and due to their solubility properties, they can be harvested as a precipitate. From Table 1 is seen solubility product constants for some of the metal sulphides studied here.

TABLE 1. Solubility product constants (K_{sp}) at 25°C for some metal sulfides. The solubility product refers to the reaction M^{2+} (aq) + H_2S (aq) ↔ MS(s) + $2H^+$ (aq) (Lide, 1995)

Metal	Formula	Solubility Product
Copper (II)	CuS	$2*10^{-16}$
Lead	PbS	$3*10^{-27}$
Zinc	ZnS	$2*10^{-04}$

MATERIALS AND METHODS

Leaching of Soil. The soil used in the experiment was taken from a contaminated site in Stockholm, Sweden. It was sieved and the fraction of particles <2 mm was used in the leaching experiments. The leaching studies were set up by mixing 1 gram of soil with 50 mL of the chelating solution. The chelators used were disodiumethylendiaminetetraacetic acid (EDTA), diethylenediaminepentaacetic acid (DTPA), trans-1,2-diaminocyclohexanetetraacetic acid (CDTA) and nitrilotriacetic acid (NTA). All chelators were bought from Merck Eurolab AB, Stockholm, and were of *pro analysi* quality. The concentrations of 0.1 and 1.0 mM were used.

SRB System. The SRB-cells were used as immobilised preparations. Two types of reactors were used: a packed bed [1 litre] (A) and a stirred tank reactor [1 litre] (B). The immobilised cells were allowed to grow on plastic carriers (Kaldnes Miljøteknologi A/S, Norway), and the amount of carriers in the stirred tank experiment filled 50% of the reactor volume. Substrate composition was the same as described elsewhere (Kolmert et al., 1997).

Each reactor was run continuously with a medium feed of 20 mL/h. The sulphide ion rich outflow (5 – 10 mM sulphide) from the reactors were mixed in a ratio of 17:1 (v/v) with a synthetic solution containing 1mM Cu 1.5 mM EDTA. Mixing the sulphide rich water with the metal containing water was done outside the SRB-reactor, since earlier experiments have shown that otherwise precipitation and eventual clogging may appear.

Precipitation. Upon mixing precipitate started to form instantaneously. The solution was held in the mixing tank for almost three hours before it passed into a settling tank where the precipitated metal sulphide was harvested. This study was carried out with synthetic mixtures as well as with real leachates from contaminated soil.

Sampling. In order to follow the progress of the reaction, samples were taken between the mixing chamber and the settling tank for analysis of metal ion content. Samples of 5 mL each were taken using a 0.45 μm filter, thereby avoiding the precipitates.

Analysis. In order to get a value of the starting concentration of heavy metals in the soil samples; 1-gram aliquots were each mixed with 20 mL of 7 M HNO_3 and then autoclaved for 30 min. at 120 °C at 2 atm. (Karstensen et al., 1998). The metal content in a filtered supernatant was analysed using an atomic absorption spectrometer (AAnalyst 800, Perkin Elmer). The aqueous leachates were analysed using the same instrument.

RESULTS AND DISCUSSION

Soil Extraction. Soil was extracted in batch mode using different concentrations of the chelating substance. From Figure 1 is seen that the concentration of the chelating substance plays an important role, whereas the time for complexation means less. The soil had an initial content of Cu of 51 μg/kg of soil, and from that approx. 20 % could be extracted.

The kinetics of the process show that 90 % of what could be extracted under the conditions used was already in solution within 30 minutes in the batch experiments. This makes the prospects good for using this method for continuous extraction when a solution of a chelating substance is pumped through the contaminated soil.

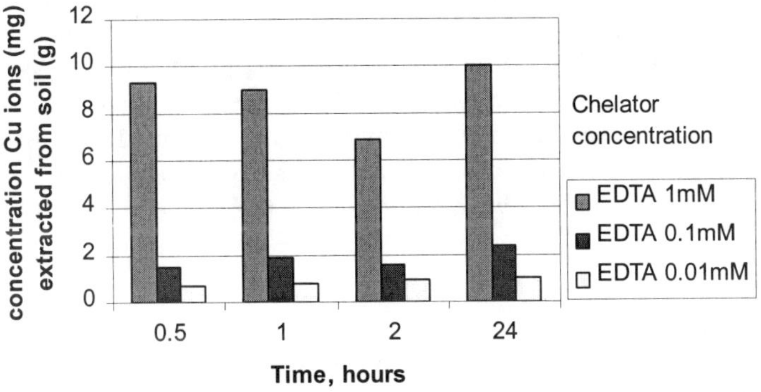

FIGURE 1. The efficiency by EDTA to remove Cu^{2+} from soil depending on the concentration

The choice of a suitable chelating substance may be critical. A range of different synthetic chelating substances were evaluated. The results from use of 4 different chelating substances in the soil extraction tests are shown in Figure 2. There is not much difference between the different chelating substances, and at the same time there is only a fraction of the total metal content that is extracted.

The extraction was run over one week and the data are relatively stable indicating that what happens will happen during the initial phase. However, when using EDTA, there is a reduction in the amount of extracted copper with time. The reason behind this observation is not clear yet. It might be that the EDTA-molecules are degraded, or that the complex is adsorbed to the soil particles.

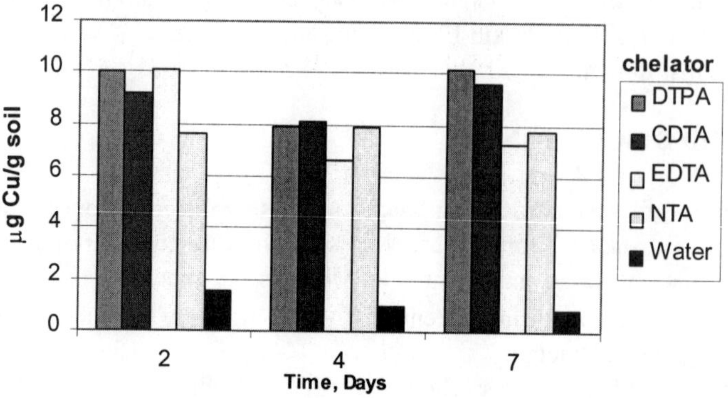

FIGURE 2. Extraction efficiency of different chelator substances at 1 mM concentration in their efficiency to extract Cu from soil to the water phase

Cu Precipitation. When extracting heavy metals as chelated complexes, it is important to show that the precipitation method that is meant to be used in order to remove the metal ions will also work when the metals are complexed. Therefore, trials were initially set up using a synthetic preparation of Cu-EDTA which was exposed to the biologically produced sulphide from the bioreactor. The metal chelate complex was formed by mixing 1 mM Cu^{2+} with 1.5 mM EDTA. Measurements were started on day 2 and the results are shown in Figure3. Approx. 95 % of the copper ions were precipitated in this system.

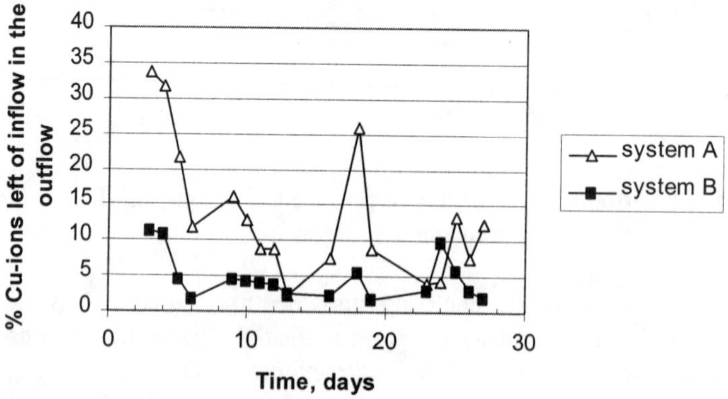

FIGURE 3. Precipitation of Cu-ions with H_2S produced by SRB. The percentage is the remaining free Cu-ions in the system

In Figure 3 there are plotted data from two different systems studied. A) represents trials using the effluent from a packed bed reactor as sulphide sources, and B) represents the sulphide produced by a stirred batch reactor with immobilised *Desulphovibrio* sp.. The higher efficiency in the latter is interpreted in terms of a better precipitation pattern. The effluent from the bioreactor contains besides the sulphide, also some bacteria. These are supposed to be used as nucleation sites in the precipitation process.

It is from the studies shown that copper ions in chelated form can be precipitated quite efficiently. A comparison with free ions shows however that there is a sluggishness in the precipitation behaviour, and also that the remaining concentration of copper ions in the complex-containing solution is higher than when the metal ion is free in solution.

This may lead to an intricate balance when choosing the chelating substance such that an efficient complexation is needed in order to get an efficient soil cleaning. At the same time one needs to strive for operation with complexes that will release the metal ions in the precipitation step. The biodegradability of the synthetic chelating substances used is an issue of interest if they are going to be used under practical remediation situations. It has in earlier studies been demonstrated that EDTA is quite efficiently degraded using specially enriched cultures or in wastewater treatment plants (Nörtemann, 1999). However, the risk evaluation of using such chelating substances has to be done before field trials can be initiated.

CONCLUSIONS

The effluent from sulphate reducing bacteria reactors can be used to precipitate heavy metals from water solutions. The metal can be either free in solution or bound in a chelate complex. Such knowledge opens up for an active soil remediation strategy where heavy metal ions will be mobilised from the site of pollution and then recovered in a designed precipitation step. The studies reported here deal with copper ions, but the strategy is applicable for a whole range of heavy metal ions that can be precipitated as metal sulphides.

ACKNOWLEDGEMENTS

This project was supported by The municipality of Stockholm and it was operated in collaboration with the COLDREM-program supported by The Swedish Strategic Fund for Environmental Research (MISTRA)

REFERENCES

Bayoumy, M. E., J. K. Bewtra, H. I. Ali and N. Biswas. 1999. “Removal of heavy metals and COD by SRB in UAFF reactor.” *Journal of environmental engineering* 125(6): 532-539.

Hallberg, R. O. 1980. "Process for precipitating heavy metals from wastewater", US4354937: 4.

Hong, A., T.-C. Chen and R. W. Okey. 1995. "Chelating extraction of copper from soil using s-carboxymethylcysteine." *Water Environment Research* 67(6): 971-978.

Karstensen, K. H., O. Ringstad, I. Rustad, K. Kalevi, K. Jörgensen, K. Nylund, T. Alsberg, K. Ólafsdóttir, O. Heidenstam and H. Solberg. 1998. "Methods for chemical analysis of contaminated soil samples - tests of their reproducibility between Nordic laboratories." *Talanta* 46(3): 423-437.

Kolmert, Å., T. Henrysson, R. Hallbeg and B. Mattiasson. 1997. "Optimization of sulphide production in an anaerobic continuous biofilm process with sulphate reducing bacteria." *Biotechnology letters* 19(10): 971-975.

Lide, D. R., Ed. 1995. "CRC Handbook of Chemistry and Physics", CRC Press Inc.

Nörtemann, B. 1999. "Biodegradation of EDTA." *Applied micorobiology and Biotechnology* 51(6): 751-759.

Steed, V. S., M. T. Suidan, M. Gupta, T. Miyahara, C. M. Acheson and G. D. Sayles. 2000. "Development of a sulfate-reducing biological process to remove heavy metals from acid mine drainage." *Water environment research* 72(5): 530-535.

METAL SPECIATION AND TOXICITY IN CHROMATED COPPER ARSENATE-CONTAMINATED SOILS

Cristina F. Balasoiu, ***Gérald J. Zagury***, and Louise Deschênes
(École Polytechnique de Montréal, Montréal, Québec, Canada)

ABSTRACT: The influence of composition and physicochemical characteristics of soils on speciation and toxicity of Cu, Cr and As in chromated copper arsenate (CCA)-contaminated soils was assessed. Physicochemical properties of synthetic soils were determined. Cu and Cr partitioning was investigated using a sequential extraction procedure while As speciation was assessed by a modified solvent extraction method. The earthworm *Eisenia fetida* survival test was used to assess the toxicity of CCA-contaminated soils. Cr and Cu present in organic soils contaminated with CCA were present in less mobile and less bioavailable forms, whereas in mineral soils, the soluble and exchangeable fraction was higher. Arsenic was present principally as As(V). The proportion of As(III) increased with the soil organic matter content. Even though organic soils were more contaminated than mineral soils, they weren't toxic for *Eisenia fetida*. It was shown that the toxicity of CCA-contaminated soils is related to the partitioning of metals.

INTRODUCTION

CCA is the most common wood preservative used in North America. There is an increasing concern about a possible environmental contamination with Cr, Cu and As from leaching losses from wood in service. Those metals are known to be toxic, but there has been little research on the effects of CCA-contaminated soils on biota.

To assess the environmental impact of contaminated soils, knowledge of the total concentration of a specific metal without considering its speciation is not sufficient. The soil composition can widely influence metal speciation. The speciation not only affects the mobility of the metal, but also affects its bioavailability and therefore its toxicity (McLean and Bledsoe, 1992).

Metals may be distributed among many components of the soil. An interesting experimental approach commonly used for studying partitioning and mobility of metals in soils is to use sequential extraction procedures (SEP).

The difficulties related to the assessment of soil quality can be minimized by combining the chemical and ecotoxicological analysis (Debus and Hund, 1997)

Objective. To our knowledge, no data is available on the partitioning, speciation, and toxicity of Cr, Cu and As in various types of CCA contaminated soils. Therefore, the objective of this study was to determine the influence of soil composition and physicochemical characteristics on partitioning and toxicity for *Eisenia fetida (E. fetida)* of Cr, Cu and As present in five synthetic CCA contaminated soils.

MATERIALS AND METHODS

Materials. Synthetic soils with different physicochemical properties were prepared. Soil components were: kaolinite, silt, sand, and organic matter.

A glacier till, from a borrow pit in Northern Quebec (Canada), was used to obtain sand and silt. The < 2 mm fraction was sieved and the silt and clay fraction was retained (< 75 μm). This fraction contained (dry weight %) 98 % of silt (< 75 μm) and 2 % of clay (< 2 μm). Sand fraction (2 mm – 75 μm) was washed with distilled water to remove the silt and clay-sized materials and then left to dry.

The kaolinite (EPK) was purchased from Sial (Quebec, Canada). According to the vendor, SiO_2 (46.2 wt %) and Al_2O_3 (37.7 wt %) were the main oxide components. Mn oxides were not detected. The organic matter source was *Spagnum* peat moss, purchased from Berger Peat Moss (Quebec, Canada). The peat was sieved and the < 1 mm fraction was retained.

Statistical Experimental Design. The constitutive proportions (wt %) of each component were : kaolinite (5 to 30 %), sand (30 to 69.5 %), organic matter (0.5 to 15 %), and silt (25 %). A fifth constraint was imposed: the sum of all components had to be 100 %.

All constraints were processed using the experimental design operating module of STATISTICA software *(Statistica for Windows. Computer program manual)*. An experimental space with five points was defined: four vertices (soil A, B, D, and E) and one centroid (soil C) (Figure 1).

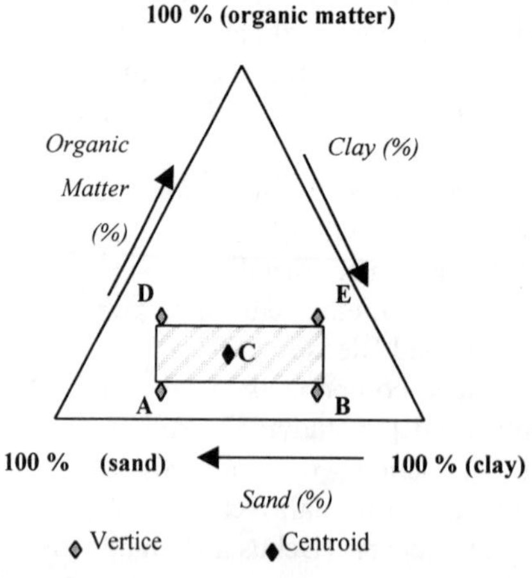

FIGURE 1. Experimental space.

Soil Synthesis. Batches of 100 g of each synthetic soil were prepared at the same time. Soil components were placed in a 500 mL polypropylene copolymerised screw cap bottle and agitated at room temperature (22 ± 1 °C) for 24 hours on a customized rotary agitator at 50 rpm.

Physicochemical Characterization. pH of sand, silt, kaolinite, and non-contaminated and contaminated soils was determined in distilled water using a soil:water ratio of 1:4. Peat pH was measured using a solid:liquid ratio of 1:16. The pH was mesured after 48 hours on triplicate samples. The pH of the synthetic soils had to be constant in order to avoid a bias of the toxicity responses during bioassays. The pH of soils C, D, and E was therefore increased to 5.5 ± 0.1 by adding powdered $CaCO_3$. All further analyses were performed onto pH adjusted soils.

Volatile solids (VS) of soil components, non-contaminated soils and contaminated soils were determined on triplicate samples at 550 °C according to Karam (1993). Cation exchange capacity (CEC) of each soil component, of synthetic soils, and of CCA contaminated soils was determined on duplicate samples according to Chapman (1965). Total Cr and Cu concentrations in non-contaminated and contaminated synthetic soils were determined after digestion with HNO_3, HF and $HClO_4$ (Clesceri et al.,1998) whereas total As content was determined after digestion with HCl according to Chappell et al. (1995). Analyses were performed using inductively coupled plasma atomic emission spectrometry (ICP-AES).

Soil Contamination Procedure. A 60 % (w/w) commercial CCA solution containing, on an oxide basis, 45.5 % CrO_3, 18.2 % CuO, and 36.3 % As_2O_5 was used to contaminate the synthetic soils. The CCA contamination level had to be realistic and compatible with Cr, Cu and As concentrations possibly found in contaminated soils close to wood treated poles in service or at timber treatment facilities. For that reason, 2573 mg of CCA (984 mg Cr, 984 mg As, and 605 mg Cu) were added per kg of dry soil. All soils were contaminated with the same concentration of CCA. Firstly, synthetic soil was humidified and allowed to stand for 24 hours (soil:solution ratio of 1:1). A known volume of CCA was then added (final soil:solution ratio of 1:3). Samples were shaken on a rotary agitator at 50 rpm for 24 hours. They were then left to stand for 120 hours. After centrifugation at 4 000 x *g* for 30 min, contaminated soil was transferred in a pyrex container and allowed to stand for 72 hours at room temperature until apparent dryness. Samples were disaggregated, passed through a 2-mm sieve, and homogenized by agitation (50 rpm) for 24 hours. Chemical and ecotoxicological analyses were carried out simultaneously.

Sequential Extraction Procedure. The basic utility of SEP is its use of appropriate chemical reagents in a manner that release different heavy metal fractions from soils by destroying the binding "agent" between the metal and the soil solids. In this study, the SEP of Tessier et al. (1979) was performed on the five CCA-contaminated soils (duplicate samples). A minor modification of this procedure was used for the determination of the residual fraction (Zagury et al., 1999). In order to evaluate the recovery of the SEP, a triplicate soil sample was simultaneously digested, following the fifth step procedure. The SEP operationally groups heavy metals into the following five fractions: F1: soluble and exchangeable (extracted with a $MgCl_2$ solution), F2: bound to carbonates or specifically adsorbed (leached by an HOAc/NaOAc solution), F3: bound to reducible Al, Fe and Mn oxides (extracted with an NH_5OH-HCl solution), F4: bound to oxidizable matter (released by HNO_3 and H_2O_2), and F5: residual metal fraction (dissolved with HNO_3, HF, and $HClO_4$).

Arsenic Speciation by Solvent Extraction. Arsenic speciation (As(V) and As(III)) was performed using an adaptation of the methods developed by Kamada (1976), Huang and Wai (1986), and Chappell et al. (1995). The method allows the recovery of all As present in the soil by multiple extractions with a 10M HCl solution, without changing its speciation. The speciation was then performed using solvent extractions (the complex ammonium pyrrolidine dithiocarbamate (APDC)-methyl isobutyl ketone (MIBK)). As(III), which is more toxic, was calculated by the difference between the results for total arsenic and As(V). Using a 25 % (v/v) HNO_3 solution, a back-extraction was performed in order to recover As(III) from the As(III)-APDC complex. Organic As species are not assessed with this method.

Ecotoxicity of CCA-contaminated Soils. The toxicity of CCA-contaminated soils was determined using the 14-day U.S. EPA (1989) earthworm survival test. Mortality of *E. fetida* was assessed on a weekly basis by lack of response to a gentle, mechanical stimulus. The toxicity of non-contaminated soils was also assessed.

RESULTS AND DISCUSSION

Soil Composition. The composition of the five synthetic soils is shown in Table 1. Soils can be classified for discussion purposes as mineral soils (A and B), moderately organic soil (C), and highly organic soils (D and E).

TABLE 1. Composition of synthetic soils (wt %).

Soil type	Peat	Kaolinite	Sand	Silt
Soil A	0.5	5	69.5	25
Soil B	0.5	30	44.5	25
Soil C	7.75	17.5	49.75	25
Soil D	15	5	55	25
Soil E	15	30	30	25

Soil Physicochemical Properties. The physicochemical characteristics of the synthetic soils are shown in Table 2. The initial pH of the soils was slightly acidic. It mainly varied as a function of peat content. In fact, the pH of sand and silt was neutral, while the pH of kaolinite (5.3 ± 0.1) and peat (4.1 ± 0.1) was acidic.

TABLE 2. Physicochemical characteristics of the non-contaminated soils.

Soil type	Initial pH[a]	CEC[b] *meq/100 g*	Volatile solids[a] *wt %*	Background level of Cr *mg/kg*	Background level of Cu *mg/kg*	Background level of As[c] *mg/kg*
Soil A	5.8 ± 0.1	4.7	1.3 ± 0.1	19.4	31.2	ND
Soil B	5.4 ± 0.2	12.0	4.2 ± 0.1	35.6	32.8	ND
Soil C	4.4 ± 0.2	24.1	9.5 ± 0.4	25.2	24.3	ND
Soil D	4.3 ± 0.1	37.8	15.1 ± 0.5	18.3	27.3	ND
Soil E	4.2 ± 0.1	42.0	18.6 ± 0.7	25.4	26.6	ND

[a] : mean values and standard deviations are calculated from three different determinations
[b] : mean values are calculated from two different determinations
[c] :ND: not detected (< 4 mg/kg)

Generally, organic matter and clay fractions are responsible for the CEC of any soil. In this study, peat had the greatest CEC (232 meq/100g), followed by kaolinite (38 meq/100g). Silt and sand had a CEC smaller than 2 meq/100g. Hence, the highly organic soils had the highest CEC. At a low organic matter content, it can be seen the small contribution of clay content to the CEC.

VS of the soil components were 96 ± 0.5 % for peat, 12 ± 0.7 % for kaolinite and < 0.5 % for sand and silt. Therefore, VS of the five soils were primarily correlated to peat content and to a lesser extent to kaolinite content. Metal background levels of the five non-contaminated soils were very low.

Characterization of Contaminated Soils. All soil characteristics were reassessed after contamination with CCA and they were very similar to the initial values.

Total Metal Content of Contaminated Soils. Metal concentrations retained in CCA-contaminated soils are shown in Tables 3, 4, and 5. Total Cu retained in soils vary from 47 to 97 %. Cu retention increases strongly with the increase of the organic matter content and with the CEC.

TABLE 3. Cu retention and partitioning in the CCA contaminated soils.

Soil type	Total Cu retained mg/kg [a]	% [b]	Fraction 1 *Soluble and exchangeable* mg/kg	%	Fraction 2 *Carbonates or specifically adsorbed* mg/kg	%	Fraction 3 *Reducible Al and Fe oxides* mg/kg	%	Fraction 4 *Oxidizable matter* mg/kg	%	Fraction 5 *Residual* mg/kg	%	Sum of fract. mg/kg
Soil A	283 ± 21	46.7	137.7	43.9	99.7	31.8	60.8	19.4	8.0	2.6	7.5	2.3	313.8
Soil B	391 ± 17	64.6	222.3	51.6	116.7	27.1	74.6	17.3	11.2	2.6	5.8	1.4	430.5
Soil C	568 ± 16	93.8	51.9	9.1	194.1	34.0	210.8	36.9	107.9	18.9	6.2	1.1	570.9
Soil D	585 ± 21	96.6	48.3	7.3	140.3	21.3	281.1	42.8	185.3	28.1	3.1	0.5	658.2
Soil E	579 ± 26	95.7	38.8	6.1	123.9	19.4	280.7	44.0	190.3	29.9	3.7	0.6	637.5

[a] : mean values and standard deviations are calculated from three different determinations
[b] : calculated as : (Total Cu / Cu added)*100

As seen in Table 4, there are considerable variations in the proportion of Cr retained in soil (from 19 to 87 %). Average Cr concentration in mineral soils is low and increases rapidly with the increase in organic matter content. As expected, kaolinite content does not increase the fraction of Cr retained.

TABLE 4. Cr retention and partitioning in the CCA contaminated soils.

Soil type	Total Cr retained mg/kg [a]	% [b]	Fraction 1 *Soluble and exchangeable* mg/kg	%	Fraction 2 *Carbonates or specifically adsorbed* mg/kg	%	Fraction 3 *Reducible Al and Fe oxides* mg/kg	%	Fraction 4 *Oxidizable matter* mg/kg	%	Fraction 5 *Residual* mg/kg	%	Sum of fract. mg/kg
Soil A	191 ± 21	19.4	54.3	18.8	39.0	13.5	160.1	55.4	20.2	7.0	15.4	5.3	289.1
Soil B	288 ± 102	29.2	60.6	18.3	46.9	14.2	165.5	50.0	25.9	7.8	32.0	9.7	331.1
Soil C	618 ± 63	62.8	3.1	0.5	76.7	11.9	373.0	58.0	162.1	25.2	28.0	4.4	643.0
Soil D	852 ± 47	86.5	1.1	0.1	113.3	14.0	410.1	50.8	259.4	32.2	21.0	2.6	805.1
Soil E	722 ± 71	73.2	1.6	0.2	55.7	7.9	360.5	51.0	260.9	36.9	28.8	4.0	707.5

[a] : mean values and standard deviations are calculated from three different determinations
[b] : calculated as : (Total Cr / Cr added)*100

Table 5 shows that total As concentrations are similar in mineral and organic soils. The retained proportion is high (from 71 to 81 %) in all soils. This is probably because the contamination was performed in very favourable conditions for arsenate retention in soils (slightly acidic pH (5.5), presence of organic matter and of kaolinite).

TABLE 5. As retention and speciation in the CCA contaminated soils.

Soil type	Total As retained mg/kg [a]	% [b]	Total As [c] mg/kg	As(V) mg/kg	As(III) calculated mg/kg	As(III) back-wash mg/kg	As(V) %	As(III) calculated %	As(III) back-wash %	Mass balance [d] %
Soil A	700 ± 38	71.1	720	684	36	34	95	5	4.7	100
Soil B	736 ± 33	74.7	751	678	73	39	90.3	9.7	5.2	95
Soil C	795 ± 29	80.7	781	637	144	118	81.5	18.5	15.1	97
Soil D	765 ± 40	77.7	742	534	208	200	71.9	28.1	26.9	99
Soil E	712 ± 11	72.3	718	493	225	174	68.7	31.4	24.2	93

[a] : mean values and standard deviations are calculated from three different determinations
[b] : calculated as : (Total As / As added)*100
[c] : mean values determined from two different determinations on arsenic extracts used for arsenic extraction
[d] : calculated as : ([As(V)]+[As(III) back-wash]/[Total As])* 100

Heavy Metal Partitioning in the CCA-contaminated Soils. The results of the SEP for Cu and Cr partitioning in soils are shown in Tables 3 and 4. The average recovery of the SEP [(sum of all fractions/total metal concentration) × 100 %] was 109 % and for Cr 124 %. There are considerable variations in the proportions of Cr and Cu in the various fractions as a function of soil composition. The only similarity is the metal proportion found in the residual fraction (F5).

In mineral soils, 47 % of Cu and 18 % of Cr are found in an easily leachable and bioavailable form (F1) that present a potential risk (see Tables 4 and 5). Moreover, Cu proportion found in F1 increases with the increase in kaolinite content and the increase in CEC (soil A and B). Very low levels of Cu are bound to oxidizable matter in these soils.

In moderately and highly organic soils, Cr and Cu are present in less mobile and less available forms for soil organisms and plants, whereas the fraction bound to organic matter is much higher. The soluble and exchangeable metal fraction is usually considered as the most hazardous (Maiz et al., 2000).

Furthermore, the level of Cr found in a reducible form (potentially mobile under strong reducing conditions) in the five soils was relatively high and constant. In short, metals present in organic soils contaminated with CCA were less mobile and less bioavailable than metals present in mineral soils. Globally, Cu was more mobile than Cr.

Arsenic Speciation in the CCA-contaminated Soils. As seen in Table 5 this element is present in two species (As(V) and As(III)). As(V) proportion decreases as the organic matter content of soils increases.

It is generally recognized that As(V) is the major species present in oxidized acidic environment, while in reducing and alkaline conditions, As(III) becomes significant (Masscheleyn et al., 1991). In this study As(III) was found in significant proportions (average of 30 % in organic soils). This is consistent with the studies of several authors who also found As (III) in large proportions in conditions theoretically favourable to the presence of As (V) (Masscheleyn et al., 1991). Some reduction of As(V) to As(III) took place in the CCA-contaminated soils, because As is present as As(V) in the wood preservative Generally, the mechanisms controlling the rate of reduction in soils are poorly understood. Reduction of As (V) to As(III) is predominantly chemical (Bowell et al., 1994), but can also occur as a result of biotic processes (Pongratz, 1998).

Ecotoxicity of CCA-contaminated Soils. It is evident that there is a difference in the toxicity of the five CCA-contaminated soils (Figure 2). The earthworm *E. fetida* didn't survive in mineral soils. However, the moderately and highly organic soils weren't toxic for the earthworm. These significant differences are mainly correlated to the soil organic matter content. No mortality was observed in all non-contaminated soils.

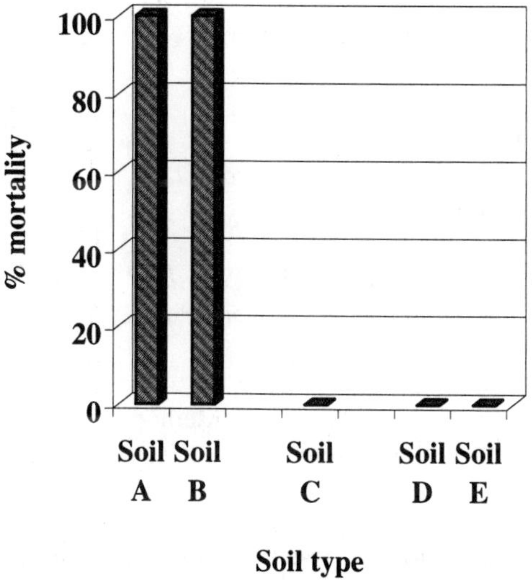

FIGURE 2. Effect of soil composition on earthworm survival.

Total metal concentration in soil is not a sufficient criterion to assess their toxic effects. It can be seen that organic soils, more contaminated than mineral soils, have no effects on the survival of the *E. fetida*.

Because As concentration is similar in all soils, but toxic effects of CCA-contaminated soils are different, it can be supposed that the toxic responses are due to the presence of Cu and Cr. It is very probable that the proportion of As(III) is not in a bioavailable form in order to cause toxic effects to earthworm *E. fetida*.

Soil composition affects significantly speciation, bioavailability, and thus, toxicity of a contaminant. The soluble and exchangeable fraction (F1) is generally considered as the fraction available for the tested organism. Tables 3 and 4 show that comparatively to organic soils, in mineral soils, the proportion of metals (Cr and Cu) in this fraction is very high. Thus, there is a direct relationship between toxic effects of CCA-contaminated soils and metal concentration present in F1: high levels of metals in F1 imply high toxicity and a greater risk for the environment.

CONCLUSIONS

In this study it was shown that soil composition has a great influence on the partitioning and toxicity of Cr and Cu present in CCA-contaminated soils. The speciation of As was also influenced by soil composition. Combination of chemical analyses and bioassays provides a better assessment of the potential risks associated with CCA-contaminated soils.

ACKNOWLEDGEMENTS

The authors acknowledge the financial support of the Chair partners: Alcan, Bell Canada, Canadian Pacific Railway, Cambior, Centre d'expertise en analyse environnementale du Québec, Gaz de France/Électricité de France, Hydro-Québec, Ministère des Affaires Municipales et de la Métropole, Natural Sciences and Engineering Research Council (NSERC), Petro-Canada, Solvay, Total Fina Elf, and Ville de Montréal.

REFERENCES

Bowell, R. J., N. H. Morley, and V. K. Din. 1994. "Arsenic Speciation in Soil Porewaters from Ashanti Mine, Ghana". *Appl Geochem. 9*: 15-22.

Chappell, J., B. Chiswell, and H. Olszowy. 1995. "Speciation of Arsenic in a Contaminated Soil by Solvent Extraction". *Talanta. 42*: 323-329.

Chapman, H. D. 1965. "Cation-Exchange Capacity". In C. A. Black (Ed), *Methods of Soils Analysis. Chemical and Microbiological Properties,* pp. 891-901. American Society of Agronomy Inc, Madison.

Clesceri, L. S., A. E. Greenberg, and A. D. Eaton (Eds.). 1998. *Standard Methods for the Examination of Water and Wastewater*. American Public Health Association, Washinghton, DC.

Debus, R., and K. Hund. 1997. "Development of Analytical Methods for the Assessment of Ecotoxicological Relevant Soil Contamination". *Chemosphere. 35*(1/2): 239-261.

Huang, Y. Q., and C. M. Wai. 1986. "Extraction of Arsenic from Soil Digests with Dithiocarbamates for ICP-AES Analysis". *Commun. Soil Sci. Plant. Anal. 17*(2): 125-133.

Kamada, T. 1976. "Selective Determination of Arsenic (III) and Arsenic (V) with Ammonium Pyrrolidinedithiocarbamate, Sodium Diethyldithiocarbamate and Dithizone by Means of Flameless Atomic-Absorption Spectrophotometry with a Carbon-Tube Atomizer". *Talanta. 23*: 835-839.

Karam, A. 1993. "Chemical Properties of Organic Soils". In M. R. Carter (Ed), *Soil Sampling and Methods of Analysis*, pp. 459-471. Lewis Publishers, Boca Raton.

Maiz, I., I. Arambarri, R. Garcia, and E. Millan. 2000. "Evaluation of Heavy Metal Availability in Polluted Soils by Two Sequential Extraction Procedures using Factor Analysis". *Environ. Pollut. 110*: 3-9.

Masscheleyn, P. H., R. D. Delaune, and W. H. Jr Patrick. 1991. "Effect of Redox Potential and pH on Arsenic Speciation and Solubility in a Contaminated Soil". *Environ. Sci. Technol. 25*: 1414-1419.

McLean, J. E., and B. E. Bledsoe. 1992. *Behaviour of Metals in Soils*. U.S. Environmental Protection Agency Technical Repport, EPA 540/S-92/018, Office of Research and Development, Ada, OK.

Pongratz, R. 1998. "Arsenic Speciation in Environmental Samples of Contaminated Soil". *Sci. Total. Environ. 224*: 133-141.

U.S. EPA. 1989. *Earthworm Survival.* Protocols for short-term toxicity screening of hazardous wastes sites. Environmental Research Laboratory. Corvallis. OR EPA. Technical report EPA/600/3-88/029, 63-68.

Tessier, A., P. G. C. Campbell, and M. Bisson. 1979. "Sequential Extraction Procedure for the Speciation of Particulate Trace Metals". *Analyt. Chem. 51*: 844-851.

Zagury, G. J., Y. Dartiguenave, and J.C. Setier. 1999. "Ex Situ Electroreclamation of Heavy Metals Contaminated Sludge : Pilot Scale Study". *J Environ. Eng. 125*: 972-978.

NAPHTHALENE DEGRADATION AND CONCURRENT Cr(VI) REDUCTION BY *PSEUDOMONAS PUTIDA* ATCC17484

Subhasis Ghoshal (McGill University, Montreal, QC, Canada)
A. Al-Hakak (McGill University, Montreal, QC, Canada)
J. Hawari (NRC Biotechnology Research Institute, Montreal, QC,Canada)

ABSTRACT: Polycyclic aromatic hydrocarbons (PAHs) and heavy metals are often present together at contaminated sites. These sites are considered difficult to clean-up using bioremediation technologies because of the toxicity of heavy metals to microorganisms. Laboratory experiments were conducted to verify if biodegradation of a model PAH, naphthalene, could be sustained over a long time-period (days to weeks) in the presence of a common heavy metal contaminant, Cr(VI). Complete reduction of Cr(VI) at concentrations up to 6.3 mg/L was achieved by *Pseudomonas putida* ATCC 17484 in the presence of naphthalene that was continuously released from an immiscible organic liquid, 2,2,4,4,6,8,8-heptamethylnonane (HMN). The HMN did not influence the growth of bacteria. The presence of Cr (VI) inhibited bacterial growth and reduced naphthalene biodegradation rates. Despite this, the microorganisms were able to degrade significant masses of naphthalene.

INTRODUCTION

Polycyclic aromatic hydrocarbons (PAHs) and heavy metals such as arsenic, chromium, copper, lead, nickel, and zinc are two classes of pollutants often found at contaminated sites. Heavy metals and PAHs are frequently found together as contaminants in soils and groundwater at sites of industrial operations such as manufactured gas plants, wood treatment, metal finishing, petroleum refining, paint and ink formulating, and automobile part manufacturing plants (Kong et al., 1995; Luthy et al., 1994; Muller et al., 1989). At such 'mixed-waste' sites, elimination of organic contaminants such as PAHs using bioremediation techniques is considered difficult because of the toxicity of heavy metals to microorganisms. However, several recent studies have reported that microorganisms can transform certain heavy metal species while carrying out its metabolic functions (Shen et al., 1996a). This suggests that it may be feasible to use bioremediation technologies to simultaneously address clean up of specific heavy metals and organic pollutants at mixed waste sites.

The overall objective of this research was to study the possibility of sustained biodegradation of a PAH compound, naphthalene, in the presence of a heavy metal, chromium. Both of these compounds are present in agents used commercially for wood treatment.

The biodegradation of naphthalene has been studied extensively in the context of bioremediation (Cerniglia, 1992; Ghoshal and Luthy, 1998). The feasibility of biotransformation of chromium by soil bacteria has only been recently reported. The most common species of chromium in the environment are its salts and ions containing chromium in the valence states +3 (trivalent or Cr (III)) and +6 (hexavalent or Cr(VI)). The trivalent chromium is the most thermodynamically stable form under ambient environmental conditions (Wittbrodt and Palmer, 1991). The trivalent form generally sorbs strongly to soil and thus its transport is retarded. Hexavalent chromium is more water soluble, is toxic to biota at relatively low concentrations, and is carcinogenic to humans. Several soil microorganisms capable of reducing soluble Cr(VI) into the less soluble Cr(III) have been identified (Ishibashi et al., 1990; Wang and Shen, 1995). Cr(VI) reduction by both gram positive and gram negative bacteria occurs both in anaerobic and in aerobic environments. While several studies have shown that Cr(VI) reduction by bacteria is feasible using easily degraded carbon sources such as glucose and acetate (Wang and Shen, 1995) a few studies have reported that chromate reduction can be coupled with the oxidation of aromatic compounds (Shen and Wang, 1995; Shen et al., 1996a; Shen et al., 1996b). The reduction of chromate coupled with the oxidation of aromatic compounds has significant applications in the simultaneous transformation of both contaminant types in the environment.

MATERIALS AND METHODS

Naphthalene mineralization and chromate reduction was evaluated in experiments carried out in 250-mL biometer flasks fitted with a side arm, and containing aqueous solutions of naphthalene and Cr(VI) (as potassium dichromate). The aqueous solution contained mineral salts to support growth of *Pseudomonas putida* ATCC 17484, a PAH degrading microorganism, that was added to the flasks at the beginning of each test. Naphthalene was dissolved into liquid 2,2,4,4,6,8,8-heptamethylnanone (HMN) at 70 mg naphthalene/mL HMN, and introduced into the biometer flasks. The naphthalene-HMN mixture served as a source-phase liquid from which the naphthalene could be released continuously into the aqueous phase as it was degraded by microorganisms. The advantage of having the naphthalene in HMN rather than in crystals is that identical initial rates of naphthalene dissolution was achievable in all test systems, because the 0.7 mL of HMN added to each system provided a constant interfacial area for mass transfer. Crystals of naphthalene, however, are irregularly shaped and thus each batch system, even when containing the same mass of naphthalene crystals, has a different interfacial area for dissolution. This makes it impossible to obtain identical initial rates of dissolution in all test systems. HMN was not toxic and not biodegraded by the bacteria used. Initial concentrations of Cr (VI) were varied in the flasks while the initial naphthalene concentration dissolved in HMN was constant in all test systems.

Radiotracer techniques methods were used to assess naphthalene mineralization. Predetermined quantities of ^{14}C-naphthalene were added to HMN along with unlabelled naphthalene. To each biometer flask, 50 mL freshly prepared MSM and 0.7 mL HMN containing 49 mg unlabelled naphthalene and a certain amount of radiolabelled naphthalene dissolved in HMN were added. Cr(VI) as potassium dichromate was added in concentrations ranging from 2.1 to 12 mg Cr(VI)/L to the biometer flasks. All flasks were incubated at 29°C in a rotary shaker set at 120 rpm. Aerobic conditions were maintained by adding air to the biometer headspace during each sampling event. Five millilitres of 2 N NaOH were added to the side arm of the flask to trap CO_2 and $^{14}CO_2$ evolved during mineralization. The side arm NaOH solution was sampled at various times and counted on a Beckman LS500TD liquid scintillation counter. At the start of the experiment, biometers were inoculated with 5 mL of active culture of *Pseudomonas putida* ATCC 17484 grown on naphthalene. The concentration of the bacteria in the culture at the time it was added to the biometer flask was measured using spread plate techniques. Standard sterile techniques were followed for all experimental procedures. At the end of each test, the remaining radioactivity in the flask was measured to determine a mass balance for ^{14}C-naphthalene.

Parallel biometer flasks with identical contents but lacking radiolabelled naphthalene were used for the monitoring Cr(VI) reduction over time. Aqueous samples withdrawn from each biometer flask at different time intervals were filtered though a 0.45 μm filter. Hexavalent chromium was determined colorimetrically by reaction with diphenylcarbazide in acid solutions as described in APHA (1998). Briefly, the filtered samples were reacted with concentrated sulfuric-nitric acid mixture then reacted with diphenylcarbazide. Samples were analyzed with a spectrophotometer set at 540 nm. Interference of salts in the growth medium was accounted for by using calibration curves obtained with the same MSM. Additional details of the methods used for all experiments are described in Al-Hakak (2001).

RESULTS AND DISCUSSION

Cr (VI) Reduction Kinetics during Naphthalene Degradation. The time-profile of Cr(VI) reduction during naphthalene degradation is presented in Figure 1. The results indicate that *Pseudomonas putida* ATCC 17484 was able to completely reduce Cr(VI) at concentrations of 2.1 and 6.3 mg/L. It took more than 550 hours to completely reduce 6.3 mg/L (0.12 mM) compared to 40 hours to reduce 2.1 mg/L (0.04 mM) of Cr(VI) respectively. Thus, the higher the Cr(VI) concentration the longer it took for its complete reduction. When inoculated biometer flasks containing 2.1 mg/L Cr(VI) had no added carbon source (naphthalene), approximately 20 % of Cr(VI) was reduced within the first 21 hours following which there was no further decrease in Cr(VI) concentration even after 240 hours of incubation. Thus in the absence of a carbon source, the microbial cells transformed some of the Cr(VI), suggesting that bacterial enzymes such as chromate reductase could be responsible for part of the reduction. Cr(VI) reduction by bacterial enzymes has been observed by other studies (Ishibashi et al., 1990; Shen and Wang, 1994; Wang and Shen, 1995). Biosorption is not suspected because Cr(VI) concentrations were similar in cell free samples and in cell suspensions extracted with a concentrated sulfuric-nitric acid mixture. Cr(VI) was not reduced in systems where naphthalene was present but no bacteria were added.

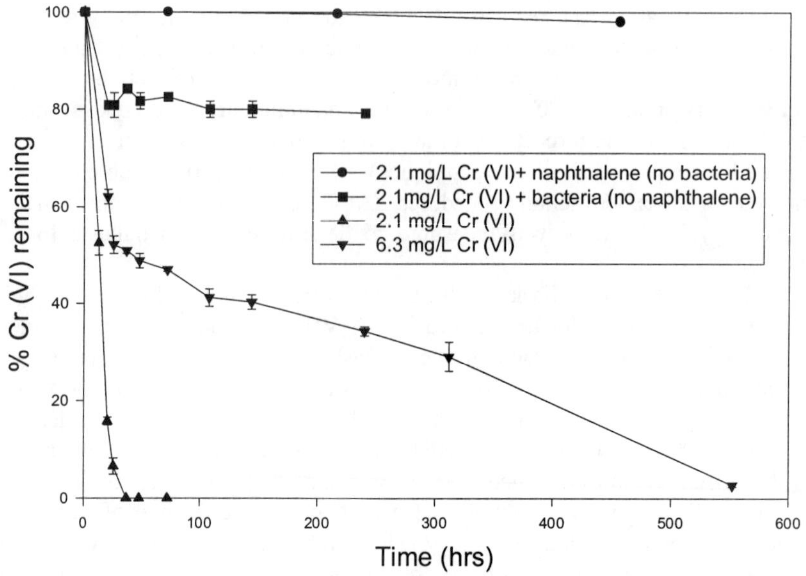

FIGURE 1. Percentage of Cr (VI) remaining in the presence and absence of naphthalene. Initial cell concentration in all inoculated reactors = 9.0 X 10^7 cfu/mL.

Naphthalene Mineralization Kinetics at Different Cr (VI) Concentrations. Naphthalene mineralization by *Pseudomonas putida* ATCC 17484 under different Cr (VI) concentrations was evaluated in biometer flasks similar to those used in the Cr (VI) reduction experiments. Figure 2 shows the effect of different concentrations of Cr(VI) on naphthalene degradation. There was a notable difference in the time at which significant mineralization commenced in the different systems. For systems containing 2.1 mg/L Cr(VI), significant naphthalene mineralization activity was observed less than 50 hours after inoculation whereas for the systems containing 6.3 mg/L of Cr(VI), significant naphthalene mineralization was observed only 400 hours after inoculation. The initial bacterial concentrations in both systems were 2.1 X 10^8 cfu/mL. The mineralization rates decreased as initial Cr(VI) concentrations were increased. In samples containing 6.3 mg/L of Cr(VI), the overall percentage of naphthalene mineralized was reduced more significantly compared to the control where no Cr(VI) was added. Systems containing Cr(VI) concentrations of 12.0 mg/L showed no significant naphthalene mineralization over 2000 hrs. This was probably due to a drastic reduction of bacterial population as a result of chromate toxicity. Losses of naphthalene due to volatilization were significant but similar (20 to 25%) for all biometer flasks. Thus the differences in mineralization patterns are likely entirely due to the effect of chromate doses.

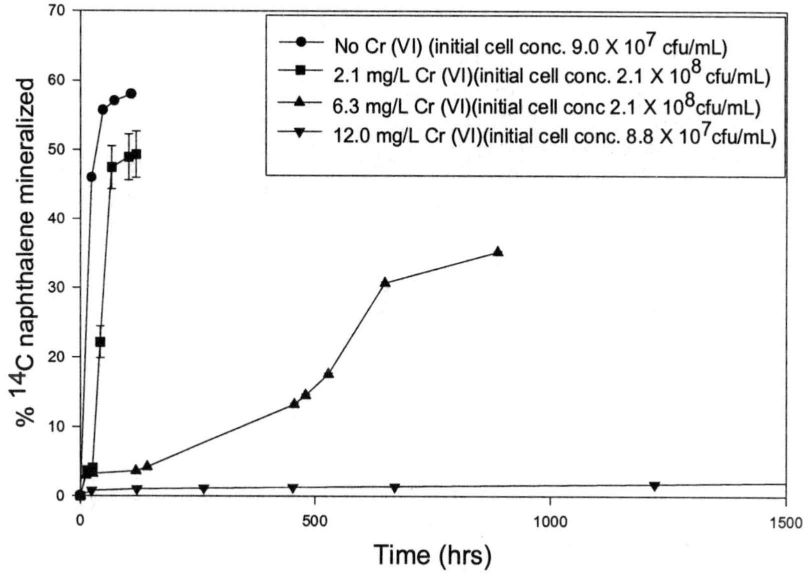

FIGURE 2: Naphthalene mineralization in the presence of different Cr (VI) concentrations.

The naphthalene degradation pattern seen in this study is different from the results reported by Shen et al. (1996b) using a mixed culture of *Bacillus sp.* K1 and *Sphingomonas paucimobilis* EPA 505. Shen et al. (1996b) observed simultaneous naphthalene degradation and Cr (VI) reduction. In that study 50% of the naphthalene was degraded within the first 7 hours followed by a decrease in mineralization rate. In this study mineralization of naphthalene was observed immediately after inoculation, but ceased thereafter during which time chromate reduction activity was noticed. After a complete reduction of Cr(VI) was attained, significant naphthalene mineralization recommenced. For the systems containing 2.1 mg/L Cr (VI), during the first thirteen hours of the experiment the mineralization rate was 0.353 mg naphthalene mineralized/h, followed by barely a detectable mineralization rate 0.007 mg naphthalene mineralized/h, and then a large increase in mineralization rate to 0.92 mg naphthalene mineralized/h. Approximately 90% of the total naphthalene mineralization activity occurred after Cr (VI) had been completely reduced. Figure 3 shows the effect of 2.1 mg/L Cr(VI) on naphthalene mineralization patterns and the Cr(VI) reduction time profile. The data suggests that as Cr(VI) enters the bacterial cells it slows down naphthalene mineralization probably due to its toxicity to bacterial culture or interaction with enzymes and its cofactors. Cr(VI) reduction during the period of negligible naphthalene mineralization activity may have been facilitated by naphthalene degradation intermediates, such as quinonic compounds, produced during the initial mineralization period.

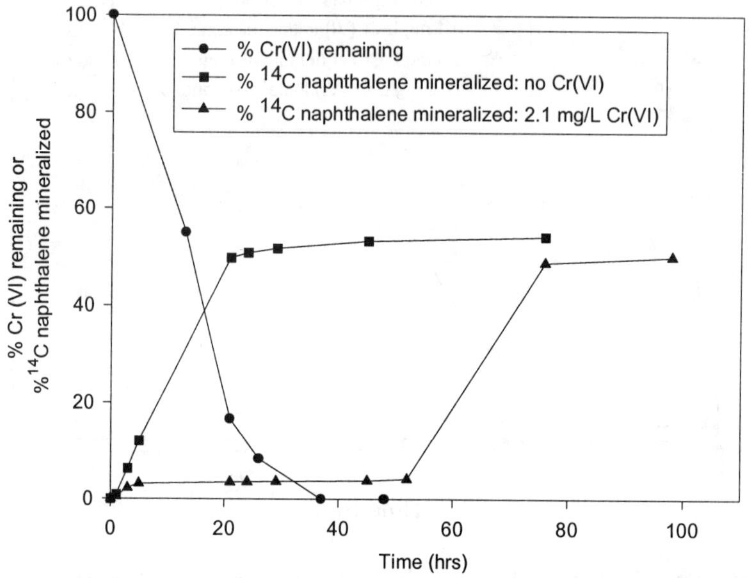

FIGURE 3: Percentage of Cr (VI) remaining and ^{14}C naphthalene mineralized for 2.1 mg/L Cr (VI). Inocula added: 8.9 X 10^7 cfu/mL.

CONCLUSIONS

Pseudomonas putida strain ATCC 17484 was able to completely reduce Cr(VI) up to a concentration of 6.3 mg/L and degrade naphthalene in systems containing both. When Cr(VI) was present at higher concentrations naphthalene was not degraded. Cr(VI) at an initial concentration of 2.1 mg/L was reduced completely in about 40 hours, however, significant naphthalene mineralization activity occurred only after Cr(VI) was completely reduced. Higher initial Cr(VI) concentrations increased the time taken for commencement of significant naphthalene mineralization. The overall naphthalene biodegradation patterns in the Cr(VI) containing systems include extended lag periods, reduced biodegradation rates, and in some cases an absence of naphthalene biodegradation.

ACKNOWLEDGEMENTS

The research was funded in part by a grant from the National Science and Engineering Research Council, Canada.

REFERENCES

Al-Hakak, A. 2001. "Degradation of Naphthalene and Concurrent Reduction of Cr(VI) by *Pseudomonas putida* ATCC 17484". M. Sc. Thesis, McGill University, Montreal, QC, Canada.

APHA. 1998. Standard Methods for the Examination of Water and Wastewater. 20th ed.

Cerniglia, C. 1992. "Biodegradation of Polycyclic Aromatic Hydrocarbons. *Biodegradation*. *3*: 351-368.

Ghoshal, S., and R.G. Luthy. 1998. "Biodegradation Kinetics of Naphthalene in Nonaqueous Phase Liquid-Water Mixed Batch Systems: Comparison of Model Predictions and Experimental Results. *Biotechnology and Bioengineering*. *57*(3): 356-366.

Ishibashi, Y., C. Cervantes, and S. Silver. 1990. "Chromium Reduction in *Pseudomonas putida*". *Applied and Environmental Microbiology*. *56*:2268-2270.

Kong, I; G. Bitton, B. Koopman, and K. Jung. 1995. "Heavy Metal Toxicity Testing in Environmental Samples". *Rev. Environ. Cont. Toxicology*. *142*: 119-147.

Luthy, R. G., D. A. Dzombak, C. A. Peters, A. Ramaswami, S. B. Roy, D. Nakles, and B. R. Nott 1994. "Remediating Tar-contaminated Soils at Manufactured Gas Plant Sites". *Environ. Sci. Technol.* *28*(6): 266A-276A.

Muller, J. G.; P. J. Chapman, and H. Prichard. 1989. "Creosote-contaminated Sites: Their Potential for Bioremediation". *Environ. Sci. Technol.* *23*(10): 1197-1201.

Shen, H., H. Pritchard, and G. W. Sewell. 1996a. "Microbial Reduction of Chromium (VI) During Anaerobic Degradation of Benzoate". *Environ. Sci. Technol.* *30*(5): 1667-1674.

Shen, H., H. Pritchard, and G. W. Sewell. 1996b. "Kinetics of Chromate Reduction During Naphthalene Degradation in a Mixed Culture". *Biotechnology and Bioengineering*. *52*(3): 357-363.

Shen, H. and Y-T Wang. 1994. "Biological Reduction of Chromium by *E. Coli*". *Journal of Environmental Engineering*. *120*(3): 560-572.

Shen, H. and Y-T Wang. 1995. "Simultaneous Chromium Reduction and Phenol Degradation in a Coculture of *Escheria Coli* ATCC 33456 and *Psuedomonas Putida* DMP-1. *Applied and Environmental Microbiology*. *61*(7): 2754-2758.

Wang, Y-T., and H. Shen, H. 1995. "Bacterial Reduction of Hexavalent Chromium". *Journal of Industrial Microbiology*. *14*: 159-163.

Wittbrodt, P. R., and C. D. Palmer. 1996. "Effect of Temperature, Ionic Strength, Background Electrolytes, Fe(III) on the Reduction of Hexavalent Chromium by Soil Humic Substances". *Environ. Sci. Technol.* *30*(8): 2470-2477.

SELECTION AND DEVELOPMENT OF CADMIUM RESISTANCE IN BACTERIAL CONSORTIA

Heather M. Knotek-Smith, Lee A. Deobald, Martina Ederer Ph.D.,
& Don L. Crawford Ph.D. (University of Idaho, Moscow, ID, USA.)

ABSTRACT: The effects of Cadmium (Cd) toxicity on bacterial consortia originating from anaerobic sewage sludge and cultivated under a variety of physiological conditions were studied. Cultures were enriched in minimal media developed specifically for Cd stress studies. At inoculation all Cd was soluble in free ion or chelated form. Electron donors and acceptors were varied to obtain each physiological enrichment type. Adaptation leading to higher levels of Cd resistance of the consortia over time was observed under all physiological conditions. Initial and increased Cd tolerances were greatest under multiphysiological (MPH) followed by sulfate reducing (SRB) and methanogenic (MET) conditions. Fermentative (FRM) enrichments had the least ability to tolerate and adapt to Cd. The Cd remained soluble as free Cd in MPH and FRM conditions and was precipitated in SRB and MET conditions. The 16S rRNA profiles of the consortia under SRB, MPH, and FRM conditions were followed over time. The consortia underwent succession under all physiological conditions when compared with the profile of the inoculum. Microbial population diversity decreased as the consortia were subcultured. The effects of chelators in the MPH media were also evaluated. The addition of chelators transiently decreased toxicity.

INTRODUCTION

The input of Cd to land through atmospheric deposition, the use of phosphate fertilizers, and the disposal of wastes constitute a risk of either excessive accumulation of Cd in the topsoil or increased leaching of Cd (Christensen, 1985). Cd is toxic to many organisms at low concentrations, a problem that hinders bioremediation efforts at Cd contaminated sites (Said and Lewis, 1991). If microorganisms are to be used for bioremediation of contaminants in mixed waste sites, it is important to understand how the organisms adapt to toxic conditions posed by soluble metals.

Microbial and heavy metal interaction studies in the laboratory require an understanding of how Cd behaves in the medium being used and how that medium relates to environmental conditions. Cd has been reported to be more mobile than most heavy metals due to its low affinity for soil (McBride, 1994). In reality, the chemistry of Cd and it's transport in the environment is complicated. Most Cd applied through phosphoric fertilizers to sandy soils will stay mobile due to a lack of adsorption sites on soil surfaces, and hence the risk of leaching to groundwater or uptake by plants is high (Mann and Ritchie, 1995). However, organic matter, hydrous oxides of iron, aluminum, manganese, and clay minerals in soil may retain Cd in non-mobile or unavailable forms. These compounds are rarely uniform in the soil structure. Therefore, the amount of Cd that might be transported is difficult to estimate.

Sorption capacities decrease with decreasing pH (Christensen, 1984). This is of great concern, since the acidification of soils by acid rain or microbial activity can mobilize Cd and make it bioavailable (Palm, 1994). As much as 36% of Cd sorbed to soil (20 mg Cd/70 g soil) was mobilized in the presence of microorganisms and nutrients, while less than 16% was released from sterile and unsupplemented soil controls (Chanmugathas and Bollag, 1988). Interactions with chelators, such as citrate, EDTA, some buffers, and soluble organics must also be considered because they can also solvate Cd.

Leachate from municipal compost, waste incinerator slag, and sewage sludge results in reduced soil sorption capacities due to increased ionic strength, inorganic and organic ligands, competing cations such as Zn, and organics, which can coat surfaces or pull Cd into solution. The presence of waste leachates, as compared to unpolluted soil solution, decreases the distribution coefficients 30 to 250 times. As a result the relative migration velocity of Cd present in waste leachate is 80-170 times faster for the same pH in an unpolluted soil solution (Christensen, 1985).

Cadmium distribution coefficients are reduced 2 to 14 times due to competition with other metals. Mixtures of Ni, Co, and Zn or Cr, Cu, and Pb, reduced Cd sorption onto soils. Zn accounted for most of the observed competition with Cd, perhaps due to the comparable chemistry and that Zn concentrations in the soil are usually 100 to 1000 times higher than Cd concentrations (Christensen, 1987).

Only when interferences are eliminated or accounted for, can it be concluded that transformation of Cd is the result of interactions with microorganisms or their metabolites. Microbial populations generally resist Cd toxicity by three processes (Ford et al., 1995). The first involves the secretion of a polymer, protein, or other component that sequesters Cd in the extracellular medium. A second is cell surface interaction which results from specific functional groups on the cell wall that bind Cd. The third is intracellular interactions such as export and intracellular sequestration. In this study cultures were enriched in minimal media developed specifically for Cd stress studies with microbial consortia.

Objective. In the present work mechanisms by which different microbial consortia adapt to aqueous phase Cd toxicity were examined. The goal of this research was not to discover a specific organism with high tolerance but rather an enrichment media that would select for indigenous organisms possessing an intrinsic resistance or ability to become resistant to Cd.

METHODS

Culture media were prepared anaerobically by the modified Hungate method in 100 ml vials and transferred to 25 ml balsh tubes. Glucose, vitamin solution, iron supplement, and Cd were added as anaerobic sterile solutions after autoclaving. Cd, added as $CdCl_2$, remained soluble as confirmed by Inductively Coupled Plasma injection (ICP) analysis (Robinson, 1994). Physiological conditions were varied by providing different electron donors and acceptors (d/a). They included: MET (d/a: acetate/CO_2) (Oremland, 1988) SRB (d/a: lactate/sulfate) (Widdel, 1988) FRM (d/a: glucose/no electron acceptor) (Stumm, 1988), and multiphysiological (MPH) (d/a: glucose/sulfate) (Widdel, 1988; Stumm, 1988). Basal media components were (g/L): 0.5 citric acid, 0.1 $MgSO_4{\bullet}7H_2O$ (0.08 $MgCl_2{\bullet}6\ H_2O$ in

FRM), 1.0 $(NH_4)_2SO_4$ (0.8 NH_4Cl in FRM), 0.0006 ferrous sulfate•$7H_2O$ (0.0005 ferric citrate in FRM), 0.03 $CaCl_2$•$2H_2O$, 1.0 HEPES, 0.212 glycerol 2-phosphate, 1.0 ascorbic acid, electron donors (added as listed above) 2.3 glucose, 6.0 sodium lactate 60% (w/v), 1.32 sodium acetate. Head space components were added aseptically after inoculation. Vitamin solution was added per Demain and Solomon (1986).

Physiological types in each medium were confirmed by standard MPN techniques in cultures without Cd. SRB's were identified using ASTM Method D 4412-84 (SRB-MPN) (Eaton et al., 1995). Fermenting organisms were identified by the production of acid in a medium containing 3 g beef extract, 5 g peptone, 5 g glucose, 0.5 g sodium thioglycolate, 0.7 g agar, 0.25 g L-cysteine, and 1 g of bromocresol purple added to 1 L of DI water (F-MPN). Methanogens were identified by the presence of methane in the culture head space when analyzed by Mass Spectrometer Flame Ionization Detector (M-A). The results were (SRB-MPN/F-MPN/M-A): FRM (-/+/-), MPH (-/+/-), MET (+/test not run/+), and SRB (+/+/+).

Media were inoculated (2% v/v) with anaerobic digester sludge from the Moscow, Idaho sewage treatment plant. Initial inoculation cultures were allowed to incubate for one day then subcultured into a second set of media. Subculturing was performed when an absorbance of 0.2 OD_{600nm} or the apparent maximum was reached. The inoculation scheme for analysis of DGGE and siderophore interaction is shown in Figure 1. Test 1 was the control condition with no Cd; Test 2 was subcultured with constant Cd level; Test 3 was subcultured with constant Cd concentration and iron (Fe); Test 4 was subcultured with increasing Cd concentration and Fe. The tests were designed to determine if consortia development was primarily effected by media components, Cd stress, or Fe.supplementation. Samples for analysis of culture development were taken at the time of subculture inoculation.

The inoculation scheme used for Cd resistance development, lag phase comparison, and chelation analysis was similar to the tests indicated by box numbers T0→1→2 and T0→9→13 in Figure 1. However, there was no Fe supplementation and SC0→SC1 was run without Citrate or HEPES for the chelation analysis.

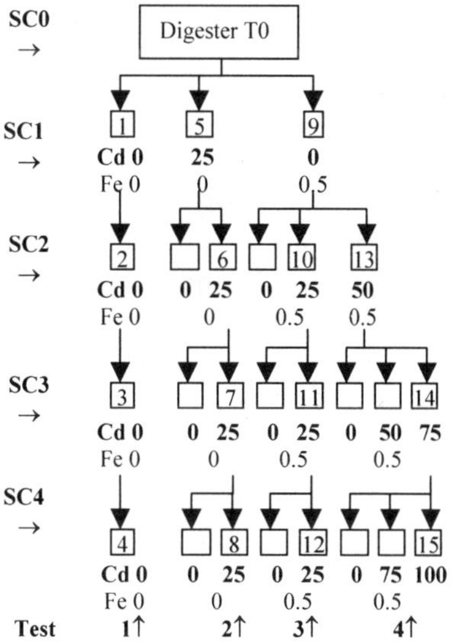

FIGURE 1. Inoculation Scheme [Cd (ppm), Fe (mM)] box numbers correspond to rRNA sample ID, SC=subculture.

Total DNA was isolated from 1.5 ml of mixed bacterial cultures or isolates by a modified method of Purdy et al. (1996). The cells were pelleted by centrifugation and resuspended in 0.5 ml cell suspension buffer (120 mM $NaPO_4$ buffer, pH 8, 1% polyvinyl polypyrrolidone). Samples were agitated twice in a Bead beater (Biospec Products, Bartlesville, OK) for 2 min at maximum speed and 4°C with 0.5 g ZrO_2/SiO_2 beads, 35 µl 20% SDS, and

0.5 ml phenol, pH 8.0. Aqueous phase was separated by centrifugation and the samples were loaded onto a hydroxyapatite spin-column equilibrated with 120 mM sodium phosphate buffer, pH 7.2 and then subjected to centrifugation at 120 x g for 2 min. After three washing steps with 0.5 ml 120 mM $NaPO_4$, the nucleic acids were eluted with 0.4 ml 1 M KPO_4 buffer, pH 7.2. The DNA was desalted on a water saturated Sephadex G-50. Sample volume was reduced by vacuum centrifugation.

The 16S rDNA was amplified using primers 907R and GC clamp-GM5F, which includes a 5' 40 base GC clamp (Muyzer et al., 1993). Conditions were 0.2 mM dNTP, 1.5 mM $MgCl_2$, 0.5 μM primers, 1% Tween-20, 5% DMSO, 1-5 μl template DNA and 1-2 units *Taq* polymerase at 50 μl volume. Samples were denatured by heating to 95°C for 5 min. Amplification was accomplished using the "touch down" program. Annealing temperature of 65°C decreased by 0.5°C each cycle for 20 cycles, to reduce amplification of non-target products, elongation was accomplished for 30 sec. at 72°C, followed by denaturation at 95°C. Twenty cycles of amplification at the last annealing temperature were added with the elongation period extended by 2 sec. each cycle. Amplification was ended by a final extension of 5 min. at 72°C.

Denaturing gradient gel electrophoresis (DGGE) was conducted on a 6% acrylamide/biacrylamide gel (19:1) with a 20–60% denaturing gradient (denaturant composed of 7M urea, 40% formamide) on a Bio-Rad DGGE apparaturs. The 10 μl of PCR product were diluted with an equal amount of loading buffer (70% glycerol, 2 mM EDTA, pH 8, 0.05% Xylene Cyanol, 0.05% Bromophenol blue) and subjected to electrophoresis at 60°C and 200V for 4 hrs. The DNA was visualized using Sybr green I (Molecular Probes, Eugene, OR).

RESULTS AND DISCUSSION

Understanding and controlling abiotic interactions of heavy metals with culture media components is indispensable when studying metal toxicity in laboratory cultures. Initial studies of Cd toxicity to anaerobic bacterial consortia were run in Starkey's Medium C (Table 1) (ATCC, 1992). These tests showed that the Cd in the medium had little or no effect on growth of cultures (unpublished data). However, screening tests of time zero samples indicated that there were numerous interactions of Cd with media components and the sludge inoculum. Nearly all Cd (150 - 500 mg/L) precipitated abiotically in the pH range of 6 to 9.

The development of a medium for studying Cd toxicity began with a medium used previously for aerobes, but known to have little or no interactions with added Cd (Roane and Kellogg, 1996). The medium, Medium 1 (Table 1), was prepared anaerobically using a modified Hungate technique. Initially, only reductant, buffer, oxygen indicator, and sodium lactate, were added to the medium formulation.

The effects of pH, reductant levels, and inoculum levels were evaluated in the enrichments. Initial pH was a significant variable with an optimum for all enrichments near neutral. Reductant levels above 0.1% (w/w) did not increase growth, and sludge inoculum levels of less than 2% (v/v) did not result in adequate growth. The carbon source requirement was also significantly higher than would be expected in an aerobically-incubated medium, as was expected based on previous literature (Tanner, 1996). Medium 1 was modified based on these initial findings, to create Medium 2. This medium didn't form the typical black precipitate of iron sulfide present in most SRB cultures. Instead a yellow precipitate was formed, which was assumed to be cadmium sulfide (CdS). There was no increase in stress in cultures between 50 ppm and 90 ppm Cd. Interaction of Cd with inorganic phosphate ($Na_4O_7P_2$) was suspected and was therefore replaced with glycerol pyrophosphate, an organic form ($C_3H_9O_6P$) to create the final medium (Trevors et al., 1985).

TABLE 1. Culture media development

Component	Starkey's	Medium 1	Medium 2	Final medium
NTA trisodium salt	0.2 g/L			
Citric Acid		0.5 g/L	0.5 g/L	0.5 g/L
Ferric Citrate			0.0006 g/L	
$MgSO_4 \cdot 7\,H_2O$	2 g/L	0.1 g/L	0.1 g/L	0.1 g/L
Na_2SO_4	1 g/L			
$(NH_4)_2SO_4$		1.0 g/L	1.0 g/L	1.0 g/L
Ferrous Sulfate				0.0006 g/L
$CaCl_2 \cdot 2\,H_2O$	0.1 g/L		0.03 g/L	0.03 g/L
NH_4Cl	1 g/L			
KH_2PO_4	0.5 g/L			
Pyrophosphate tetra sodium		.017 g/L	.017 g/L	
Glycerol 2-Phosphate				0.212 g/L
Yeast Extract	1 g/L			
Vitamin Solution				Methods section
60% Sodium Lactate	5.8 g/L	3.0 g/L	3.0 g/L	3.0 g/L
pH adjusted with	NaOH	NaOH	KOH	KOH
HEPES buffer		1.0 g/L	1.0 g/L	1.0 g/L
Cysteine HCl	0.75 g/L			
Ferrous Ammonium Sulfate	0.025 g/L			
Ascorbic acid		1.0 g/L	1.0 g/L	1.0 g/L

Enrichment experiments in the Final Medium indicated that there was Cd stress under all physiological conditions. As shown in Figure 2 for MPH, growth decreased as Cd concentration was increased, indicating that Cd in the medium was bioavailable. This was true for all enrichment conditions.

A separate test was run to evaluate the effects of chelators. Chelators bind metals and are able to keep them in solution. Some chelators will bind metals in a non-bioavailable form causing a false resistance mechanism to be deduced. As shown in Figure 3, the presence of citrate and HEPES effected growth under control and stress conditions. In controls these compounds increased overall growth.

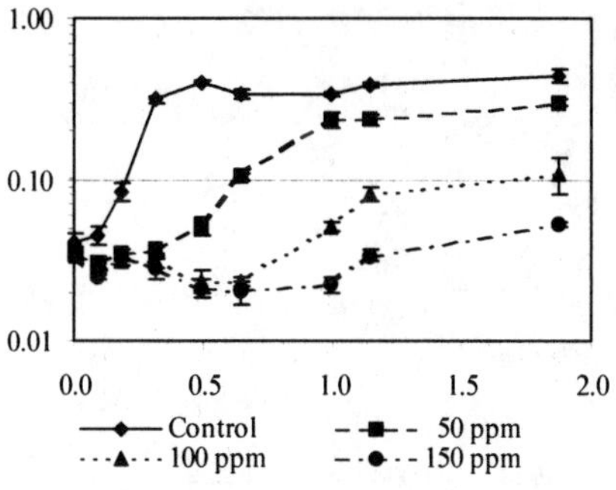

FIGURE 2. Cadmium stress in MPH SC 1. (X Axis- days, Y axis- OD$_{@\,600\,nm}$)

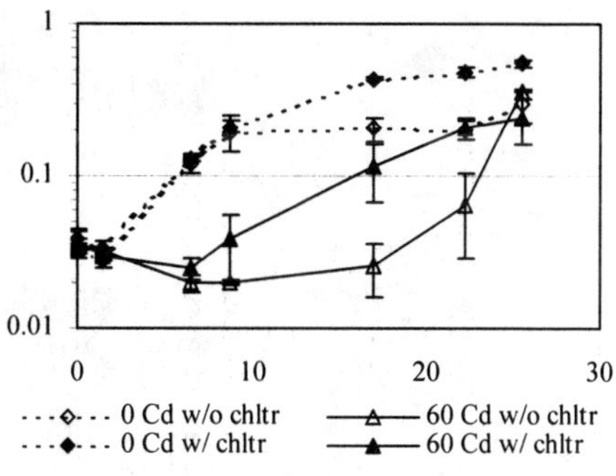

FIGURE 3. Effects of chelation molecules in MPH media. (Cd ppm, X axis- hours, Y axis- OD$_{@600\,nm}$)

The effect may have been due to the use of citrate as a carbon source or the HEPES as a buffer. Cultures with Cd and chelators had a transient decrease in Cd toxicity; however, growth of the cultures without the chelators recovered. It was also determined in these cultures that the Cd was ultimately released from the chelation molecules, as shown by ion selective probe that does not detect Cd if bound to another molecule. This finding was supported by Electrospray Ionization-MS (ESI-MS) analysis of a fraction of the supernatant separated on Bio-Rad P2 gel. The presence of Cd in this fraction was verified by ICP, however, a Cd signature was not found in the ESI-MS spectrum (data not shown). This indicates that the Cd was most likely free in solution and could not be ionized.

The level of Cd resistance of the consortia, as shown by a decreased lag phase upon subculture, increased for all physiological conditions studied. A lag phase comparison of the MPH medium is shown in Figure 4.

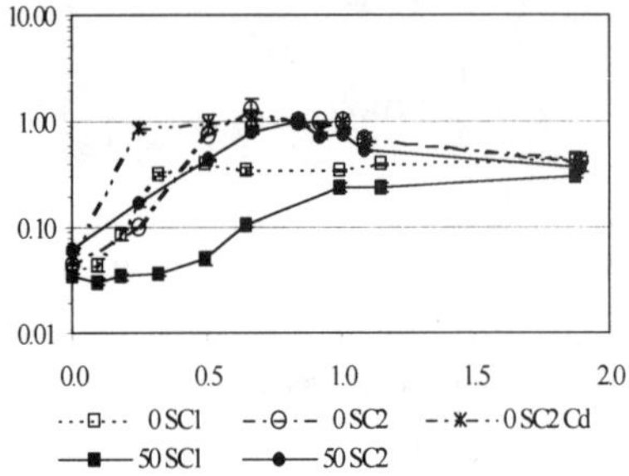

FIGURE 4 . MPH Lag phase comparison of SC1 to SC2 (Cd ppm, X axis- days, Y axis- OD $_{@\ 600\ nm}$)

The 16S rRNA profiles showed that the consortia underwent succession under all physiological conditions when compared to the profile of the inoculum. Succession stabilized by the fourth cycle of subculturing. Under SRB conditions (not shown), the presence of Cd led to four predominant bands in the DGGE analysis independent of the iron concentration. In the control (no Cd) only one predominant band emerge after four subculturing cycles. Cd resistant organisms enriched under SRB conditions include *Veillonella* sp., *Klebsiella* sp., and *E. coli.*, based on 16S rRNA gene sequencing.

Culture development under MPH conditions is shown in Figure 5(A). Under these culture conditions the medium was not the primary selection influence, as shown by the differences in band development in lanes 1-3. The addition of iron (lanes 9-12), allowed a variety of species to prosper in addition to the species selected without iron enrichment (lanes 5-8). These lanes also indicate that resistance was not siderophore production due to the presence of the same organism with and without Fe. Cd resistant organisms enriched under MPH conditions in-

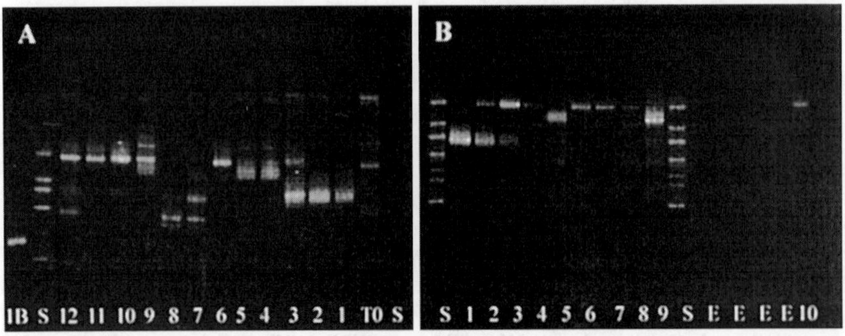

FIGURE 5. (A) MPH (B) FRM Culture Development
[Lane numbers indicate sample number (see Figure 1.) S=standard, E=empty, 1B=isolate *Enterbacter*

cluded *Klebsiella sp*., and *Enterbacter* sp.. As shown in Figure 5(A), *Enterbacter* (1B) was isolated from the MPH culture but was not a dominant member in the enrichment. It should be noted that the organisms isolated from Test 1 were able to tolerate Cd at low concentrations in all but MET. This indicates that low level Cd resistance was intrinsic in these organisms. FRM conditions in Figure 5(B) selected for the same species under control (lanes 1-4), Cd (lanes 5-8), and increasing Cd with iron (lanes 9-10). This shows that culture adaptation was due primarily to media formulation rather than Cd stress.

CONCLUSIONS

This work describes development of a medium specifically for Cd stress studies. Cd resistance increased with subculturing for all physiological conditions studied. However, the more effective resistance mechanisms seemed to be due to fortuitous selection. In fact, some physiological conditions especially those that left the Cd free in solution seemed to be dependent on the enrichment type rather than Cd selection as demonstrated by the existence of cadmium resistance in control cultures. This work also demonstrated that development Cd toxicity resistance does not necessarily require that the metal be removed from solution and the resistance mechanism may be dependent on media formulation.

ACKNOWLEDGMENTS

This research was supported by grant # DE-FG03-98ER62688 of the US Department of Energy NABIR Program, and by the University of Idaho Agricultural Experiment Station. We thank Alisa Perez-LaPlath for technical assistance.

REFERENCES

ATCC. 1992. "Media 207: Modified Starkey's Medium C", *American Type Culture Collection Catalogue of Bacteria & Phages*, pp. 424,

Chanmagathas, P & JM Bollag. 1998. "A Column Study of the Biological Mobilization & Speciation of Cadmium in Soil". *Arch. of Env. Cont. & Tox.*. 17:229-237.

Christensen, TH 1984. "Cadmium Soil Sorption at Low Concentrations: I Effect of Time, Cadmium Load, pH, & Calcium". *Wat., Air, & Soil Poll.* 21:105-114.

Christensen, TH 1985. "Cadmium Soil Sorption at Low Concentrations: III Prediction & Observation of Mobility & IV Effect of Waste Leachates on Distribution Coefficients ". *Wat., Air, & Soil Poll.* 26:255-274.

Christensen, TH 1987. "Cadmium Soil Sorption at Low Concentrations: V Evidence of Competition by Other Heavy Metals & VI A Model for Zinc Competition". *Wat., Air, & Soil Poll.* 34:293-314.

Demain, AL, & NA Solomon. 1986. *Manual of Industrial Microbiology & Biotechnology*, pg. 89, ASM, Washington D.C.

Eaton, AD., LS Clesceri, & AE Greenberg (ed.). 1995. *Standard Methods for the Examination of Water & Wastewater*, 19th ed, Vol. Am. Pub. Health Assoc., Washington, DC.

Ford, T, J Maki, & R Mitchell. 1995. "Metal-microbe Interactions", In CC Gaylarde & HA Videla (ed.), *Bioextraction & Biodeterioration of Metals*, pp. 1-23, Cambridge University Press, Boston.

Mann, SS, & GSP Ritchie. 1995. "Forms of Cadmium in Sandy Soils After Amendment with Soils of Higher Fixing Capacity". *Env. Poll.* 87(1): 23-29.

McBride, MB 1994. "Trace & Toxic Elements in Soils", *Environmental Chemistry of Soils*, pp. 318-325, Oxford University Press Inc., New York.

Muyzer G, EC DeWaal, AG Uitterlinden. 1993. Profiling of complex microbial populations by denaturing gel gradient electrophoresis analysis of polymerase chain reaction-amplified genes coding for 16S rRNA. *Appl. Env. Microbiol.* 59:695-700.

Oremland, RS 1988. "Biogeochemistry of Methanogenic Bacteria", In A. J. G. Zehnder (ed.), *Biology of Anaerobic Microorganisms*, pp. 651, John Wiley & Sons, Inc., Toronto.

Palm, V 1994. "Model for Sorption Flux & Plant Uptake of Cadmium in Soil Profile: Model Structure & Sensitivity Analysis". *Wat., Air, & Soil Pol.* 77:169-190.

Purdy, KJ, TM Embly, S Takii, DB Nedwell. 1996. Rapid extraction of DNA and RNA from sediments by a novel hydroxyapatite spin-column method. *Appl. Environ. Microbiol.* 62(10): 3905-3907.

Roane, TM, & ST Kellogg. 1996. "Characterization of bacterial communities in heavy metal contaminated soils". *Can. J. Micro.* 42:593-603.

Robinson, JW 1994. "Emission Spectrography, Inductively Coupled Plasma Emission (ICP), & ICP-Mass Spectroscopy", *Undergraduate Instrumental Analysis*, 5 ed, vol. pp. 474-475, Marcel Dekker, Inc., New York.

Said, WA., & DL Lewis. 1991. "Quantitative Assessment of the Effects of Metals on Microbial Degradation of Organic Chemicals". *Appl. & Env. Micro.* 57(5): 1498-1503.

Stumm, W 1988. "Geochemistry & Biogeochemistry of Anaerobic Habitats", In A. J. G. Zehnder (ed.), *Biology of Anaerobic Microorganisms*, pp. 16-20, 29-33, John Wiley & Sons, Inc., Toronto.

Tanner, RS 1996. "Cultivation of Bacteria & Fungi", In C. J. Hurst (ed.), *Manual of Environmental Microbiology*, pp. 54-55, ASM, Washington, D.C.

Trevors, JT, KM Oddie, & B. H. Belliveau. 1985. "Metal Resistance in Bacteria". *FEMS Microbiology Reviews.* 32:39-54.

Widdel, F 1988. "Microbiology & Ecology of Sulfate- & Sulfur-Reducing Bacteria", In A. J. G. Zehnder (ed.), *Biology of Anaerobic Microorganisms*, pp. 493, 500-502, John Wiley & Sons, Inc., Toronto.

MICROBIALLY MEDIATED REDUCTION AND IMMOBILIZATION OF URANIUM IN GROUNDWATER AT KÖNIGSTEIN

Werner Lutze (University of New Mexico, Albuquerque, NM 87131)
Zhu Chen, Danielle Diehl, Weiliang Gong, H. Eric Nuttall (University of New Mexico, Albuquerque, NM 87131)
Günter Kieszig (WISMUT GmbH, 09117 Chemnitz, Germany)

ABSTRACT: Experiments have been conducted to determine whether *in situ* bioremediation could be applied at Königstein, Germany. There, an aquifer located on top of a former uranium mine might be contaminated after flooding of the mine. The flooding water will contain uranium and other contaminants. On site chemical conditions were closely simulated in the laboratory. Water was amended with lactate. The most likely chemical reactions leading to microbially mediated mineralization of uranium were identified and then experimentally verified. The results showed that indigenous microbes reduced U(VI) to U(IV) that precipitated as uraninite UO_2. A large excess of mackinawite [$Fe_{(1-x)}S$] was also precipitated by microbially mediated reactions and was shown to provide long-term protection against oxidative dissolution of uraninite.

INTRODUCTION

The large number of sites and volumes of contaminated groundwater and soil call for innovative and economically attractive remediation technologies. To date, pump-and-treat is the most widely used technology. Frequently, pump-and-treat has been ineffective in permanently lowering contaminant concentrations in groundwater (Travis and Doty, 1990). A recent study by Quinton et al. (1997) showed that groundwater cleanup technologies such as pump and treat, permeable reactive barriers with zero-valent iron, and bio-barriers are more expensive than *in situ* bioremediation.

Microorganisms can reduce uranium indirectly by producing H_2S or H_2 in the course of other processes (abiotic reduction) or directly using enzymes (enzymatic reduction). The first microorganisms identified to enzymatically reduce U(VI) were the dissimilatory Fe(III)-reducing microorganisms, *Geobacter metallireducens* and *Shewanella putrefaciens* (Lovley et al., 1991). These microorganisms used uranium as an electron acceptor and H_2 or acetate as an electron donor to support growth and tolerated U(VI) concentrations as high as 8 mM. Several authors studied the enzymatic reduction of U(VI) by various pure or mixed cultures of microorganisms, including metal- and sulfate-reducing bacteria (a summary of previous work can be found in Abdelouas et al. 1999a). These authors reviewed the literature on microbial reduction of uranium and the significance of biogeochemical processes related to uranium mining, tailings and groundwater remediation.

In summary, the literature on microbially mediated reduction of U(VI) showed that U(VI) can be reduced to U(IV) by enzymatic activity of microorganisms including metal- and sulfate-reducing bacteria; U(VI) can be reduced either by pure cultures or by mixed indigenous cultures; U(IV) precipitates as uraninite (UO_2); complexation of U(VI) with organic and inorganic ligands can inhibit its reduction by microorganisms; complexation of U(IV) may inhibit its precipitation.

Objective: The objective of this work was to identify and understand microbially mediated redox reactions after amending an aquifer located above the uranium mine at Königstein, Germany. Experiments were conducted in the laboratory to determine whether an *in situ* bioremediation process could effectively separate uranium from the groundwater by biomineralization, should the aquifer become contaminated by water from the flooded mine.

Site Description: A sketch of the site is shown in Figure 1. The mine is located in sandstone in aquifer G4, about 260 m below the surface. Uranium was mined by *in situ* leaching the sandstone with sulfuric acid. The mine will be flooded in the framework of remediation actions. After flooding, the open space is filled with groundwater contaminated with uranium and other toxic species. Clay layers (dashed zones in Figure 1) separate layers of sandstone. Each layer of sandstone constitutes an aquifer, including G3, a drinking water reservoir. A vertical fault provides a potential path for flooding water to reach and contaminate aquifer G3. Creation of a bioremediation zone is envisaged near the point of entry of G4 water (Figure 1). Figure 2 shows the details of the bioremediation zone. Water compositions are given in Table 1. Major constituents in the acid (pH<3) water G4 are dissolved iron and sulfate. Arrows indicate flow directions. Amendment (e.g. sodium lactate solution) can be delivered from the surface.

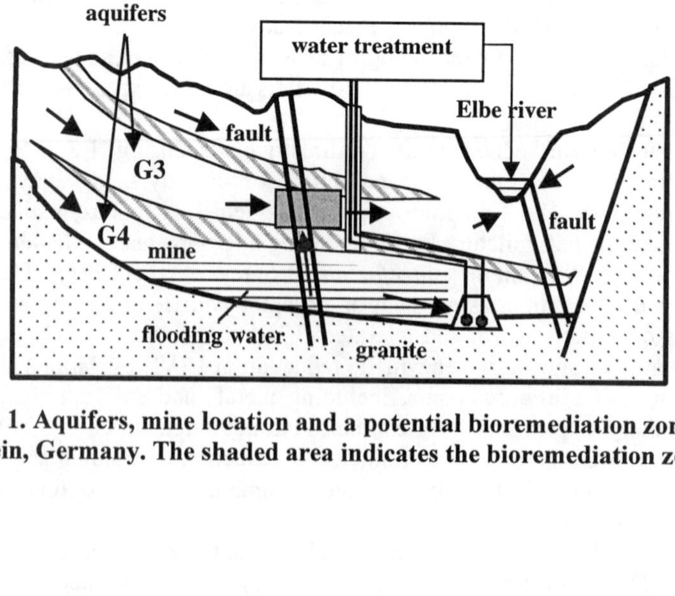

FIGURE 1. Aquifers, mine location and a potential bioremediation zone at Königstein, Germany. The shaded area indicates the bioremediation zone.

Table 1. Groundwater composition in aquifers G3 and G4.

Constituents	Aquifer G3 [mg/L]	Aquifer G4 [mg/L]
U^{6+}	0.002	17 (0.07mM)
SO_4^{2-}	39	1450 (15mM)
Fe^{3+}	1.12	200 (3.5mM)
O_2	8.7	3
pH	6.2	2.7
Zn	0.1	40
Cd	<0.0001	1
Ni	<0.02	4
Co	<0.01	3

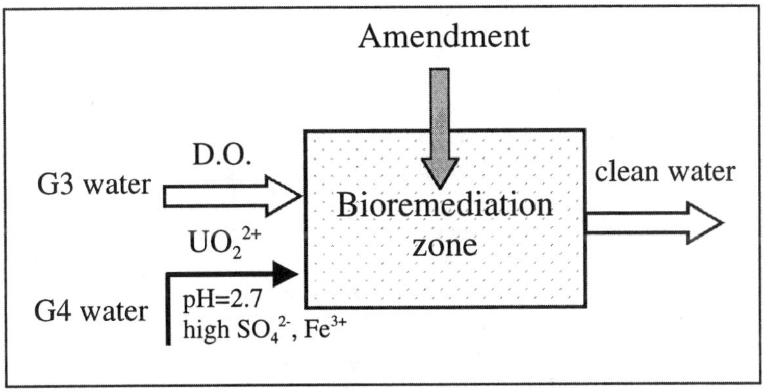

FIGURE 2. Bioremediation zone.

Expected Chemical Reactions in the Bioremediation Zone: Dilution of G4 by G3 water increases the pH and iron oxyhydroxide, e. g. goethite precipitates:

$$Fe^{3+} + 3H_2O = FeO(OH) + 3H^+ \quad (1)$$

Uranium binds to the surface of FeO(OH):

$$FeO(OH) + UO_2^{2+} = FeOO\text{-}UO_2^+ + H^+ \quad (2)$$

Addition of an organic amendment (e.g. lactate) stimulates growth of indigenous microbes (if present) in aquifer G3 and aerobes mediate reduction of dissolved oxygen:

$$O_2 + 4H^+ + 4e^- = 2H_2O \quad (3)$$

Nitrate (may be part of the amendment as a nitrogen source for microbial growth) is reduced to N_2 by various microbial species (Abdelouas et al., 1998; Lu et al., 1999):

$$NO_3^- + 6H^+ + 5e^- = 1/2N_2 + 3H_2O \quad (4)$$

U(VI) is reduced to U(VI) by sulfate and iron reducing microbes

$$UO_2^{2+} + 2H_2O + 2e^- = U(OH)_4 \quad (5a)$$

and precipitates as UO_2, uraninite (Abdelouas et al., 1998):

$$U(OH)_4 = UO_2 + 2H_2O \quad (5b)$$

followed by reduction and dissolution of FeO(OH):

$$FeO(OH) + 3H^+ + e^- = Fe^{2+} + 2H_2O \quad (6)$$

Sulfate is reduced to sulfide:

$$SO_4^{2-} + 9H^+ + 8e^- = HS^- + 4H_2O \quad (7)$$

The amendment (lactate) is oxidized:

$$CH_3CH(OH)COO^- + 3OH^- = 3CO_2 + 9H^+ + 12c^- \quad (8)$$

Iron reacts with sulfide to form the mineral mackinawite $FeS_{(1-x)}$ ($x \approx 0.1$) (Abdelouas et al., 1999b):

$$Fe^{2+} + (1-x)HS^- = FeS_{(1-x)} + (1-x)H^+ \quad (9)$$

Reactions (3) to (8) are sequential or overlap as the redox potential in solution decreases (Abdelouas et al., 1998).

In the presence of microbes the sulfur content of mackinawite increases with time and phase transformations produce a series of minerals, e.g. greigite, smythite and pyrrhotite, $FeS_{(1+x)}$, with $x<1$ (McNeil and Little 1990). The thermodynamically stable phase in a reducing natural environment is pyrite, FeS_2. Pyrrhotite and pyrite were detected upon formation of mackinawite (Abdelouas et al., 2000).

When the newly formed minerals in the bioremediation zone are exposed to dissolved oxygen, e.g. after termination of amendment delivery, iron sulfides oxidize forming sulfate and iron oxyhydroxide, e.g. goethite (Abdelouas et al. 2000).

Oxidative dissolution is expected to take place in the sequence of increasing redox potential:

$$FeS_{(1-x)} + (1-x)O_2 = Fe^{2+} + (1-x)SO_4^{2-} \quad (10)$$

$$2Fe^{2+} + 1/2O_2 + 3H_2O = 2FeO(OH) + 4H^+ \quad (11)$$

$$UO_2 + 1/2O_2 + 2H^+ = UO_2^{2+} + H_2O \quad (12)$$

Formation of crystalline goethite was verified experimentally (Abdelouas et al. 1999b). The experiments described in the next section were conducted to provide evidence that reactions of the kind described in equations (1, 2, 5, 6, 7, 9, 10) take place. It is expected that reaction (12), oxidation of uranium, is delayed because the large abundance of mackinawite buffers the redox potential at $E_H < -200$ mV. UO_2 oxidizes at >+100 mV.

MATERIALS AND METHODS

Water samples were taken from aquifer G3 (Figure 1) and from the currently available acid water (G4) in the mine (Table 1). The concentration of the constituents in the G4 water is likely to decrease upon flooding of the mine but the water will still be acid and rich in dissolved iron and sulfate. Sandstone core samples were obtained from aquifer G3. The material was crushed to obtain sand. The following composition is typical of the sandstone: SiO_2 (90), Al_2O_3 (5.8), MgO (3.9), CaO (0.16), TiO_2 (0.12), in weight percent, and Fe (600), Sr (25), Zr (55), Pb (10), Sn (8), in mg/kg, and trace elements at lower concentrations.

Microbial Activity Test: 5 g sandstone were added to 100 mL G3 or G4 water amended with 0.09 g sodium lactate and 2 mg sodium trimetaphosphate (TMP). The mixture was stored in sterilized and sealed serum bottles at 13°C (the water temperature in aquifer G3). In G3 water the sandstone turned black within three weeks, indicating microbially mediated reduction of sulfate (in solution) and iron (in the sandstone) followed by precipitation of iron sulfide. Mixtures without lactate did not turn black. There was no microbial activity in the acid water G4.

Reduction of Uranium: It was assumed that leakage of G4 water into aquifer G3 would be small. Leakage was simulated by diluting G4 water 1:50 with G3 water. As with microbial activity tests, sandstone and amendment were added and the experiments were conducted at 13 and 24°C. The lids of the serum bottles contained a self-sealing rubber stopper. To measure reaction progress, samples were taken from solution by penetrating the stopper with a sterilized syringe. Blanks were run whenever the experimental conditions changed. Experimental details can be found in Abdelouas et al. (1998). U(VI) was analyzed using a laser fluorescence analyzer (Scintrex UA-3). Anions were measured by ion chromatography (DIONEX DX 500 HPLC). Solid reaction products were investigated by transmission electron microscopy (TEM), selected area electron diffraction (SAED), and energy dispersive x-ray spectroscopy (EDX), using a JEM-2010 TEM with an Oxford Link ISIS EDS system. The TEM was operated at 200 KeV.

Oxidation of Uranium: Experiments in sandstone columns (25 cm long, 2.5 cm wide) and in serum bottles were conducted at 24°C to measure how much uranium is oxidized by dissolved oxygen as a function of pore volumes passed through the columns. Groundwater with uranium, sulfate and iron was amended as described above and pumped into the column. The column was sealed to avoid access of oxygen. After three months the sandstone was black and all uranium and sulfate was reduced, as determined in batch experiments. Then, 15 pore volumes of pristine, oxygen saturated water from aquifer G3 were pumped through the column and the effluent was analyzed for sulfate and uranium.

RESULTS

Our results are shown in Figures 3 to 5. Without amendment, a large decrease in U(VI) concentration was observed in the first day followed by a slow further decrease and approaching a constant final value of about 15 μg/L (Figure 3, circles). Iron oxyhydroxide precipitated after mixing groundwater G3 and G4, adsorbing UO_2^{2+}; see reactions (1) and (2) in the introduction. With amendment, the first decrease in U(VI) concentration was the same (Figure 3, square symbols). However, after one month, microbial activity became significant reducing sulfate to sulfide (reaction (7)) and Fe^{3+} to Fe^{2+}, evidenced by precipitation of iron sulfide (reaction (9)). Reduction of Fe^{3+} led to disappearance of iron oxyhydroxide (reaction 6) as indicated by release of adsorbed UO_2^{2+} into solution between 30 and 40 days (Figure 3). Then the concentration of U(VI) in solution decreased until all uranium was reduced and precipitated as UO_2 (reactions 5a, b). The products mackinawite and uraninite were identified by TEM, SAED and EDX. Figure 4 shows the decrease of SO_4^{2-} and U(VI) concentration with time, indicating quantitative reduction of S(VI) and U(VI) in about 80 days. Figure 5 shows the evolution of sulfate and uranium concentration upon elution of a sandstone column with G3 water. The molar excess of iron relative to uranium in the column was about 50:1 (Table 1, last column), but was about 500:1 when the iron content in the G3 sandstone (about 100 mg) was included. There was enough

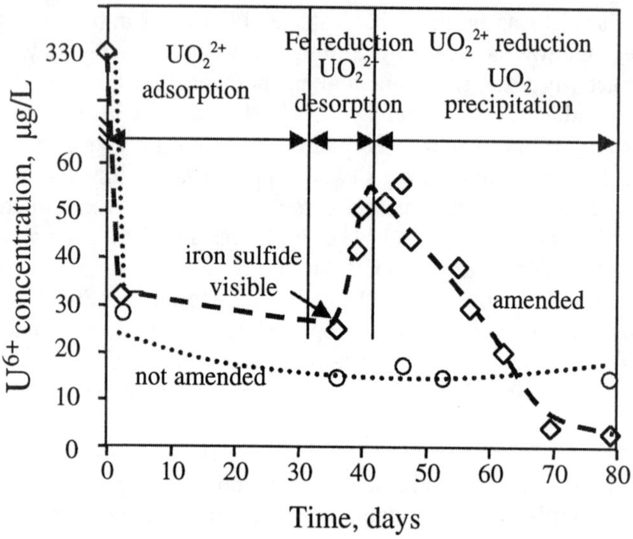

FIGURE 3. Adsorption of UO_2^{2+} on ferric hydroxide followed by microbially mediated reduction to UO_2 (13°C; groundwater mixed G4:G3=1:50).

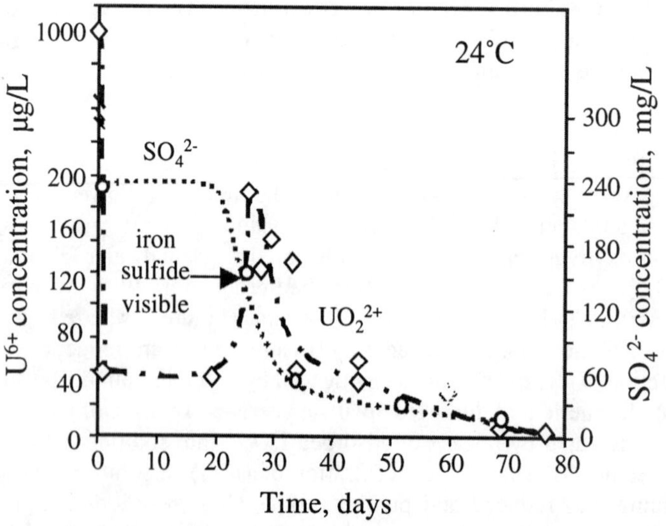

FIGURE 4. Microbially mediated reduction of SO_4^{2-} to HS^- and U^{6+} to U^{4+}.

sulfate in the column (pore volume 60 mL) and thus enough sulfide produced to precipitate all the iron in the column as mackinawite. We know from previous work that the iron in sandstone is bioavailable and reacts with sulfide in solution (Abdelouas, et al., 1999b). Upon elution oxygen was consumed by sulfide

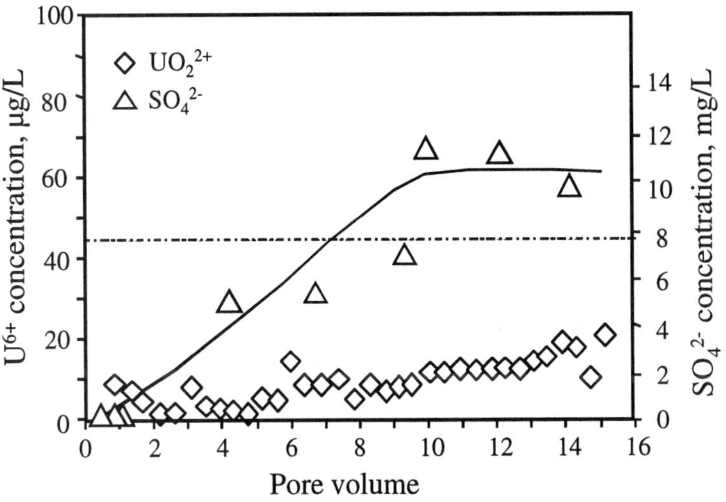

oxidation (reaction 10) as indicated by the increase in sulfate concentration (Figure 5). Only a small fraction (3%) of the uranium inventory in the column was released (oxidized) after passing 15 pore volumes of G3 water through the column. During elution the U(VI) concentration was less than 20 μg/L.

FIGURE 5. Concentrations of UO_2^{2+} and SO_4^{2-} vs. throughput of oxygen-saturated groundwater G3 (24°C). Dotted line marks the US-EPA concentration limit of 44μg/L.

DISCUSSION

We have shown that the fate of U(VI) can be confirmed experimentally, if described by the chemical reactions (1-12). Indigenous microbes mediated the chemical reactions essential for *in situ* biomineralization of uranium and protection against oxidation. At 13°C, reactions were complete within 80 days. Groundwater in aquifer G3 travels about 3 m in 80 days. Hence, in the field the length of the bioremediation zone can be on the order of 10 m. Oxidation experiments were particularly encouraging as they have shown that the large excess of iron sulfide over uraninite provides effective protection against oxidative dissolution of uraninite. A preliminary estimate showed that the 15 pore volumes passed through the column in one day translate into 400 years in the field. The concentration of U(VI) in these 15 pore volumes was less than 20 μg/L, a factor of two less than the US-EPA groundwater protection standard (Federal Register 1995). The sulfate concentration increased to its maximum value of 12 mg/L after 9 pore volumes. The sulfate concentration is limited by the oxygen concentration in the water (Table 1). The slow increase in the sulfate concentration within the first 9 pore volumes is probably due to consumption of oxygen by oxidation of biomass.

Beside uranium the final concentrations of the metals Zn, Cd, Ni, and Co, listed in Table 1, were all on the order of 1 μg/L or less. Precipitation of sulfides is the most likely process lowering their concentrations.

There are still important questions to answer, e.g. what is the effect of phase transformations of mackinawite into pyrite on the stability of uraninite and other precipitated or co-precipitated contaminants. Engineering effective mixing of amendment with the groundwater to establish a bioremediation zone is another issue to be addressed. However, the database acquired in this study shows that further investigation of *in situ* bioremediation of a potentially contaminated aquifer at Königstein is justified.

REFERENCES

Abdelouas, A., Y. Lu, W. Lutze, and H. E. Nuttall. 1998. "Reduction of U(VI) to U(IV) by Indigenous Bacteria in Contaminated Groundwater." *J. Contam. Hydrology 35*, 217-233.

Abdelouas, A. W. Lutze, and H. E. Nuttall. 1999a. Chapter 9: Uranium Contamination in the Subsurface: Characterization and Remediation." In *URANIUM: MINERALOGY, GEOCHEMISTRY AND ENVIRONMENT*, P. C. Burns and R. Finch, eds; Reviews in Mineralogy, Vol. 38, P. H. Ribbe, ed., 433-473.

Abdelouas, A., W. Lutze, and H. E. Nuttall. 1999b. "Oxidative Dissolution of Uraninite Precipitated on Navajo Sandstone." *J. Contam. Hydrology 36*, 353-375.

Abdelouas, A. W. Lutze, W. Gong, H. E. Nuttall, B. A. Strietelmeier, and B. J. Travis. 2000. "Biological Reduction of Uranium in Groundwater and Subsurface Soil." *Sci. Total Environ. 250*, 21-35.

Federal Register. 1995. Environmental Protection Agency, CFR 40 Part 192. "Groundwater Standards for Remedial Actions at Inactive Uranium Processing Sites." Table 1, p. 2866, Nov. 1995.

Lu, Y., H. E. Nuttall, and W. Lutze. 1999. "Denitrification Kinetics: Model and Comparison to Experimental Data." In: *Bioremediation of Metals and Inorganic Compounds*. A. Leeson and B. C. Alleman, eds. Battelle Press, Columbus Richland, 65-71.

McNeil, M. B. and B. J. Little. 1990. "Mackinawite Formation during Microbial Corrosion." *Corrosion 46*: 599-600.

Quinton, G. E., R. J. Buchanan, D. E. Ellis and S. H. Shoemaker. 1997. "A Method to Compare Groundwater Cleanup Technologies." Remediation, Autumn 1997, 7-16.

Travis, C. C. and C. B. Doty. 1990. "Can Contaminated Aquifers at Superfund Sites be Remediated?" *Environ. Sci. Technology 24*, 1464-1466.

Lovely Dr, Phillips EJP, Gorby Y, Landa E. 1991 Microbial reduction of uranium. Nature 350:413-416.

URANIUM SEQUESTRATION BY MICROBIALLY INDUCED PHOSPHORUS BIOAVAILABILITY

Leigh G. Powers (Georgia Institute of Technology, Atlanta, Georgia)
Heath J. Mills (Georgia Institute of Technology, Atlanta, Georgia)
Anthony V. Palumbo (Oak Ridge National Laboratory, Oak Ridge, Tennessee)
Chuanlun L. Zhang (University of Missouri, Columbia, Missouri)
Patricia A. Sobecky (Georgia Institute of Technology, Atlanta, Georgia)

ABSTRACT: Traditional approaches to the remediation of metals and radionuclides typically utilize dissimilatory reduction processes. However, in oxygenated environments such as the vadose zone, dissimilatory reduction can be problematic. Thus, new technologies are needed to broaden the scope of available bioremediation approaches. The goal of this study was to test the feasibility of a procedure that could immobilize contaminants, e.g., uranium, in aerobic environments by modifying soil microbes to constitutively overproduce the enzyme alkaline phosphatase. The substrate of alkaline phosphatase, organic P, is highly mobile in porous media, e.g., subsurface soils. Overproduction of alkaline phosphatase was achieved by introduction of plasmid pJH123 containing a *pglA-phoA* hybrid gene encoding a fusion protein. Plasmid pJH123 can be transformed into a number of subsurface pseudomonad isolates; each selected for their potential in field-scale delivery systems. In the presence of selection, plasmid pJH123 was stably maintained in the majority of the subsurface hosts tested. In addition, the plasmid conferred significantly higher levels of alkaline phosphatase enzyme activity. We hypothesize that an increase in phosphate levels due to alkaline phophatase activity may result in precipitation of uranium from solution and possibly soil. Studies are ongoing to determine the applicability of this uranium immobilization bioremediation strategy.

INTRODUCTION

Subsurface environments contaminated with mixed wastes are a consequence of the disposal of nuclear industry by-products (Riley et al., 1992) and represent a challenge to the growth and activity of subsurface microbes. These wastes are ill-defined mixtures of metals, radionuclides and organics. Much of this waste is in contact with surrounding geologic media allowing the unimpeded migration of contaminants into the surrounding soils and groundwater. Presently, techniques being employed to remediate waste from soil and groundwater include leaching (Hutchins et al., 1986), biosorption (Hu et al., 1996), acidification (Fortin et al., 1994), phytoremediation (Robinson et al., 1999), and bioaugmentation (El Fantrousi et al., 1999). Dissimilatory reduction processes have also shown promise in reducing heavy metal contamination in shallow, surface environments. Microbial reduction of heavy metals, such as uranium, has been previously shown to be coupled to electron transport processes involved with microbial growth (Lovley et al., 1991). Although reductive precipitation is a promising method for immobilization of metals and radionuclides, an alternative approach is phosphate mineral formation (Macaskie et al., 1994), which is more relevant to oxygenated environments, e.g., vadose zones. The precipitation of phosphorus with divalent cations and ferric iron has been shown to occur in many environments (Atlas and Bartha, 1987). Robertson and Alexander (1992) have reported Ca and Fe can decrease biodegradation rates by limiting availability of phosphorus. Low mobility of inorganic phosphorus limits the applicability of direct addition of inorganic P for immobilization of uranium. Therefore, the greatest potential for immobilization of U may be the use of organic phosphorus compounds coupled to bacterial enzymatic cleavage.

In the present study, we propose an innovative bioremediation treatment strategy to immobilize uranium using microbially produced inorganic phosphate. Previously, Sobecky et al. (1996b) have shown that marine bacteria can be genetically modified to overproduce alkaline phosphatase. Increased alkaline phosphatase activity persisted for up to a month in microcosms. Although results varied, addition of 10 μM glycerol-3-phosphate resulted in final PO_4^{3-} concentrations of greater than 5 μM from original concentrations of less than 1.0 μM (Sobecky et al., 1996b).

Our proposed method utilizes intact cells representing indigenous subsurface natural microorganisms. The rationale for use of native organisms is based on their ability to compete and survive in subsurface environments (Balkwill et al., 1998). Accordingly, several indigenous *Pseudomonas* species were selected and transformed with an alkaline phosphatase gene expressing plasmid, pJH123. Our results indicate stable maintenance of plasmid pJH123 that confers significantly higher levels of alkaline phosphatase activity in two of the three pseudomonad hosts tested relative to parental control strains. Our future objectives are to demonstrate the utilization of alkaline phosphatase overproduction by subsurface pseudomonads to precipitate radionuclides, e.g., uranium, from contaminated soils.

MATERIALS AND METHODS

Bacterial strains and plasmids used in study. The bacteria and plasmids used in this study are listed in Table 1. The subsurface isolates were kindly provided by Dr. David Balkwill through the Department of Energy Subsurface Microbial Culture Collection (SMCC). The strains were incubated in either nutrient broth (Difco) or basal salts minimal media (Schell et al., 1988) amended with glycerol-3-phosphate (G3P) (Sigma Chemical Co.) (30°C, 200 rpm). We selected spontaneous mutants of *Pseudomonas* species resistant to rifampicin (37 μg mL^{-1}) (Table 1). Broad-host-plasmids pRK404 and pJH123 where transferred from *Escherichia coli* to rifampicin-resistant *Pseudomonas* species as described previously (Ditta et. al., 1980). Tetracycline was added to a final concentration of 30 μM mL^{-1} (Table 1). Plasmid pJH123 is a derivative of pRK404 containing the *pglA-phoA* hybrid gene (Huang and Schell, 1990). This gene encodes a fusion protein comprising the signal sequence and first 177 residues of the extracellular

PglA polygalaturonase of *Pseudomonas solanacearum* fused to residue 14 of mature PhoA (Huang and Schell, 1990).

Determination of Plasmid Stability. Plasmid stabilization assays were conducted as previously described (Sobecky et al., 1996a). Briefly, overnight liquid cultures of *Pseudomonas* strains containing the plasmids of interest were obtained with antibiotic selection. An aliquot of cells was diluted 10^4- to 10^6-fold into pre-warmed nutrient broth and minimal media amended with 9.3 mM G3P and grown to mid-log phase. An aliquot of this dilution was plated. The remaining stationary-phase cells where used to isolate plasmid DNA to verify the presence and integrity of the DNA. All plates were incubated at 30°C, and the percentage of cells maintaining the plasmids was determined by replica plating at least 200 colonies from cells plated at the various sampling intervals onto the same medium amended with tetracycline (30 μg mL^{-1}). Percentage plasmid loss per generation was calculated as previously described (Roberts and Helinski, 1992) and averaged over at least three trials.

Table 1. Bacterial Strains and Plasmids

Strain or Plasmid	Relevant Characteristics	Reference or Source
Strain		
Pseudomonas veronii		Balkwill et al., 1998
P. veronii	Rif^r	This Study
P. rhodesiae		Balkwill et al., 1998
P. rhodesiae	Rif^r	This Study
Pseudomonas. sp. strain 150		Balkwill et al., 1998
Pseudomonas. sp. strain 150	Rif^r	This Study
Plasmid		
pRK404	Tc^r	Ditta et al., 1985
pJH123	PglA-phoA, Tc^r	Huang and Schell, 1990

Alkaline Phosphatase Activity Determination. Alkaline phosphatase activity is in nanomoles of *p*-nitrophenylphosphate hydrolyzed per minute per 10^9 cells, measured as described in Huang and Schell (1990). All measurements were taken from intact cells grown to stationary phase in basal salts minimal media (BSM) amended with 9.3 mM G3P. Cell counts were determined by direct microscopy and viable plating (Porter and Feig, 1980).

RESULTS AND DISCUSSION

The low solubility of phosphate minerals (e.g., log solubility product for meta-autunite ($Ca(UO_2)_2(PO_4)_2 \cdot 4H_2O$) = -48.5) (Faure, 1991) suggests that phosphate may be used to extract U from groundwater. The addition of calcium phosphate to a synthetic uranium-contaminated waste water decreased the concentration of uranium by more than 99% (Morrison and Spangler, 1992). Uranyl phosphate (meta-autunite) was one of the major uranium phases identified in soils (Buck et al., 1996). Geochemical calculations indicate that uranyl phosphates will control uranium concentrations in groundwater when $[HPO_4^{2-}]/[HCO_3^-] > 10^{-3}$ and that uranyl phosphates will precipitate if a high enough phosphate concentration exists. Unfortunately, the phosphate concentration in most natural waters limits the potential for uranyl phosphate precipitation to occur. In this study, the potential for microorganisms to liberate phosphate that would facilitate the precipitation of low solubility U-phosphates has been examined.

The stability and expression of a plasmid-encoded alkaline phosphatase was determined for three indigenous subsurface *Pseudomonas* sp. (Table 1). In the absence of antibiotic selection, plasmid pJH123 and the control vector pRK404 were not maintained in the three isolates during incubation in nutrient broth (Table 2). In contrast, *P. veronii* and *Pseudomonas* sp. strain 150 maintained pJH123 in the absence of selection (92% and 82% respectively; Table 2). These results are consistent with earlier findings that demonstrated that plasmid maintenance by bacteria in natural systems does not necessarily require selection (Sobecky et al., 1992). One explanation for the lack of plasmid pJH123 retention is that this IncP derivative lacks plasmid-encoded stabilization and partitioning loci found shown to promote the stable maintenance and retention of the parental plasmid RK2 in the absence of selection (Roberts and Helinski, 1992; Roberts et al., 1994). RK2 has been shown to contain the *parCBA* system, a putative multimer resolution system (Sobecky et al., 1996a) and *parDE*, a toxin/antitoxin cell killing system that either inhibits or kills plasmid-free daughter cells that arise during cell division (Roberts et al., 1994).

Table 2. Plasmid Retention

Isolate	Nutrient Broth	BSM[a]
P. veronii (pRK404)	0.0%	ND[b]
P. veronii (pJH123)	2.0%	92.0%
P. rhodesiae (pRK404)	5.0%	ND[b]
P. rhodesiae (pJH123)	1.0%	6.0%
Pseudomonas. sp. isolate 150 (pRK404)	0.0%	ND[b]
Pseudomonas. sp. isolate 150 (pJH123)	3.0%	82.0%

[a] BSM amended with 9.3 μM G3P
[b] ND - not done

All three pJH123-containing subsurface isolates exhibited elevated extracytoplasmic alkaline phosphatase production ranging from 110 to >240-fold relative to that by the parent and parent containing pRK404. The level of alkaline phosphatase production was 2.7- to 6-fold greater than was observed with the marine isolates (Sobecky et al., 1996b). One explanation may be the *pglA-phoA* hybrid gene isolates introduced into the three pseudomonads directed considerably more effective extracellular alkaline phosphatase export.

Specifically, the *pglA* gene from the plant pathogen *Pseudomonas solanacearum* encodes a 52 kD, extracellular polygalacturonase protein, one of several excreted extracellular enzymes (Huang and Schell, 1990). Thus, *pglA* and its promoter may have been transcribed more efficiently, e.g., the codon utilization typical of a *Pseudomonas* gene (Nakai et al., 1983; Normore, 1976), and exported to the periplasm of the pseudomonads used in this study relative to the marine bacterium previously studied.

Table 3. Alkaline Phosphatase Activity

Isolate	Plasmid	Alkaline Phosphatase Activity[a] ($nM\ 10^9\ cells^{-1}\ min^{-1}$)
P. veronii	-	0.05 (0.02)
P. veronii	pRK404	0.03 (0.01)
P. veronii	pJH123	11.4 (0.05)
P. rhodesiae	-	0.02 (0.01)
P. rhodesiae	pRK404	0.01 (0.01)
P. rhodesiae	pJH123	16.4 (0.06)
Ps. sp. isolate 150	-	0.07 (0.02)
Ps. sp. isolate 150	pRK404	0 (0.002)
Ps. sp. isolate 150	pJH123	7.5 (0.09)

[a] Standard deviation given in parentheses.

Preliminary studies have indicated that as much as 250 $\mu g\ L^{-1}$ phosphate accumulates in batch cultures with *Pseudomonas* isolates containing pJH123 relative to the same isolates with or without pRK404 (data not shown). Cell-free supernatant collected from the batch cultures complexed/precipitated as much as 15 μM of uranyl acetate (data not shown). These results indicate that overproduction of microbially-produced phosphate precipitates uranium from solution. Efforts are underway to determine the feasibility of this intact, non-immobilized cell approach in soil microcosms. A longer-term goal of this research is to develop the technology for use in the field. Our present approach relies on the formation of low solubility phosphate minerals produced from the (bacterial) liberation of phosphorus from added organic phosphorus compounds. To circumvent the need for recombinant microorganisms, acid and alkaline phosphatase enzymatic activities in naturally occurring subsurface bacteria are being characterized. An approach that couples the introduction of organic phosphorous compounds (biostimulation) and native bacteria with suitable phosphate liberating activities (bioaugmentation) may be a feasible *in situ* means to control the migration of contaminants and stabilize contaminant plumes via bacterially mediated co-precipitation of low-solubility solid phases.

ACKNOWLEDGEMENTS

We thank Kelly Delaney for providing excellent technical assistance. Research was supported by the Department of Energy, grant DE-AC05-96OOR22464.

REFERENCES

Atlas, R.M., and R. Bartha. 1987. *Microbial Ecology. Fundamentals and Applications.* Benjamin/Cummings Publishers: Menlo Park, CA.

Balkwill, D.L., E.M. Murphy, D.M. Fair, D.B. Ringelberg, and D.C. White. 1998. "Microbial Communities in High and Low Recharge Environments: Implications for Microbial Transport in the Vadose Zone." *Microbiol. Ecol. 35*:156-171.

Buck, E.C., N.R. Brown, and N.L. Dietz. 1996. "Contaminant Uranium Phases and Leaching at the Fernald Site in Ohio." *Environ. Sci. Technol. 30*:81-88.

Ditta, G., T. Schmidhauser, E. Yakobson, P. Yu, Y.-W. Liang, D.R. Finlay, D. Guiney, and D.R. Helinski. 1985. "Plasmids Related to the Broad Host Range Vector pRK290 for Genetic Cloning and for Monitoring Gene Expression." *Plasmid. 13*:149-153.

Ditta, G., S. Stanfield, D. Corbin, and D.R. Helinski. 1980. "Broad-Host-Range DNA Cloning System for Gram-Negative Bacteria: Construction of a Gene Bank of *Rhizobium meliloti*." *Proc. Natl. Acad. Sci. USA. 77*:7347-7351.

Faure, G. 1991. *Principles and applications of Geochemistry*. 2nd ed. Prentice Hall Publishers: Upper Saddle River, NJ.

Fortin, D., G. Southam, and T.J. Beveridge. 1994. "Nickel Sulfide, Iron-Nickel Sulfide, and Iron Sulfide Precipitation by a Newly Isolated *Desulfotomaculum* Species and its Relation to Nickel Resistance." *FEMS Microbiol. Ecol. 14*:121-132.

El Fantroussi, S., M. Belkacemi, E.M. Top, J. Mahillon, H. Naveau, and S.N. Agathos. 1999. "Bioaugmentation of a Soil Bioreactor Designed for Pilot-Scale Anaerobic Bioremediation Studies." *Environ. Sci. Technol. 33*:2992-3001.

Hu, M. Z.-C., J.M. Norman, B.D. Faison, and M.E. Reeves. 1996. "Biosorption of Uranium by *Pseudomonas aeruginosa* Strain CSU: Characterization and Comparison Studies." *Biotech. Bioeng. 51*:237-247.

Huang, J., and M.A. Schell. 1990. "DNA Sequence Analysis of *pgl*A and Mechanism of Export of its Polygalacturonase Product from *Pseudomonas solanacearum*." *J. Bacteriol. 172*:3879-3887.

Hutchins, S.R., M.S. Davidson, J.A. Brierley, and C.L. Brierley. 1986. “Microorganisms in Reclamation of Metals.” *Annu. Rev. Microbiol. 40*:311-336.

Lovley, D.R., E.J.P. Phillips, Y.A. Gorby, and E.R. Landa. 1991. “Microbial Reduction of Uranium.” *Nature. 350*:413-416.

Macaskie, L.E., K.M. Bonthrone, and D.A. Rouch. 1994. "Phosphatase-Mediated Heavy Metal Accumulation by a *Citrobacter* sp. and Related Enterobacteria." *FEMS Microbiol. Lett. 121*:141-146.

Morrison, S.J., and R.R. Spangler. 1992. "Extraction of Uranium and Molybdenum from Aqueous Solutions: A Survey of Industrial Materials for Use in Chemical Barriers for Uranium Mill Tailings Remediation." *Environ. Sci. Technol. 26*:1922-1931.

Nakai, C., H. Kagmiyama, M. Nozaki, T. Nakazawa, S. Inouye, Y. Ebina, and A. Nakazawa. 1983. "Complete Nucleotide Sequence of the Metapyrocatechase Gene on the TOL Plasmid *Pseudomonas putida* mt-2." *J. Biol. Chem. 258*:2923-2928.

Normore, N.W. 1976. “Guanosine-Plus-Cytosine (GC) Composition of the DNA of Bacteria, Fungi, Algae, and Protozoa.” In G.D Fasman (Ed.), *Handbook of Biochemistry and Molecular Biology*, pp. 65-240. CRC Press: Boca Raton, FL.

Porter, K.G., and Y.S. Feig. 1980. "The Use of DAPI for Identifying and Counting Aquatic Microflora." *Limnol. Oceanogr. 25*:943-948.

Riley, R.G., J.M. Zachara, F.J. Wobber. 1992. *Chemical Contaminants on DOE Lands and Selection of Contaminant Mixtures for Subsurface Science Research.* DOE/ER-0547T. US Department of Energy, Washington, DC.

Roberts, R.C., and D.R. Helinski. 1992. “Definition of a Minimal Plasmid Stabilization System from the Broad-Host-Range Plasmid RK2.” *J. Bacteriol. 174*:8119-8132.

Roberts, R.C., A.R. Strom, and D.R. Helinski. 1994. “The *parDE* Operon of the Broad-Host-Range Plasmid RK2 Specifies Growth Inhibition Associated with Plasmid Loss.” *J. Mol. Biol. 237*:35-51.

Robertson, B.K., and M. Alexander. 1992. “Influence of Calcium, Iron and pH on Phosphate Availability for Microbial Mineralization of Organic Chemicals.” *Appl. Environ. Microbiol. 58*:38-41.

Robinson, B.H., R.R. Brooks, and B.E. Clothier. 1999. “Soil Amendments Affecting Nickel and Cobalt Uptake by *Berkheya coddii*: Potential Use for Phytomining and Phytoremediation.” *Ann. Bot. 84*:689-694.

Schell, M.A., D.P. Roberts, and T.P. Denny. 1988. "Analysis of the *Pseudomonas solanacearum* Polygalacturonase Encoded by *pgl*A and its Involvement in Phytopathogenicity." *J. Bacteriol. 170*:4501-4508.

Sobecky, P.A., M.A. Schell, M.A. Moran, and R.E. Hodson. 1992. "Adaptation of Model Genetically Engineered Microorganisms to Lakewater: Growth Rate Enhancements and Plasmid Loss." *Appl. Environ. Microbiol. 58*:3630-3637.

Sobecky, P.A., C.L. Easter, P.D. Bear, and D.R. Helinski. 1996a. "Characterization of the Stable Maintenance Properties of the *par* Region of Broad-Host-Range Plasmid RK2." *J. Bacteriol. 178*:2086-2093.

Sobecky, P.A., M.A. Schell, M.A. Moran, and R.E. Hodson. 1996b. "Impact of a Genetically Engineered Bacterium with Enhanced Alkaline Phosphatase Activity on Marine Phytoplankton Communities." *Appl. Environ. Microbiol. 62*:6-12.

BIOREMEDIATION OF NITRATE CONTAMINATED WASTEWATER

P. C. Mishra and N. Behera
(Sambalpur University, ORISSA, INDIA)

ABSTRACT: IDL Industries Limited, with its explosive manufacturing factory at Rourkela, India, produces mining explosives of a water gel type. The major raw materials used for manufacturing these explosives are nitrates of ammonia, calcium, and sodium in addition to powder of aluminium, thickeners, and common salts. Being a nitrate-based factory, the trade effluent contains a significantly high quantity of dissolved nitrates, sometimes exceeding 500 mg/L in wastewater. The present practice of disposal of such wastewater is to allow the water to enter an impervious-lined large lagoon. However, during monsoon season, the overflow water contaminates the soil and water bodies outside the premises. The statutory requirement for disposal of nitrate contaminated wastewater is 45 mg/L. As the conventional treatment methods are cost-prohibitive, time consuming, and frequently incomplete, biological methods of treatment with hydrophytes like water hyacinth (*Eichhornia crassipes*) alone and in combination with water fern (*Salvinia molesta*) in the presence of cow dung as a source of denitrifying bacteria were tried in the laboratory as well as in open lagoons. In the laboratory study, these hydrophytes reduced the nitrate load by 90% within a period of 30 days. In the field study, water hyacinth alone in the presence of a bacteria source reduced the nitrate load from 118 to 15.5 mg/L within 15 days. However, in combination with water fern, the nitrate load could be reduced from 390 to 35 mg/L (93.5%) within 30 days indicating the potential of these hydrophytes in removing the nitrate load from wastewater.

INTRODUCTION

In India, some 8,000 large and medium industrial units are responsible for the pollution of water bodies, but their contribution to water pollution load is 10 to 15%. The rest are of domestic origin. However, the industrial wastewater poses a greater threat due to the presence of toxic chemicals, nutrients, and non-biodegradable organic matter. Although the conventional chemo-mechanical methods of wastewater treatment involving physical, chemical, and biological methods are by far the best methods of treatment, these methods are cost-intensive in developing countries owing to constraints like paucity of funds for initial investment, lack of adequate foreign exchange to import spare parts, and want of facilities for the repair of sophisticated instruments. Therefore, an appropriate technology that may be less expensive and require minimally technically trained personnel has been in demand by municipalities and industrial establishments in the developing countries. Bioremediation, a low-cost method of treatment, encompasses biological methods for cleaning up contaminated areas and involves establishing the conditions in contaminated environments so that appropriate microorganisms flourish and carry out the metabolic activities to detoxify the contaminants (Jogdand, 1995). This paper tries to evaluate the efficacy of bioremediation involving hydrophytes and bacteria in removing nitrate load in wastewater.

Study Area. The study was undertaken in IDL Industries Limited at Rourkela, India. It is an explosives-manufacturing company producing commercial mining explosives of water gel type. The major raw materials used are nitrates of ammonium, calcium, and sodium in addition to powders of aluminum, thickeners, common salts etc. As the explosives manufactured are nitrate-based, the trade effluent contains high quantities (200 to 500 mg/L) of dissolved nitrates. The Central Pollution Control Board (CPCB) of India has prescribed a limit of 45 mg/L nitrate in wastewater for discharge to open water bodies, sewage drains, and soil (CPCB, 1993).

It is common knowledge that the current methods for removing nitrate are either ineffective or impractical. A process that uses only live microorganisms to denitrify water is slow, frequently incomplete, and difficult to set up and maintain. Against this background the emerging technology of using aquatic macrophytes in combination with denitrifying bacteria which enables cost-effective treatment assumes great significance and holds great promise (Boyd, 1970; Wolverton et al., 1976; Wolverton, 1987). Among the various types of aquatic macrophyte-based treatment systems, pond systems containing floating macrophytes such as water hyacinth (*Eichhornea crassipes)* and water fern (*Salvinia molesta*) are most commonly used in tropical and subtropical regions (Abbasi, 1987; Abbasi and Nipaney, 1994; Tchobanoglous et al., 1989; Wood and McAtamney, 1994)

Methodology. To assess the potentiality of hydrophytes in combination with microorganisms in reducing nitrate load from IDL wastewater, laboratory and field studies were conducted. Laboratory studies were made in aquariums (30 × 12 × 10 cm size) with a known concentration of nitrate-enriched water and grown with water hyacinth and water fern independently and in combination. Field studies were conducted on IDL wastewater containing 300 to 500 mg/L nitrate in small lagoons with water hyacinth, water fern, sewage sludge, and fresh cow-dung. Sewage sludge serves as a food source for microorganisms and fresh cow-dung as a source of denitrifying bacteria. The nitrate content was analyzed by phenol disulphonic acid method (Michael, 1984).

RESULTS AND DISCUSSION

Table 1 shows the characteristics of wastewater generated from IDL, Rourkela. Table 2 shows the reduction in nitrate content of contaminated water in the aquarium grown with water hyacinth and water fern independently and in combination. Figure 1 shows percent reduction in nitrate content over the previous day.

TABLE 1. Quality of IDL wastewater.

Parameter	Content (mg/L except pH)
pH	7.5 – 8.5
Suspended solids	60 – 75
Total dissolved solids	1,500 – 1,800
Chloride	60 – 80
Nitrate	300 - 500

TABLE 2. Nitrate reduction in laboratory study.

Plants/days	0	5	10	15	20	25	30
Water hyacinth	100	82	73	50	35	18	10
Water fern	100	87	76	54	42	23	13
Hyacinth + fern	100	74	65	47	31	12	3

Within a period of 30 days, there was around 90% reduction in nitrate content with a maximum reduction of 97% in water grown with both hyacinth and fern (Table 2). However, percent reduction was highest after 20 days of growth. The weight/surface area of plants grew 2.5 times during 30 days indicating that both plants could survive in nitrate stress and possibly reduce the nitrate load.

Based on laboratory studies, field studies were designed with small impervious lagoons sized 2 m × 2m × 1m (l × b × h). Dried sewage sludge collected from sewage drain was put into the bottom of the lagoon (6"). Effluent from IDL containing 300 mg/L nitrate was poured into the lagoon up to a height of 75 cm, and strands of young water hyacinth collected from nearby fresh water source were fixed to the bottom sediments in the periphery of the lagoon covering one-fourth of the water surface. Approximately 1 kg of fresh cow-dung was also put into the lagoon as a source of denitrifying bacteria and the nitrate content was monitored. Table 3 shows a significant reduction in nitrate content in five such lagoons from an average initial load of 118 to 15.45 mg /L (87.3%).

The results obtained from the small-scale field experiment prompted the authors to construct another larger impervious lagoon of the size 6 × 6 × 1.5 m. Both water hyacinth and water fern were grown covering one-third of the surface area and 5 kg of fresh cow dung were added to the lagoon. Table 4 reveals a reduction in nitrate content from an initial level of 390 to 35 mg/L with an increase in NO_3^-N content in leaf sample from 14 to 38 and 21 to 43 mg/L in water hyacinth and water fern respectively within a period of 30 days. Yoeh (1993) enhanced the final discharge quality of some agrochemicals with removal of 53.7% ammoniac nitrogen and 68.8% total Kjeldahl nitrogen. High productivity of *Salvinia* and its resilience make it suitable to recover nutrients in the tertiary treatment stage. The dense cover makes the water relatively anaerobic that, in turn, favors denitrification.

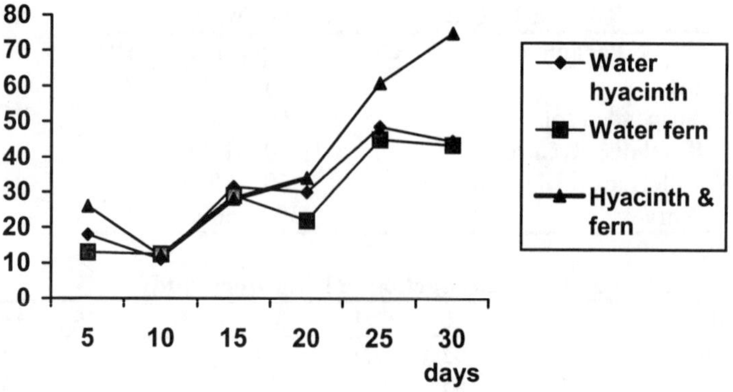

FIGURE 1. Per cent reduction in nitrate over previous day in laboratory study.

TABLE 3. Nitrate reduction (mg /L) in small lagoons treated with water-hyacinth (X ± SD).

Lagoon	**Days**		
	0	**7**	**15**
Lagoon-1	90	13.2 ± 1.4	7.7 ± 1.4
Lagoon-2	120	44.2 ± 20.5	26.2 ± 9.1
Lagoon-3	110	25.9 ± 23.2	27.8 ± 0.8
Lagoon-4	130	13.4 ± 1.0	8.6 ± 1.54
Lagoon-5	140	16.9 ± 2.3	9.46 ± 1.05
Average (X ± SD)	118 ± 19	22.72 ± 3.1	15.45 ± 1.5

TABLE 4. Nitrate reduction in wastewater (figures in parentheses show per cent reduction from 0 day).

Days	**Nitrate content (mg/L)**	**NO_3-N in leaf of water hyacinth (mg/g)**	**NO_3-N content in leaf of water fern (mg/g)**
0	390	14	21
5	285 (27)	15	23
10	210 (46)	22	23.5
15	143 (70)	24	24.5
20	95 (75.5)	26	25
25	64 (86)	32	37
30	35 (93.5)	38	43

Thus, in the absence of any technology or when the existing technology becomes cost-intensive, the removal of nitrate from wastewater through bioremediation may be the only choice for India and other developing countries.

ACKNOWLEDGEMENTS

Financial support from IDL, Rourkela, and laboratory facilities in the Department of Environmental Sciences, Sambalpur University is gratefully acknowledged.

REFERENCES

Abbasi, S.A. 1987. "Aquatic plants-based water treatment systems in Asia." K.R. Reddy and W.H. Smith (Eds.). *Aquatic Plants for water treatment and resource recovery.* Magnolia Publ. Inc. Orland, 175-98.

Abbasi, S. A. and P.C. Nipaney.1994. " Potential of aquatic weed *Salvania molesta (*Mitchell) for water treatment and energy recovery ". *Ind. J. Chem. Tech. 1*: 204-13.

Boyd, C.E. 1970. "Vascular aquatic plants for mineral nutrient removal from polluted water". *Econ. Bot. 23*: 95-103.

CPCB. 1993. "Standards for wastewater effluent discharge." *The Gazetteers of India,* No. 174.

Jogdand, S.N. 1995. *"Environmental Biotechnology".* Himalaya Publishing House, Delhi, 235 p.

Michael, P. 1984. "*Ecological methods for field and Laboratory investigations.*" Tata McGraw Hill. 404 P.

Tchobanoglous, G., P. Maitski, K Thompson, and T.H. Chadwick. 1989. "Evaluation and performance of the city of San Diego pilot scale aquatic wastewater treatment system using water hyacinth." *J .Water Pollution Control Fed. 61*(11/2): 1625-35.

Wolverton, B.C. 1987. "Artificial marshes for wastewater treatment." K.R. Reddy and W.H. Smith (Eds.). *Aquatic Plants for water treatment and Resource Recovery*. Pp. 13-19. Magnolia Publishing Inc. Orlando, FL.

Wolverton, B.C., R.M. Barlow, and R.C. McDonald. 1976. "Application of vascular aquatic plants for pollution removal, energy and food production in a biological system." J. Tourbier and R.W. Pierson, Jr. (Eds.). *Biological control of water pollution.* Univ. of Pennsylvania Press, USA. 141.

Wood, B. and McAtamney.1994. "The use of macrophytes in bioremediation." *Biotechnology Advances 12*(4): 653-62.

Yeoh, B.G. 1993. "Use of water hyacinth (*Eichhornea crassipes)* in upgrading small agroindustrial wastewater treatment plants". *Water Science Technology. 28*: 207-13.

NATURAL ATTENUATION OF NITRATE IN THE BIG DITCH WATERSHED, ILLINOIS

Shawn Shiffer and Robert Sanford (University of Illinois at Urbana-Champaign, Illinois)
Tania Matos (University of Puerto Rico at Rio Piedras, Puerto Rico)
Edward Mehnert, Donald A. Keefer and William S. Dey (Illinois State Geological Survey, Champaign, Illinois)
Thomas R. Holm (Illinois State Water Survey, Champaign, Illinois)

ABSTRACT: The focus of this research was to determine the natural attenuation capacity for nitrate in the groundwater throughout the Big Ditch Watershed. Eleven sites were sampled within the watershed and sediment samples were taken from several depths at each site. The natural attenuation capacity of the watershed was determined by enumeration of denitrifiers by most probable number (MPN) technique and by determining the intrinsic denitrification rate using a modified denitrifying enzyme activity (DEA) method. Enumeration results showed that the microorganisms were less abundant in sediment samples taken from the middle depths of the saturated zone. Three separate total carbon amendments of 2 mM, 8 mM and 20 mM were used in the denitrification rate determination. The low carbon (2mM) amendment resulted in a significantly lower denitrification rate (29.84 ± 0.25 mg N/kg soil-day) than the other carbon amendments (8mM: 132.63 ± 1.17 mg N/kg soil-day, 20mM: 193.98 ± 3.11 mg N/kg soil-day) and varied little with depth. In contrast, the denitrification rate increased with the depth of the sample for the 8 mM C and 20 mM C (193.98 ± 3.11 mg N/kg soil-day) amendments. The data shows that denitrifier abundance and activity in the groundwater are dependent on the spatial variables of depth and location, the dissolved oxygen, and on the availability of carbon.

INTRODUCTION

Nitrate is a significant groundwater contaminant throughout the world. When nitrate is ingested it is converted to nitrite, which can cause methemoglobinemia in infants. As a result, the U.S. Environmental Protection Agency (EPA) has set the maximum contaminant level (MCL) under the Safe Drinking Water Act at 10 mg/L nitrate-nitrogen or 44 mg/L nitrate. Agricultural fertilizer and animal wastes leaching into the groundwater are the major causes of nitrate contamination (Spalding and Exner, 1993). Lee and Nielson (1989) used information on aquifer vulnerability, nitrogen fertilizer usage, and maps of groundwater nitrate concentrations (Madison and Brunett, 1985) to determine the nitrate contamination potential for aquifers in the United States. Approximately 20% or more of wells sampled in Iowa, Nebraska and Kansas exceed the MCL for nitrate (Spalding and Exner, 1993). Many wells studied in Ohio were deemed vulnerable to nitrate contamination; however, only 2.7% of the wells sampled exceeded the MCL (Baker et al., 1989). There is evidence of stratification in nitrate concentration with depth in aquifers in Ohio and Illinois. At other vulnerable sites, the nitrate discharged to the surface does not even reach the groundwater. These discrepancies between the actual contamination of the groundwater and the amount predicted have been attributed to denitrification, rapid nitrogen uptake by plants, and tile drainage that discharges nitrate contaminated water to surface waters (Spalding and Exner, 1993).

Natural attenuation of nitrate occurs through denitrification. Nitrate is removed from the groundwater by denitrifying microorganisms that reduce nitrate to nitrogen gas. The formation of nitrogen gas serves as an inert nitrogen sink, where the other forms of nitrogen in groundwater will persist and continue to contaminate the groundwater. The denitrification pathway of nitrate to nitrogen gas is shown below (Tiedje, 1994). Some bacteria only convert nitrate to nitrite

$$NO_3^- \rightarrow NO_2^- \rightarrow NO \rightarrow N_2O \rightarrow N_2$$

or dissimilatorily reduce nitrate to ammonia. When these latter two pathways occur, nitrogen is not removed from the groundwater. The significance of these alternative pathways is usually small in oligotrophic (low carbon) groundwater ecosystems, where denitrification is predominant (Tiedje, 1994). Most denitrifying bacteria are facultative, heterotrophic anaerobes and therefore denitrification can only occur in the absence of significant dissolved oxygen (DO) (Korom, 1992). Organic carbon is required for heterotrophic growth, so the amount of dissolved organic carbon (DOC) and the amount of carbon in the sediment are important factors determining the extent of natural attenuation of nitrate. Spalding and Exner (1993) determined that soils with high organic carbon protected vulnerable aquifers from nitrate contamination by stimulating denitrification. Ohio and other corn producing states like Illinois and Indiana have organic rich soils and subsoils that are favorable for denitrification and thus the incidence of groundwater contamination is lower than expected. Other areas where this has been observed are the riparian woodlands in the Southwest (Spalding and Exner, 1993) and parts of Illinois (Risatti and Mehnert, 1998), where the soils also have a high organic carbon content.

Since the fate and attenuation of nitrate in groundwater is not well defined in Illinois, the purpose of this study was to quantify the activity of denitrifying microorganisms in groundwater throughout the Big Ditch Watershed and to determine the nitrate attenuation capacity in the aquifer. The groundwater in the watershed consists of a shallow unconfined aquifer and a deeper confined aquifer. Denitrifying activity by microorganisms in the watershed was analyzed using two experimental approaches. First, denitrifying microorganisms in the aquifer sediment were enumerated. Second, the rate of denitrification was measured for different samples from different areas of the watershed.

METHODS

Sediment Core Collection. Eleven sites were selected within the Big Ditch Watershed, with shallow and deep wells installed at 9 of the 11 sites, giving a total of 20 different wells. These sites were selected for a complete representation of the watershed and to determine the variability of denitrification rates across it. Samples of core material were taken at different depths at each site to determine the vertical variability of denitrification rates in the watershed. In order to prevent contamination, the sediment samples were taken from the center of the core material retrieved from drilling boreholes at each site. Aseptic technique was used when collecting the samples to prevent cross-contamination. The samples were stored at 4°C for up to 3 months until the analyses were completed. Most probable number (MPN) analyses were started within one week of sample collection.

Enumeration. The sediment samples were enumerated using a modified denitrifier MPN method (Tiedje, 1994). The MPN determinations were completed using R2A broth media with two different nitrate concentrations, 0.5 mM and 10 mM. A three-tube MPN test was used. The samples were incubated for two weeks and then the number of positive and negative tubes were counted for each sample. The positive tubes containing 10 mM nitrate media were indicated by gas bubble formation in inverted Durham tubes and a diphenylamine indicator test was used to detect residual nitrate or nitrite (Tiedje, 1994). The positive MPN tubes for the 0.5 mM nitrate media were indicated by the diphenylamine indicator test and terminal dilution tubes were verified using HPLC analysis.

Rate Determination. The soil denitrification rate was measured by using a modification of the standard denitrifying enzyme activity (DEA) assay for soil (Tiedje, 1994). In this assay, 160-mL serum bottles, with ten grams of soil and phosphate buffer up to a total volume 100 mL were used. The headspace in these bottles was 60 mL of 90% nitrogen gas and 10% acetylene. The original assay was modified by changing the carbon source and increasing the nitrate concentration to 2 mM. In order to determine the effect of available carbon, the N_2O production rate of the soil was measured for three separate total carbon amendments: 2 mM, 8 mM and 20 mM. Acetate and glucose were added in equimolar amounts as the source of carbon. This means that the 2 mM total carbon amendment consisted of 0.25 mM acetate and 0.25 mM glucose, the 8 mM amendment consisted of 1 mM of acetate and 1 mM of glucose, and the 20 mM amendment consisted of 2.5 mM acetate and 2.5 mM glucose. Chloramphenical was not used because it has been shown to underestimate the actual denitrification rates (Pell et al., 1995) and because N_2O production was undetectable when it was used. Instead of the usual 2-hour DEA test used for surface soils, headspace gas samples were taken at 12, 36, 40, 44, 48, 60, 64, 68, and 72 hours. The rate was determined from the slope of the line between 36 and 48 hours.

Analysis. The headspace samples for the DEA assay were analyzed for the N_2O concentration using a Varian 3800 Gas Chromatograph equipped with a Thermal Conductivity Detector (TCD) and Poropak Q column. The nitrate and nitrite concentration of liquid samples was analyzed with High Performance Liquid

Chromatograph (HPLC) analysis using a Waters 486 Tunable Absorbance Detector and a Partisil SAX 10 micron column.

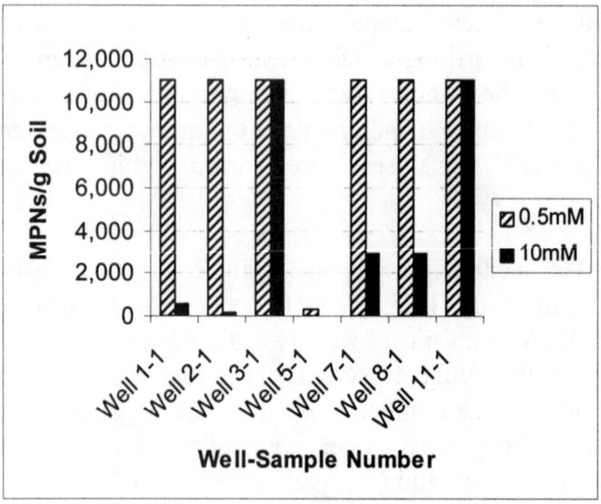

FIGURE 1: Comparison of MPN results for 0.5mM and 10mM nitrate media. 11,000 is the upper MPN limit of this assay.

RESULTS

MPN Enumeration. Denitrifier MPNs observed were consistently higher with the 0.5 mM nitrate concentration than with the 10 mM nitrate concentration (Figure 1). Wells 3 and 11, however, did not show a difference related to the nitrate concentration in the MPNs. This lack of difference may be because the limit of quantification for the MPN assay was 11,000/g soil, which is due to the limits of the highest dilution used. The abundance of denitrifying microorganisms in sediment samples decreased down to sample depths of 15-20 ft and then increased again at lower depths (Table 1).

TABLE 1: MPN results (0.5mM nitrate) compared with depth.

Depth (m)	Average MPN/g soil
< 3	11,000[a]
3-4.6	9,380
4.6-6.1	4,726
6.1-7.6	8,583
7.6-10.7	8,975

[a] 11,000 is the upper MPN limit of this assay.

DEA Rate Analysis. In all samples tested, the low carbon (2 mM) amendment resulted in significantly lower denitrification rates (21.04 to 45.34 mg N/kg soil-day [$\bar{x} = 29.84 \pm 0.25$]) than observed with the 8 mM (11.26 to 216.61 mg N/kg soil-day [$\bar{x} = 132.63 \pm 1.17$]) and the 20 mM carbon amendments (11.74 to

337.67 mg N/kg soil-day [$\bar{x}$ = 193.98 ± 3.11]) (Figure 2). Spatially, denitrification rates varied with the depth of the sample taken for each carbon concentration (Figure 3). The 2 mM carbon amendments had low denitrification rates at all depths. In contrast, the denitrification rate increased with the depth of the sample for the 8 mM and 20 mM carbon amendments.

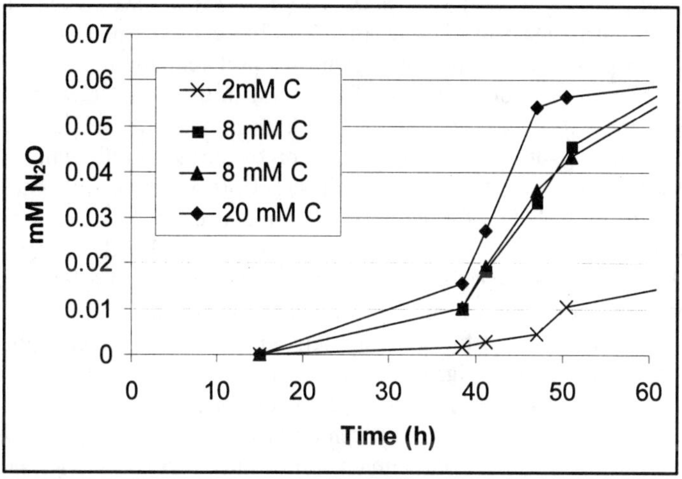

FIGURE 2: N_2O production rate for Well 1 at a depth of 27 feet. Lines indicate the different carbon amendments.

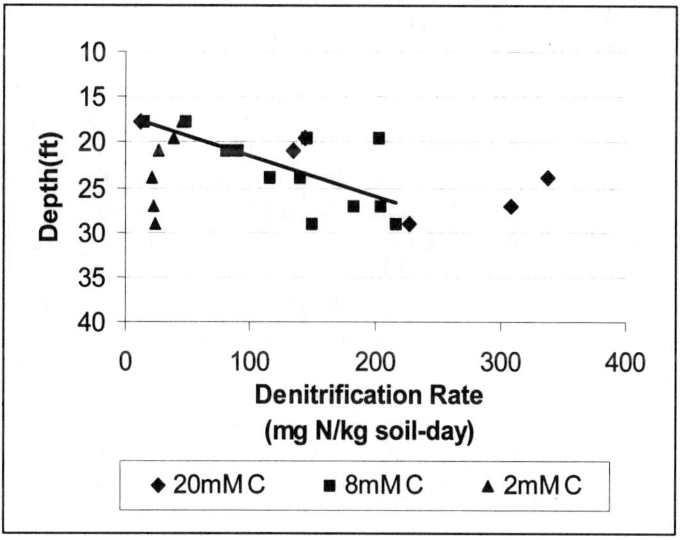

FIGURE 3: Comparison of denitrification rates with depth. The line indicates a linear regression line for the 8mM carbon amendment.

DISCUSSION

The low variability of denitrification rates for the low carbon amendment compared to the high variability with depth of the other two higher amendments was unexpected. Studies of denitrification rates in aquifers have shown that carbon is a limiting nutrient (Starr and Gillham, 1993); however, in this study the amount of carbon in all of the amendments was in excess of the amount required for complete conversion of NO_3^- to N_2O. For the lower (2 mM C) concentration, however, the rates appear to be first order with respect to carbon. One possible reason for this could be that the microorganisms preferentially use acetate before glucose. If the acetate concentration of 0.25 mM in the 2 mM C amendment is the only carbon used for denitrification, then the carbon available from acetate is not sufficient to support the complete reduction of nitrate to nitrogen gas. Thus the amount of available carbon could limit the affect of depth on denitrification rates. Figure 3 shows that for the lower carbon concentration there is also very little variability in the denitrification rate with depth. It is slightly higher near the surface, but this small variability is probably due to the higher number of denitrifiers near the surface. This low variability of the denitrification rate with depth may also be expected for other aquifers with low carbon availability. The carbon concentrations in the groundwater throughout the Big Ditch Watershed varies between 0.067 and 0.525 mM C, with the majority of the wells having around 0.08 mM C and the highest concentrations observed in the shallow wells. Based on these carbon levels, the attenuation of nitrate in the aquifer should be higher closer to the surface and decrease in the deeper parts of the aquifer due to the limiting effect of carbon. This is in contrast with what occurs when carbon was increased in the DEA experiments, where denitrification rates increased with depth. These observations are perhaps related to the intrinsic characteristics of the aquifer material in the Big Ditch Watershed.

Although 10 mM nitrate is typically used for enumeration of denitrifiers, the results show that the number of denitrifiers are underestimated when this concentration is used. This is probably due to the concentration of nitrate in the groundwater being closer to 0.5 mM than to 10 mM and that denitrifying bacteria from the aquifer are not adapted to high nitrate concentrations. The apparent lack of difference in MPNs between wells 3 and 11 for both nitrate concentrations might be because both samples are from a shallow depth where typically higher numbers of microorganisms are observed. The upper quantification limit of 11,000 MPN/g soil is also not high enough to determine if there is a real difference in the number of microorganisms between the two nitrate concentrations.

The middle levels of the core material have lower MPNs than shallow or deep levels (Table 1). The reason for this is unclear, however the types of nitrate reducing bacteria may differ with depth. Another reason for the difference of MPNs with depth might be related to the nature of the aquifer material. For example, some layers might be richer in organic carbon than others, thus leading to different microbial populations. The MPN assay uses a measure of nitrate removal to indicate activity. The assay does not differentiate between the microorganisms that dissimilatorily reduce nitrate to ammonia and the denitrifiers.

It is assumed that these dissimilatory ammonifying microorganisms are only a small fraction of the total number of nitrate reducers in the system because the groundwater is oligotrophic. Also such organisms would not produce significant N_2O in the DEA assay, so it is not likely that they are significant.

According to the results of this study denitrifier abundance and activity in the groundwater are dependent on the spatial variables of depth and location, the DO, and the available organic carbon. The spatial variability can be used to compare the different geological characteristics of the site to the attenuation of nitrate in the groundwater of the watershed. The denitrification variability in the aquifer will probably depend more on the concentration of carbon and DO rather than the geological characteristic variability. This is because no matter what the geological characteristics of the sediments are, if there is not enough available carbon and there are significant amounts of DO then denitrification will not occur. In an experiment at the Rodney Site (Starr and Gillham, 1993) in situ rates of denitrification of 2.4×10^{-5} g N/L-h were measured. Of the two sites studied, the Rodney Site has available carbon concentrations similar to the Big Ditch Watershed. When carbon as glucose was added the rate increased to 1.4×10^{-4} g N/L-h (Starr and Gillham, 1993). This glucose amended rate is similar to the average rate for the 2 mM carbon amendment of $1.16 \times 10^{-4} \pm 0.009 \times 10^{-4}$ g N/L-h (29.84 ± 0.25 mg N/kg soil-day) observed in this study. Thus, it is probable that because the available carbon concentrations are similar the in situ rate of denitrification in the Big Ditch Watershed is similar to the unamended denitrification rate found at the Rodney Site (Starr and Gillham, 1993).

The rate of denitrification will probably be higher in the shallower wells and decrease with the depth of the wells due to the carbon concentration. It is also possible that the nitrate applied to the surface of the watershed could be attenuated before it reaches the deeper aquifer as was seen at the Rodney Site (Starr and Gillham, 1993). More studies on the watershed need to be completed before the in-situ denitrification rates can be accurately determined. It is clear, however that the attenuation capacity is significant at the site. Rate studies are planned on additional core samples and rates will be determined for samples unamended with carbon at each site. The geological characteristics of the site, and the variable DO, DOC, and nitrate concentrations will be factored into the analysis of the natural attenuation potential.

ACKNOWLEDGEMENTS

We would like to acknowledge The Council for Food and Agricultural Research (C-FAR) for funding this project and the Illinois State Geological Survey for technical support.

REFERENCES

Baker, D.B., L.K. Wallrabenstein, R.P. Richards, and N.L. Creamer. 1989. "Nitrate and pesticides in private wells of Ohio: A state atlas." *The Water Quality Laboratory*, Heigelberg College, Tiffin, OH.

Korom, S.F. 1992. "Natural denitrification in the saturated zone: a review." *Water Resour. Res.* 28 (6): 1657-1668.

Lee, L.K., and E.G. Nielson. 1989. "Farm chemicals and groundwater contamination." *In* J.R. Nelson and E.M. McTernan (ed.) *Agriculture and groundwater quality – examining the issue.* p. 2-10. University Center for Water Resources, Oklahoma State University, Stillwater, OK.

Madison, R.J., and J.O. Brunett. 1985. *Overview of the occurrence of nitrate in groundwater of the United States.* U.S. Geological Survey Water Supply Paper. 2275.

Pell, M., B. Stenberg, J. Stenström, and L. Torstensson. 1996. "Potential denitrification activity assay in soil – with or without chloramphenicol?" S*oil Biol.Biochem.* 28 (3): 393-398.

Risatti, J.B., and E. Mehnert. 1998. "Nitrate attenuation by a riparian woodland: injection test experiments." *Proceedings of Eighth Annual Conference of the Illinois Groundwater Consortium*, April 1-2, Makanda, IL.

Spalding, R.F., and M.E. Exner. 1993. "Occurrence of nitrate in groundwater – a review." *J.Environ. Qual.* 22 (3): 392-402.

Starr, R.C., and R.W. Gillham. 1993. "Denitrification and organic carbon availability in two aquifers." *Ground Water* 31 (6): 934-947.

Tiedje, J.M. 1994. "Denitrifiers." *In Methods of soil analysis. Part 2. Microbiology and Biochemical Properties.* p. 245-267. Soils Science Society of America, Madison, WI.

IN SITU EVALUATION OF ENBEDDED CARRIER IN SOIL TO REDUCE NITRATE LEACHING FROM CROPLAND

Teruo Higashi (University of Tsukuba, Tsukuba, Japan)
Takahiro Oshio and Tsuyoshi Kawakami (University of Tsukuba, Tsukuba, Japan)
Bio-Consortium (organized by The Japan Research Institute Ltd., Tokyo, Japan)

ABSTRACT: In situ experiments using no-bed lysimeters were carried out during two years to evaluate the efficiency of embedded carriers in soil to reduce nitrate leaching from upland agricultural area in Japan where volcanic ash soils (Umbric Andosols) are widely distributed. Two carriers embedded in the soils are a polyethyleneimine-coated cellulose with continuous micropore structure and a calcined fine particle of diatomaceous earth. Under the fertilizer application rate of 300kg nitrogen per ha without crop cultivation, the values of nitrate concentration and relative abundance of ^{15}N of the water samples taken from the soils at different depths in the lysimeters and also the number of vital count of denitrifying bacterias revealed that denitrification processes in the carriers as well as in the soils were positively enhanced by the embedded carriers in the soils.

INTRODUCTION

Soil surface nitrogen balance, proposed as one of the environmental indicators for agriculture (OECD, 1999) showed that the agricultural lands of Japan have recieved annual surplus of about 150kg nitrogen per ha from 1985 to 1997. This figure is one of the highest among the countries encountering groundwater pollution with nitrate. Reflecting such circumstance, nitrate leaching from cropland has become a serious problem in Japan since 1980's, and the concentration of over 10mg nitrogen per liter has been frequently observed, as recently reviewed by Kumazawa (1999) and Takeuchi (1997). This has been caused by the intensive fertilizer and livestock waste application for various vegetables, fruit crops and tea trees cultivated in the upland agricultural areas in these 20 years.

Besides the recommended agroenvironmental techniques including reduced application rate and efficient use of fertilizers and manure, we tried here to reduce nitrate leaching from cropland in situ by the use of the embedded carriers in soil. Since many preceeding studies have revealed that the incorporation of the organic substances with high C/N ratio into soils makes dramatic decrease in nitogen concentration in leachate from the soil column experiments (e.g. Lens et al., 1994; Robertson and Cherry, 1995; Kihou, 1999, 2000; Kawanishi et al, 1997), it is of value to study whether the same results as observed for soil column is obtained or not by the use of embedded organic carriers under field soil conditions.

Study Site and Soil Description. Study site at Miyakonojo-city, Miyazaki prefecture, located on Kyushu Island of southwestern Japan, was selected through the preliminary survey on the distribution of unconfined groundwater having higher nitrate concentration. The study site is covered with volcanic ash soils, classified as Umbric Andosols according to World Reference Base for Soil Resources (ISSS/ISRIC/FAO, 1998), having a very thick A horizons of about 85cm overlying the pumice layer. This kind of soils occupies about 40% of the upland agricultural areas of Japan, being rich in humic substances which are responsible for their high porosity and water holding capacity. Below the pumice layer there is a pyroclastic flow layer of tuffecious coarse-textured materials of about 15 to 20m thick which is so-called "Shirasu deposits". Due to the presence of the latter two layers upward capillary movement of soil water would be intercepted, although the area recieves higher annual precipitation of about 2500mm and mean annual temperature of about 18 ℃.

MATERIALS AND METHODS

Two carrier materials were embedded to provide the suitable habitat for autochtonous denitrifying bacterias. Their growth would be favored by its continuous micropore strucuture of a polyethyleneimine-coated cellulose (Bm-Aquacel, Biomaterial Co., Ltd.), and by its enhanced water holding capacity of a calcined fine diatomaceous earth (CG1C: Isolite Kogyo Co., Ltd.).

Five different sylindrical no-bed lysimeters with a diameter of 1.5m and 1.3m depth which is made of polyethylene were installed under the ground to the depth of 1.0m, where excavated soils were refilled inside the lysimeters to keep the original order of stratification of each soil horizons. As shown in Figure 1., besides a lysimeter without the carrier materials (Plot 1), used as the control in this experimets, two carrier materials were embedded in different manner at the depth of 45 to 55cm from the surface: only cellulose layer (Plot 2), double layered with the two materials (Plot 3), throughly mixed single layer (Plot 4), and the mixed single layer with manure above the carrier (Plot 5) were prepared. In each lysimeter water sampler using porous ceramic cups were installed at the depth of

20, 40, 60, and 100cm from the surface and soil water was sampled periodically from each depths. The experiment was conducted from April, 1999 to September, 2000. Fertilizer application rate of 300kg nitrogen per ha was carried out on April, 1999 and April and June, 2000, without crop cultivation. This application rate is about double compared with the conventional one in the areas. In addition, 0.4g of soudium tetrafluoropropionate was applied as a herbicide.

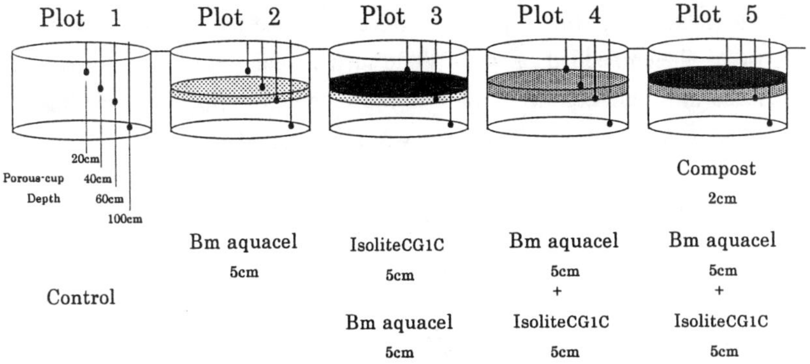

FIGURE 1. Scheme of embedded carriers in 4 different lysimeters (from Plot 2 to 5) and the control (Plot 1 with no embedded carriers).

Analyses of water samples were done as follows: nitrate ion was determined using ion chromatography, and the relative abundance of ^{15}N was measured for the selected water samples obtained in May 1999 by ANCA-Mass (Automatic Nitrogen and Carbon Analyzer-Mass Spectrometer, Europe Scientific Co., Ltd.). All water samples were filtered with cellulose acetate membrane filter (0.2 μm).

Refilled soils in the lysimeters were re-excavated at the end of the experiment, and soil profile were re-examined to observe the changes of the nature of embedded carrier materials and soils. Soil samples were taken from the soil profiles at the 20cm interval from the surface in addition to the carriers. Undisturbed soil core samples were analysed for several physical properties at the time of sampling, and also measurement of vital count of denitrifying bacteria was done by conventional MPN method using PYN medium (peptone, yeast powder and pottasium nitrate) where the bubbles in Durham tube were detected.

RESULTS AND DISCUSSION

Percolation Rate and Nitrate Concentration of Soil Water. In 1999, as shown in Figure. 2 (illustrated for the Plots 1 and 4), percolation rate of soil water was clearly detected by the front of increased nitrate concentration at different depths when plotted against the amount of cumulated rainfall after fertilizer application. This shows that the percolation of nitrate moved downward by about 10cm in depth with every 100mm rainfall. Profile distribution of nitrate concentration in Plot 4 showed a distinct decrease in nitrate concentrations at 40 and 60cm (just above and below the carrier materials), compared with that at 20cm (above the carrier materials), suggesting a positive effect to reduce the nitrate concentration by the use of the embedded carriers. While such a clear trend for percolation rate and decrease in nitrate concentration was not obtained during the experiment in 2000, probably due to the irregular heavy rainfall in studied area.

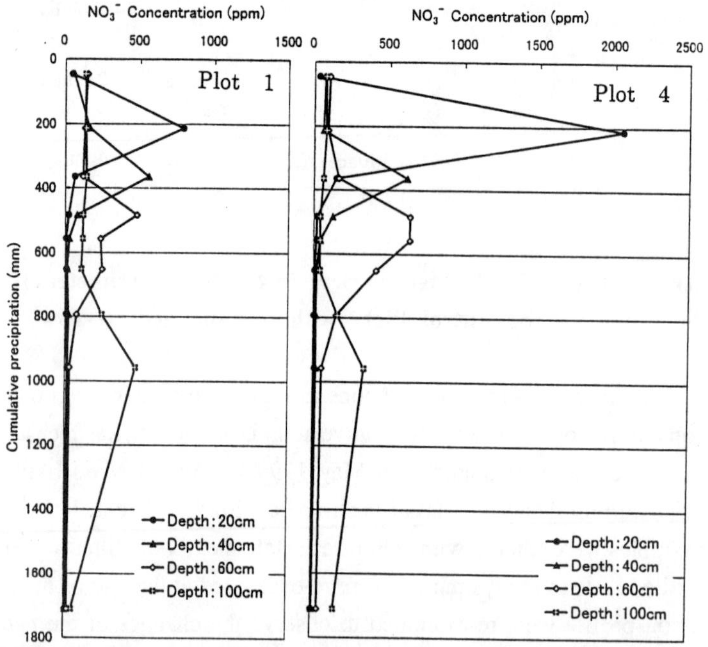

FIGURE 2. Nitrate concentration at different depths against cumulated rainfall in Plots 1 and 4 from April to July in 1999.

Amounts of Nitrate Leached and the Removal % by Embedded Carriers. Amounts of nitrogen (NO_3-N) leached in each plots in Table 1 were calculated as follows: first, the sum of products between the concentration of nitrate (NO_3-N) and the amounts of soil water sampled at the depth of 100cm were calculated over whole sampling dates in 1999 and also in 2000 (untill June), supposed soil water is sampled from the soil in distance of 4.5cm around popous ceramic cup. Then these sum of the products were converted to the scale of lysimeter in each Plots.

In 1999, compared with the the nitrogen leached from the control (Plot 1), about 40% of the control was removed in Plots 2 and 4 by the presence of the embedded carriers, and about 23% in Plot 3, but application of manure on the carrier as a proton donner in Plot 5 had negative effect, probably due to the mineralization of manure itself. When the same calculation is applied for the period of heavy rainy season from the end of June to early in July , this tendency was more enhanced to about 64% in Plot 4, 52% in Plot 1 and about 20% in Plots 3 and 5. In 2000 the figures were about 46% in Plot 1, 36% in Plot 2 and 43% in Plots 4 and 32% in Plot 5. Thus, the embedded carrier materials showed a positive role to reduce nitrate leaching and their removal % was different among each plots and increased within the two years.

TABLE 1. Efficiency of embeded carriers to reduce nitrate in each Plots.

Plot Number	N leached (g/lysimeter) 1)		Removal % against Plot 1	
Year	1999	2000 2)	1999	2000
Plot 1	24.07 (14.43) 3)	8.01	-	-
Plot 2	14.41 (6.99)	4.31	40.14 (51.54)	46.19
Plot 3	18.44 (11.68)	5.09	23.37 (19.05)	36.45
Plot 4	15.05 (5.17)	4.57	37.46 (64.20)	42.95
Plot 5	36.40 (11.24)	5.49	-51.21 (22.13)	31.46

1) calculated by the radius of lysimeter and that of soil area around porous cup.
2) to compare the concentration more precisely a volume of water samples is normalized to 100ml using conical flask under 40cmHg pressure.
3) figures in the parentheses denote the values in the heavy rainy season.

Relative Abundance of ^{15}N of Soil Water. Measurement of ^{15}N of the nitrogen fertilizer used in the present study showed the value of 0.1(‰). However, the values obtained for soil water samples were distinctly higher, ranging from 0.97 to 2.79. As it is clear from Table 2 the values of ^{15}N increase when the soil water pass through the embedded carriers in Plots from 2 to 5, but decrease in the control of Plot 1. Difference in the values between the water samples from 40 cm depth (just above the carrier) and those from 60cm (just below the carrier) suggests the higher efficiency of denitrification processes in Plots 2 and 4, compared with Plots 3 and 5. Moreover the values of ^{15}N seems to be correlated with the nitrogen removal % in each plots in Table1. Since the denitrification processes generally raise the value of ^{15}N (Mariotti et al., 1988), the data obtained here strongly suggest the increased colonization of denitrifying bacterias in the carriers, especially in the mixed single layer of the two materials in Plot 4.

TABLE 2. Abundance of ^{15}N and vital count of denirifying bacterias.

Plot Number	Abundance of ^{15}N 1) (‰) 40cm	60cm	Difference	Vital count2) (40-80cm depth)
Plot 1	1.60	1.03	- 0.57	8.01×10^{8}
Plot 2	1.05	2.08	1.03	3.01×10^{9}
Plot 3	1.08	1.26	0.18	1.17×10^{9}
Plot 4	1.15	2.79	1.65	1.76×10^{9}
Plot 5	0.97	1.65	0.68	1.23×10^{10}

1) measured for the soil water samples obtained from 40cm depth during 25 to 28th May, and from 60cm depth during 9 to 10th, July , 1999.
2) expressed on lysimeter volume basis.

Numbers of Vital Count of Denitrifying Bacterias. Numbers of vital count of denitrifying bacterias shown in Table 3 are expressed in three different ways: on 1g dry carrier or dry soil basis, on 1mL carrier or soil volume basis and on each carriers or soil layers basis. Numbers based on volume and layer basis are calculated by the use of the values of bulk density of soils or carriers and the thickness of the relevant layers. It is clear that the number of vital counts on dry matter basis were higher for Plots 3, 4 and 5, compared with those for the control Plot 1, especially for the carriers which showed the higher values than the corresponding soils just above (A4) and below (A5) the carriers in the same Plots. However, somewhat it is not the case in Plot 2. Moreover, it is worth to mention that the value for the soils of A5 in Plot 5 was fairly high, probably resulting from the supply of available carbon and electron from manure which positively affects on the bacterial growth. When calculated on volume and layer basis, this trend mentioned above bacame less conclusive. When it is calculated for each lysimeter between 40 to 80cm in depth from the surface (shown in Table 2), that is obtained from the sum of the each layers in each Plots, the number of vital counts were slightly higher for Plots 2 and 4 compared with Plot 3, and was the highest for Plot 5. However these values are significantly higher copmpared with those in Plot 1. Thus the denitrifying bacterias are to be colonized in the embededd carries in Plots from 2 to 5, especially in Plot 4 compared with Plot 1 with no carriers .

TABLE 3. Number of vital counts of denirifying bacterias by YPN method.

Plot Number		Number of Vital Count DM (1g) basis	VM (1mL) basis	Layer basis
Plot 1	A4	1.60×10^{2}	1.22×10^{2}	1.03×10^{7}
	A5	1.84×10^{2}	1.45×10^{2}	2.05×10^{7}
Plot 2	A4	2.69×10^{3}	1.10×10^{4}	2.93×10^{9}
	Carrier	1.72×10^{2}	1.20×10^{1}	1.28×10^{6}
	A5	5.95×10^{2}	2.56×10^{2}	8.59×10^{7}
Plot 3	A4	1.68×10^{3}	7.75×10^{2}	1.37×10^{8}
	Carrier (Iso)	1.51×10^{3}	6.80×10^{2}	6.01×10^{7}
	Carrier (Bm)	8.85×10^{4}	6.19×10^{3}	3.28×10^{8}
	A5	3.21×10^{3}	1.67×10^{3}	6.48×10^{8}
Plot 4	A4	3.15×10^{3}	1.45×10^{3}	3.84×10^{8}
	Carrier	2.80×10^{4}	7.28×10^{3}	9.01×10^{8}
	A5	3.19×10^{3}	1.50×10^{3}	4.76×10^{8}
Plot 5	A4	2.05×10^{3}	9.86×10^{2}	2.96×10^{8}
	Manure	7.91×10^{3}	2.45×10^{3}	1.30×10^{8}
	Carrier	6.49×10^{4}	1.69×10^{4}	1.79×10^{9}
	A5	7.96×10^{4}	4.06×10^{4}	1.00×10^{10}

Physical Properties of Undisturbed Soil Core Samples after the Experiment. Although there was a slight increase in capillary water holding capacity of the soils in Plot 2 and also in water contents of the soils at the time of sampling in Plot 4, no other apparent changes was observed among the plots. It is suggested that soil physical properties does not change much within two years of the experiment.

CONCLUSION

The use of embedded carriers in soils, consisting of a polyethyleneimine-coated cellulose with continuous micropore structure supported by a calcined fine particles of diatomaceous earth, showed a fairly positive role to reduce nitrate leaching from volcanic ash soils, suggested by the values of removal %, relative abundance of ^{15}N and the number of vital count of denitrifying bacterias.

ACKNOWLEDGEMENT

The authors wish to express the sincere thanks to the following persons who have assisted in every aspects of the present study ; Dr. K. Kimura, Maesawa

Kogyo Co. Ltd., Mr. M. Nishimura and Mr. D. Maruoka, The Japan Research Institute Ltd., Mr.M. Koyama, Isolite Kogyo Co. Ltd., and T. Hirayama, Kokusai Kogyo Co. Ltd.

REFERENCES

Kawanishi, T., Z. Jiang, M. Inagaki, N. Shimizu, and Y. Hayashi. 1997. "The Way Rice Straw Input into Soil Affects the Nitrogen Removal from Soil". *Jpn. J. Water Environment.* 20(3):347-351. (in Japanese with English summary)

Kihou, N. 1999. "Toward the Development of New Technologies to Reduce the Nitrate Leaching from Croplands". *Jpn. J. Farm Work Research.* 34. Extra Issue (2): 20-29. (in Japanese)

Kihou, N. 2000. "Organic Amendments to Reduce Nitrate Leaching from Croplands." *Farmlands and Soils.* 32(6):18-23. (in Japanese)

Kumazawa, K. 1999. "Present State of Nitrate Pollution in Groundwater". *Jpn. J. Soil Sci. and Plant Nutrition.* 70(2):207-213. (in Japanese)

Lens, P. N., P. M. Vochten, L. Splees, H. Verstraete. 1994. "Direct Treatment of Domestic Wastewater by Percolation over Peat, Bark, and Woodtips". *Water Research.* 28(1):17-26.

Mariotti, A., A. Landreau, and B. Simon. 1988. "^{15}N Isotope Biogeochemistry and Natural Denitrification Process in Groundwater: Application to the Chalk Aquifer of Northern France". *Geochimica et Cosmochimica Acta* 52:1869-1878.

OECD. 1999. "Environmental Indicators for Agriculture: Methods and Results-The Stocktaking Report." *COM/AGR/CA/ENV/EPOC(99)(80)*. 95pp.

Robertson, W. D. and J. A. Cherry. 1995. "In Situ Denitrification of Septic System Nitrate Using Reactive Porous Media Barriers: Field Trials" . *Ground Water.* 33:99-111.

Takeuchi, M. 1997. "Nitrate and Phosphate Outflow from Arable Land". *Jpn. J. Soil Science and Plant Nutrition.* 68(6):708-715.

AUTOTROPHIC DENITRIFICATION OF BANK FILTRATE USING ELEMENTAL SULFUR

Hee Sun Moon, ***Kyoungphile Nam***, Jae Young Kim
(Seoul National University, Seoul, The Republic of Korea)

ABSTRACT: The present study was designed to develop a permeable reactive barrier (PRB) as a means of *in situ* bioremediation of nitrate-contaminated bank filtrate. As a preliminary bench-scale study, transformation of nitrate to nitrogen gas was conducted in an upflow column by using autotrophic denitrifiers containing *Thiobacillus denitrificans* and elemental sulfur as an electron donor. The results indicate that nitrate was almost completely transformed to nitrite in the first four days of column operation and nitrite accumulation was observed. After two days of accumulation, however, nitrite concentration slowly decreased and the compound was detected less than 0.5 mg/L in 14 days. When the influent nitrate concentration was 30 mg-N/L and hydraulic retention time was 12 hours, sulfate concentration in the effluent was 70~90 mg-S/L and pH was maintained around 7.5. When sampled at 17 cm from the bottom of the column, the effluent showed the highest nitrite concentration, and nitrate concentration decreased rapidly to the point of 33 cm from the bottom. The sulfur-oxidizing activity increased continuously to the point of 33 cm and then it was maintained or slightly decreased. Further study to determine kinetic parameters and optimal wall thickness is underway.

INTRODUCTION

Bank filtration, utilizing dunes in natural river streams, has been practiced in Europe for a long time as a pre-treatment for water intake system and can be successfully applied in areas where the permeability of water is well established by the development of alluvial formation. In Korea, such geological formation is found along the Nakdong river and thus, bank filtration has been considered for an alternative method for improvement of water quality and development of water intake system.

It is generally considered that the water quality of bank filtrate is better than that of surface water. But, in Korea, bank filtrate has become increasingly contaminated with nitrate originating from the use of fertilizers and stock breeding near rivers.

Nitrate is highly mobile in soil and diffuses easily in the subsurface environment, resulting groundwater contamination. Nitrate level in drinking water is strictly regulated in Korea and the requirement is 10 mg-N/L.

Conventional nitrate treatment technologies including ion exchange, reverse osmosis, electrodialysis, chemical precipitation, and distillation have low removal efficiency and heterotrophic denitrification as biological treatment also needs an external organic carbon source such as methanol or ethanol (Lampe and Zhang, 1996). Biological process using autotrophic denitrification process and element sulfur can be an alternative because this process requires no external carbon source and produces low amounts of biomass (Batchelor and Lawrence, 1978). Autotrophic denitrification using S^0 as an electron donor occurs by sulfur-oxidizing bacteria such as *Thiobacillus denitrificans* and *Thiomicrospira denitrificans*, and stoichiometric equation can be expressed as follows (Flere and Zhang, 1999).

$$\mathbf{55S + 20CO_2 + 50NO_3^- + 38H_2O + 4NH_4^+ \rightarrow 4C_5H_7O_2N + 25N_2 + 55SO_4^{2-} + 64H^+}$$

In this study, we attempted to implant autotrophic denitrification into the permeable reactive barrier system to improve the water quality of bank filtrate. For the purpose, column experiments were performed. (1) to measure the basic kinetic parameters of autotrophic denitrification reaction; (2) to determine the effects of nitrate concentration, hydraulic retention time on the removal efficiency of nitrate.

MATERIALS AND METHODS

Microorganisms. A bacterial consortium containing *Thiobacillus denitrificans* used in this study was isolated from a tidal flat near Inchon, Korea. The culture was enriched and monitored in a liquid medium containing 2g/L KNO_3, 5g/L $Na_2S_2O_3 \cdot 5H_2O$, 2g/L K_2HPO_4, 1g/L $NaHCO_3$, 0.5g/L NH_4Cl, 0.5 g/L $MgCl_2 \cdot 6H_2O$, 0.02 g/L $FeSO_4 \cdot 7H_2O$ at 30°C.

Column Reactor. The column reactors for autotrophic denitrification using elemental sulfur were made of Pyrex with inside diameter of 70 mm and height of 700 mm (FIGURE 1). They were packed with granular elemental sulfur with 2 mm diameter and limestone with 2~5 mm diameter at the ratio of 3 to 1(Flere and Zhang, 1999). Limestone was added to maintain the pH of the system. The enriched consortium described above was introduced into the packed bed reactors for a period of time three days to attach bacteria on sulfur particles. Influent artificially contaminated with nitrate at the levels of 30 to 60 mg-N/L was introduced into the column reactors from the bottom.

At intervals, samples were collected from various points of the column through sampling ports. Experiments were performed using an upflow column reactor with flow rate of 1.0 mL/min to represent the flow of bank filtrate in field. All experiments were conducted at 20°C.

Analytical Method. The concentrations of nitrate, nitrite, and sulfate were determined by using an ion chromatography (Dionex DX500). Alkalinity and pH of the system were also monitored because sulfate accumulation may inhibit the denitrifying activity of the consortium.

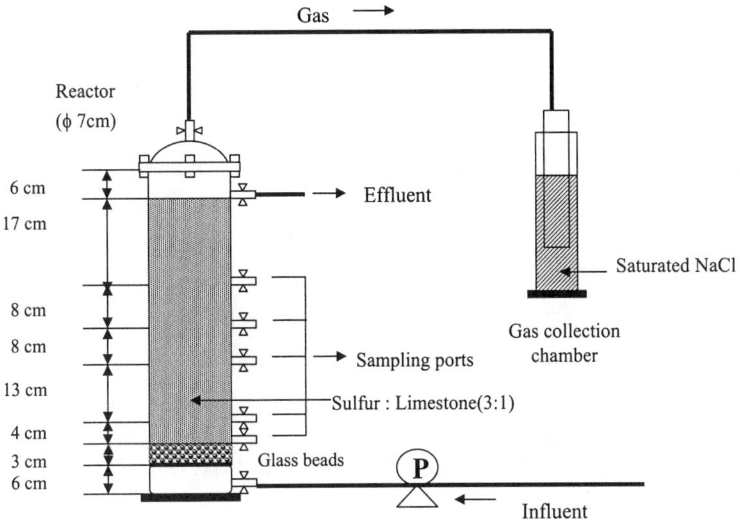

FIGURE 1. Schematic of column reactor.

RESULTS AND DISCUSSION

Effect of Nitrate Concentration. To evaluate the effect of influent nitrate concentration on the removal efficiency, influent nitrate concentrations varying from 30 mg-N/L to 60 mg-N/L were tested. Figures 2 and 3 show overall profiles of effluent nitrate, nitrite, and sulfate concentrations with time at different influent nitrate concentrations and a hydraulic retention time of 12 hr. At the level of 30 mg-N/L, nitrate was almost completely transformed to nitrite in the first four days of column operation and nitrite accumulation was observed.

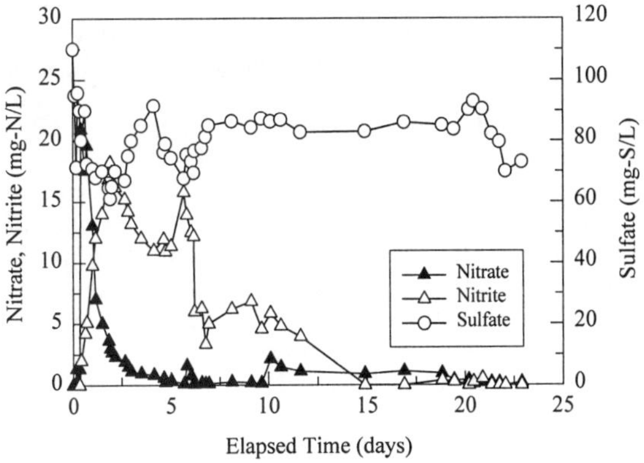

FIGURE 2. Profiles of nitrate, nitrite, and sulfate concentration with time at influent nitrate concentration of 30 mg-N/L and HRT of 12hr.

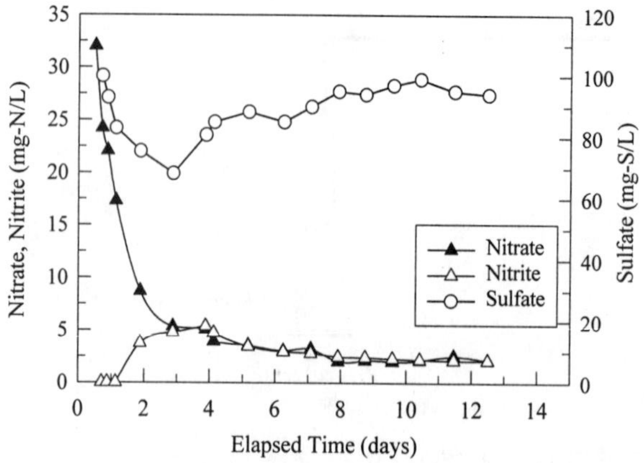

FIGURE 3. Profiles of nitrate, nitrite, and sulfate concentration with time at influent nitrate concentration of 40 mg-N/L and HRT of 12hr.

After two days of accumulation, however, nitrite concentration slowly decreased and the compound was detected less than 0.5 mg-N/L in 14 days (FIGURE 2). Nitrite accumulation of initial period also reported previously (Furumai et al., 1996). Therefore, nitrite control seems to be an important factor for this operation.

After 23 days of operation at the level of 30 mg-N/L, nitrate concentration increased to 40 mg-N/L maintaining the other conditions identical. As shown in Figure 3, during the initial two days of operation, nitrate rapidly decreased and detected less than 5 mg-N/L. For the first four days, effluent nitrite concentration increased and then decreased. Compared to the case of influent nitrate concentration of 30 mg-N/L, nitrate reduction rate increased. It is probably because the metabolic activity of the consortium tested was more stabilized or enhanced during the column operation. When 60 mg-N/L of nitrate concentration was inflowed, the pattern of nitrate removal was very similar (data not shown). The effluent nitrate concentration was less than 3 mg-N/L after three days and then further decreased. However, nitrite was not detected after one day of operation. This also might be due to the successful adaptation of nitrite conversion activity of the consortium, but it cannot be proved at this time.

Characteristics of Effluent. Figure 4 shows the change of pH, sulfate concentration, and alkalinity in effluents with time. Sulfate is the end product of sulfur-based denitrification. As shown in Figure 4, sulfate concentration in the effluent was the 70~90 mg-S/L when the influent nitrate concentration was 30 N-mg/L and hydraulic retention time was 12 hr. In Korea, the standard level of sulfate concentration for drinking water is 67 mg-S/L (200 mg-SO_4^{2-}/L), which is a little lower than the level obtained from this study. Also, even though the nitrate-contaminated groundwater is treated through the sulfur-limestone permeable reactive barrier, there will be no significant problem due to the increased sulfate concentration because the treated groundwater will be mixed with a large volume of water from river. Alkalinity and pH were maintained constantly after four days of operation. pH was found 7.5 and alkalinity was 85 90 mg-$CaCO_3$/L. Based on the stoichiometric equation for autotrophic denitrification, the theoretical value of alkalinity consumed is 4.57 mg-$CaCO_3$/mg NO_3^--N removed. The constant pH value was acceptable for treated bank filtrate.

In spite of sulfate production, constant pH maintenance was probably due to the granular limestone provided.

Spatial Distribution of Reactants. Figure 5 shows the spatial distribution of nitrate, nitrite, and sulfate throughout the column. When sampled at 17 cm from the bottom of the column, the effluent showed the highest nitrite concentration, and nitrate concentration decreased rapidly to the point of 33 cm from the bottom. The sulfur-oxidizing activity increased continuously to the point of 33 cm and then it was maintained or slightly decreased. Data showed that the nitrite reduction occurred between 17 cm to 33 cm from bottom, indicating that the activity of nitrite reductase, responsible for the conversion of nitrate to nitrite, increased in this area. In terms of stoichiometric mole ratio of NO_3^- to SO_4^{2-}, we obtained the experimental value of 0.83, which was close to the theoretical value of 0.91 by stoichiometric equation described previously.

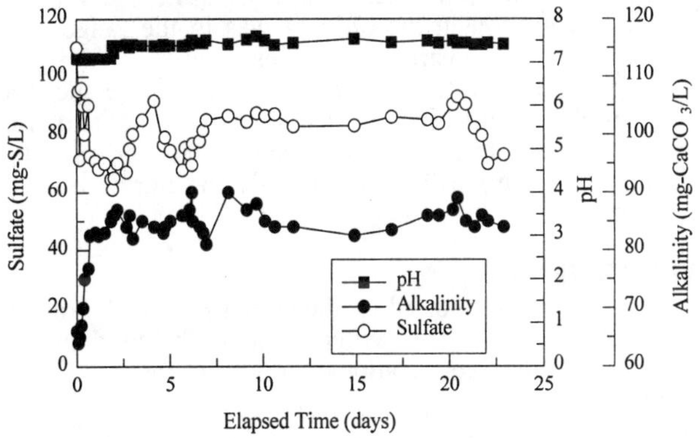

FIGURE 4. Profiles of the effluent sulfate concentration, pH, and alkalinity with time at influent nitrate concentration of 30 mg-N/L and HRT of 12 hr.

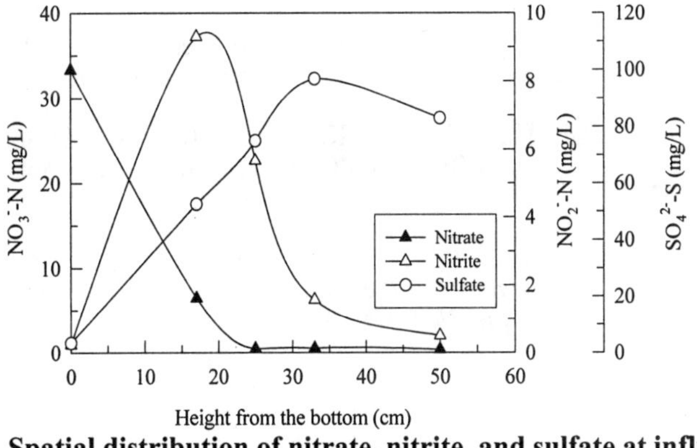

FIGURE 5. Spatial distribution of nitrate, nitrite, and sulfate at influent nitrate concentration of 30 mg-N/L and HRT of 12 hr.

Further Direction. Further study is being conducted to determine the spatial distribution of denitrification process throughout the column and the effects of various initial concentrations of nitrate on denitrification efficiency as well. We will also evaluate kinetic parameters and effects of environmental conditions on the denitrification efficiency of the system and develop a mathematical model for autotrophic denitrification using sulfur and limestone. An extended study will be performed to optimize the column-operating conditions to scale-up the system and to provide the design manual for the field application.

CONCLUSIONS

1. Nitrate was almost completely transformed to nitrite range from 30 mg-N/ to 60 mg-N/L of nitrate concentration at HRT of 12 hr.
2. Initial nitrite accumulation was observed, however, nitrite concentration slowly decreased for the nitrate concentrations tested. Effluent concentration of nitrite was less than 2 mg-N/L.
3. Sulfate concentration in the effluent was in the range of 70~90 mg-S/L and pH was maintained around 7.5 constantly.
4. Data suggest that nitrate in the bank filtrate can be successfully removed by autotrophic denitrification using the sulfur and limestone. An appropriate wall thickness of reactive barrier for the autotrophic denitrification seems to be 30 cm when the nitrate concentration is less than 60 mg-N/L.

ACKNOWLEDGMENTS

This research was supported by the Green Korea 21 Project and the Brain Korea 21 Project in 2000. The writers would like to thank the Institute of Engineering Science at the Seoul National University for the technical assistance.

REFERENCES

Batchelor, B., and Lawrence, A.W. 1978. "Autotrophic Denitrification using Elemental Sulfur". *J. Water Pollution Control Federation*, *50*(8): 986-2001.

Flere, J. M., and Zhang, T. C. 1999. "Nitrate Removal with Sulfur-Limestone Autotrophic Denitrification Process". *J. Environ. Engineering*, *125*(8): 721-729.

Furumai, H., Tagui, H., and Fujita, K. 1996. "Effects of pH and Alkalinity on Sulfur-Denitrification in a Biological Granular Filter". *Water Sci. Technol.*, *34*(1-2): 355-362.

Lampe, D. G. and Zhang, T. C. 1996. "Evaluation of Sulfur-Based Autotrophic Denitrification", *Proceedings of the 1996 HSRC/WERC Joint Conference on the Environment, Albuquerque, New Mexico.*

HIGH-RATE DENITRIFICATION IN BIOFILM-ELECTRODE REACTOR COMBINED WITH MICROFITRATION

Michal Prosnansky (Gunma University, Gunma, JAPAN)
Yutaka Sakakibara (Waseda University, JAPAN)

ABSTRACT: In this study, a novel multi-cathode biofilm-electrode reactor (BER) combined with micro-filtration (MF) was investigated using a laboratory-scaled experimental apparatus for on-site remediation of nitrate-contaminated groundwater. In BER, autotrophic denitrification takes place by applying electric current. Extended surface area of multi-electrodes enables BER to operate at high electric currents with low current densities. MF membranes with plate modules and pore size of 0.2 μm were chosen for the final rejection of suspended solids (SS) escaping from BER. Experimental results demonstrated that it was possible to operate the multi-cathode BER with high denitrification rates and low hydraulic retention time (HRT = 20 min). The overall performance was enhanced by 10 to 60 times in comparison with those in former studies. MF membranes successfully rejected bacteria escaping from BER, so that effluent concentration of SS was below 1 mg-SS/L throughout the experiment. The present BER/MF process is considered applicable for on-site remediation of nitrate-polluted groundwater.

INTRODUCTION

Groundwater is one of the principal sources of water supply. However, nitrate contamination has become an increasing problem in many parts of the world. The main sources of the contamination are nitrogen fertilizers and discharge of wastes (Env. Eng. Research Council of ASCE, 1990; Mackay and Cherry, 1989).

Biological denitrification seems to be an effective process in the treatment of nitrate, since nitrate is converted to non-toxic N_2 gas. Either heterotrophic or autotrophic technologies have been used for nitrate removal (Mateju et al., 1992).

Recently, processes using H_2-utilizing bacteria have been studied extensively. H_2 gas can be supplied by external injection or internal production by the electrolysis of water (e.g., Flora et al., 1993; Sakakibara and Kuroda, 1993) or chemical reaction with water (DeJournett and Alvarez, 1999). Since H_2 is only slightly soluble in water (e.g., 0.4 mM at 1 atm, 25°C), bacteria can utilize H_2 more effectively by the electrochemical or chemical H_2 production than by the external feeding of H_2.

Biofilm-electrode reactor (BER) consisting of mono-polar electrodes, where denitrifying bacteria were cultured on cathode and H_2 gas was produced by the electrolysis of water, has been proposed by Sakakibara and Kuroda (1993). The advantage of this process is easy operation and maintenance. However, denitrification rate was slow and longer hydraulic retention time (HRT = 10 hours to several days) was required to achieve complete denitrification (Cast and Flora,

1998; Feleke et al., 1998; Flora et al., 1993; Islam and Suidan, 1998; Sakakibara and Kuroda, 1993; Sakakibara et al., 1995).

This study presents a novel high-rate multi-cathode BER combined with microfilters. The process is designed for small scaled and dispersed on-site remediation of nitrate-polluted groundwater, because of its easy control, operation and maintenance.

DENITRIFICATION IN BER

In the multi-cathode BER used in this study, autotrophic H_2-utilizing denitrifying bacteria are attached on the surface of each cathode. Direct current (DC) is applied to produce H_2 via electrolysis of water, as expressed by Eq. 1. Granular activated carbon (GAC) was packed to each cathode to enlarge surface area of electrodes and to attach bacteria quickly and firmly. On the surface of the counter electrode (anode), O_2 gas is produced according to Eq. 2.

$$2\,H_2O + 2e^- \rightarrow H_2 + OH^- \tag{1}$$
$$2\,H_2O - 4e^- \rightarrow O_2 + 4H^+ \tag{2}$$

The overall reaction of denitrification can be expressed by (Feleke et al., 1998; Sakakibara and Kuroda, 1993);

$$NO_3^- + 5e^- + 3H_2O \rightarrow \tfrac{1}{2}\,N_2 + 6OH^- \tag{3}$$

MATERIALS AND METHODS

The scheme of the experimental apparatus is shown in Fig. 1. The reactor is composed of two compartments: multi-cathode BER and membrane compartments. Synthetic groundwater (the composition is shown in Table 1) was fed into cathodic zone and leaving BER through anodic zone. Micro-filtration

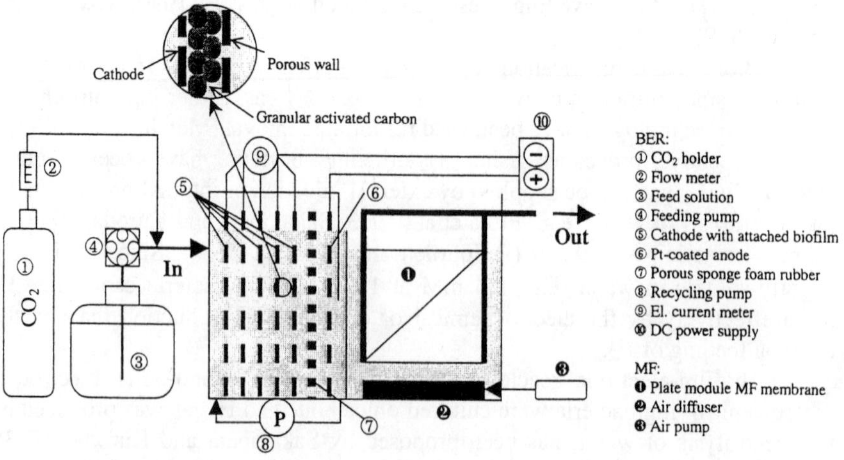

FIGURE 1. Experimental apparatus used for remediation of nitrate-polluted groundwater.

(MF) membranes completed final purification. The experimental conditions are shown in Table 2. Details of each compartment are explained as follows;

Multi-Cathode BER Compartment. This compartment consists of 5 porous GAC electrodes acting as multiple-cathodes and an inert anode (platinum coated titanium). GAC electrodes were made by contacting GACs with expanded metals (stainless steel) as shown in Figure 1 and were connected in series with the anode. Electric current was controlled independently for each cathode. Porous sponge-foam rubber was placed between cathodic and anodic zones to prevent from mixing of H_2 and O_2 bubbles produced by electrolysis of water. BER was operated in a flow-through mode, where water flows through pores of multi-electrodes with pore diameter of roughly about 3 mm.

TABLE 1. Composition of synthetic groundwater.

$NaNO_3$	15 ~ 40 mg-N/L
K_2HPO_4	1.76 mg/L
KH_2PO4	2.08 mg/L
NaCl	0.96 mg/L
$MgSO_4.7H_2O$	4.00 mg/L
$CaCl_2$	1.12 mg/L
$FeCl_3.6H_2O$	1.92 mg/L

TABLE 2. Experimental conditions in this study.

Run No.	HRT (hr)	Flow (L/h)	Current (mA)	NO_3^--N (mg/L)	CO_2 injection (mL/h)
1	6.00	0.10	50	40	60
2	2.88	0.21	60	25	72
3	1.66	0.36	70	20	96
4	1.00	0.60	90	15	108
5	0.50	1.20	180	15	240
6	0.33	1.82	300	15	438
7	1.00	0.60	90	15	108
8	2.00	0.30	52.5	15	65

Membrane Compartment. MF membrane made of poly-sulfon with 0.2 μm pore size was inserted behind BER (Fig. 1). Four sheets of the membrane were set in the compartment by 2 cm in parallel with each other. Total membrane area was 374 cm^2. Membrane module was operated in an interval mode: 8 min of suction followed by a pause of 2 min. Aeration was used for cleaning of membrane. Air was injected into the bottom of membrane compartment. A stone diffuser was used to attain uniform air distribution.

Analytical Methods. Analyses were performed on daily basis. TCD gas chromatograph (GC-3BT and GC-8A, Shimadzu) was used to analyse gas composition. Nitrate and nitrite concentrations were measured by an ion chromatograph (IC 7000 Series II, Yokogawa Analytical Systems). ORP and pH were measured by an ORP/pH meter (UC 23, Central Kagaku). Dissolved oxygen was analysed by a DO meter (UC 12, Central Kagaku) and conductivity by a conductivity meter (SE 12, Horiba). Furthermore, permeate flux; trans-membrane pressure and air-bubbling rate were monitored. Also, total suspended solids (TSS) were measured according to the Standard Methods (APHA et al., 1995).

RESULTS AND DISCUSSION

BER performance. The BER/MF reactor was operated under continuous mode with eight runs (Table 2). Figure 2 shows influent and effluent concentrations of nitrate and nitrite escaping from BER and applied electric current at different hydraulic retention time (HRT). At the beginning of the experiment, the desired effluent concentration of NO_3^- (NO_3^- < 10 mg-N/L, WHO guideline for drinking water) was not reached in Run 1 (HRT = 6 h, current = 50 mA and influent NO_3^- = 40 ~ 46 mg-N/L). The reason was high pH (pH = 9.0 ~ 11.9) due to the production of OH^- (Eq. 1), which is undesired condition for biological denitrification. Feeding of CO_2 was thereby applied to correct pH around neutrality. It resulted in improved denitrification rate in Run 1 with average effluent NO_3^- = 5.0 mg-N/L (Fig. 2).

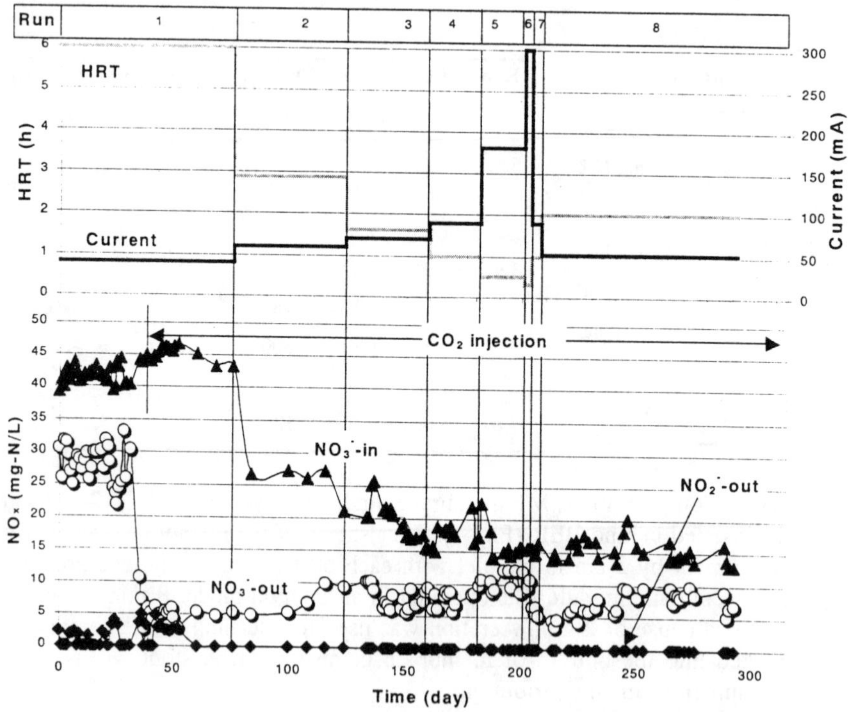

FIGURE 2. Influent and effluent concentration of NO_x^- at different applied current and HRT.

In runs 2 to 8, net denitrification rates increased with increasing electric current. Nitrite was not detected during the experiment, except Run 1 (Fig. 2).

Table 3 shows the maximum denitrification rate (Run 6) compared with former studies dealing with BER (mono-polar electrode system). As can be seen from the table, the denitrification rate was significantly increased by 8 ~ 61 times.

However, with an increase of electric current, the effect of electric field became gradually significant in this study. Since NO_3^- is an ionic constituent, part of NO_3^- migrated towards anode, and escaped from BER without sufficient treatment. Current-denitrification efficiency based on Eq. 3 was about 70 ~ 90 % in Runs 1 to 4 and Runs 7 to 8 (for electric currents 40 ~ 90 mA) and in Runs 5 and 6 around 30 ~ 60 % (for currents 80 ~ 300 mA). As this decline of current efficiency might bring about an increase in electric energy consumption, further investigation is necessary to eliminate migration effect at high currents. A different electrode configuration, such as flow-by configuration may be an effective alternative.

TABLE 3. The comparison of BER performances in former studies with this study.

NO_3^--in mg-N/L	HRT h	Denitrification rate, R_i mg-N/h L	Ratio of denitrification rate (-)**	Operating mode	References
10	10.5	0.82	1/20	Batch	Sakakibara and Kuroda (1993)
20	13.4	1.46	1/11	Continuous	Flora et al. (1993), Islam and Suidan (1998)
28	96	0.27	1/61	Batch	Sakakibara et al. (1995)*
20	72	0.28	1/59	Batch	Cast and Flora (1998)
20	10	2.00	1/8	Continuous	Feleke et al. (1998)
15	0.33	16.4	1	Continuous	This study

* In-situ denitrification

**calculated from $R_i/R_{in\ this\ study}$

Dissolved oxygen present in influent (DO ≅ 5.0 mg/L) also affected the denitrification efficiency. DO is reduced according to electrochemical and biological reactions, (4) and (5), respectively.

$$½O_2 + 2e^- + H_2O \rightarrow 2OH^- \tag{4}$$

$$½O_2 + H_2 \rightarrow H_2O \tag{5}$$

These reactions cause the loss of electron (Eq. 4) or the consumption of H_2 (Eq. 5). In this study, the decline was estimated to be from 8 to 22 % of the total applied current.

MF Membrane Performance. Microorganisms escaping from biofilm are rejected at microfilters. The influent and effluent concentration of SS is shown in Figure 3. During the observation (120 days), the influent concentration of suspended solids increased in the membrane compartment from 5.0 to 28 mg/L with average 11.3 mg/L. The concentration of SS in the effluent did not exceed 1.0 mg/L, having average value SS = 0.21 mg/L. This demonstrates good rejection ability of applied MF membrane.

To see the effect of permeate flux and aeration on MF membrane performance, variation of flux (F = 0.063 ~ 0.200 m/d) and airflow (q_{air} = 0.34 ~ 2.02 L/m^{2}min) were applied. In case of higher fluxes (F = 0.130 ~ 0.200 m/d) even high aeration (q_{air} = 2.02 L/m^{2}min) was not sufficient to remove cake

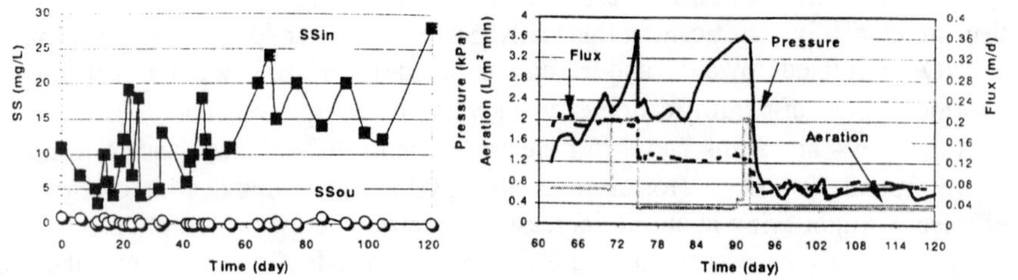

FIGURE 3. MF membrane performance.

formation. After applying F = 0.063 m/d, only q_{air} = 0.34 L/m^2min was necessary to reach quasi-steady conditions. Trans-membrane pressure was kept around constant value ($P = 0.7 \pm 0.2$ kPa) throughout the rest of the experiment.

CONCLUSIONS

This study demonstrates enhanced denitrification and SS removal in multi-cathode BER combined with MF membrane. In comparison with former BER studies, denitrification rate was increased about 8 ~ 60 times. No accumulation of more toxic nitrite was detected, except initial phase. The concentration of suspended solids in the effluent did not exceed 1.0 mg-SS/L throughout the experiment, attaining average value 0.2 mg-SS/L. Sludge cake formed on MF membrane was efficiently removed by air bubbling.

This novel combined process is suitable for enhanced treatment of dilute solution such as groundwater. It is designed for small scaled and dispersed on-site remediation of nitrate-polluted groundwater.

ACKNOWLEDGEMENT

This study was funded in part by the grant-in-aid for the development of Scientific Research (# 11 650 556) from the Ministry of Education, Science and Culture of Japan.

REFERENCES

APHA, AWWA and WEF. 1995. "Standard methods for examination of water and wastewater." *The 20th edition*, 2-57~2-58.

Cast, K. L. and Flora, J. R. V. 1998. "An evaluation of two cathode materials and the impact of copper on bioelectrochemical denitrification." *Wat. Res.* **32**(1), 63-70.

DeJournett, T. D. and P. J. J. Alvarez. 1999. "Combined microbial-Fe(0) system to treat nitrate-contaminated groundwater." *Proceedings from the Fifth International In Situ and On-Site Bioremediation Symposium* **5**, 79-84.

Environmental Engineering Research Council of ASCE. 1990. "Ground-water protection and reclamation." *J. Environ. Eng.* **116**(4), 654-662.

Feleke Z., Araki, K., Sakakibara, Y., Watanabe, T. and Kuroda, M. 1998. "Selective reduction of nitrate to nitrogen gas in a biofilm-electrode reactor." *Wat. Res.* **32**(9), 2728-2734.

Flora, R. V., M. T. Suidan, S. Islam. 1993. "Numerical modelling of a biofilm-electrode reactor used for enhanced denitrification." 2nd *International Specialized Conference on Biofilm Reactors*, 613-620.

Islam, S. and M. T. Suidan. 1998. "Electrolytic denitrification: Long term performance and effect of current intensity." *Wat. Res.* **32**(2), 528-536.

Mackay, D.M. and J.A. Cherry. 1989. "Groundwater contamination: Pump-and-treat remediation. *Environ. Sci. Technol.* **26**(6), 630-636.

Mateju, V., S. Cizinska, J. Krejci and T. Janoch. 1992. "Biological water denitrification – A review. *Enzyme Microb. Technol.* **14**, 170-183.

Sakakibara, Y and Kuroda, M. 1993. "Electric prompting and control of denitrification." *Biotechnol. Bioeng.* **42**, 535-537.

Sakakibara, Y., Flora, J. R., Suidan, M. T. and Kuroda, M. 1994. "Modeling of electrochemically-activated denitrifying biofilms." *Wat. Res.* **28**(5), 1077-1086.

Sakakibara Y., T. Tanaka, K. Ihara, T. Watanabe and M. Kuroda. 1995. "An in-situ denitrification of nitrate-contaminated groundwater using electrodes." *Proc. of Environ. Eng. Research* **32**, 407-415.

NITRITE MAY ACCUMULATE IN DENITRIFYING WALLS WHEN PHOSPHATE IS LIMITING

W. J. Hunter (USDA-ARS, Fort Collins, Colorado, USA)

ABSTRACT: Permeable *in situ* denitrifying walls or barriers have been investigated as a means of removing nitrate from flowing groundwater. Denitrifying walls function by providing a microbial energy source in a highly permeable matrix. They usually are constructed by back-filling a trench or excavated area with a mixture of sand and fine gravel, which provide a porous matrix, and a substrate that serves as an energy source or electron donor for microbial denitrifiers. The present laboratory scale study demonstrates a need for adding phosphate when remediating some natural waters and shows that nitrite may accumulate in large amounts when phosphate is limiting. Denitrifying columns, 2.5 x 30 cm sand filled glass tubes infused with a vegetable oil substrate, were used as a scale model of a denitrifying wall. The water used was a natural groundwater that contained 16 to 18 mg/L nitrate-N and ~ 0.010 mg/L phosphate-P. When this low phosphate groundwater was pumped through the denitrifying columns little denitrification was observed and nitrite accumulated. When supplemental phosphate at 0.040 and 0.080 mg/L-P was used denitrification was still retarded and nitrite still accumulated. Increasing phosphate to 0.160 mg/L-P (N/P = 100) resulted in a rapid decrease in nitrate and only a brief accumulation of nitrite in the column effluents. The addition of solid rock phosphate or Biofos™ to the denitrifying columns at the time of packing provided adequate phosphate for denitrification over an extended period of time. These results illustrate the importance of assuring that adequate phosphate is available in denitrifying walls.

INTRODUCTION

Nitrate in groundwater is both a health and an environmental hazard. Permeable denitrifying barriers or walls are an emerging new technology that shows great promise as a method for protecting groundwater from nitrate contamination and for the *in situ* remediation of groundwaters that are contaminated with nitrate. In recent years several labs have investigated the use of permeable *in situ* denitrifying walls or barriers as a means of removing nitrate from flowing groundwater (Hunter, 1999; Hunter et al., 1997; Robertson and Anderson, 1999; Robertson and Cherry, 1995; Schipper and Vojvodic-Vukovic, 1998).

Denitrifying bacteria are ubiquitous in groundwaters but usually are inactive because energy sources are inadequate. Denitrifying walls function by providing a microbial energy source in a highly permeable matrix. Walls may be horizontal or vertical and usually are constructed by back-filling a trench or excavated area with a mixture of sand, fine gravel, and substrate. The sand and gravel provide a porous matrix and the substrate serves as an energy source or electron donor for microbial denitrifiers. Sawdust, crop residues, newspaper, cotton, vegetable oils, inorganic sulfur etc. have been used or proposed as substrates for denitrification reactions (Blowes et al., 1994; Flere and Zhang, 1999; Hunter et al., 1997; Lampe and Zhang, 1997; Volokita et al., 1996a; Volokita et al., 1996b).

Objective. This laboratory scale study investigated the denitrifying activity of a vegetable oil based denitrifying wall supplied with well water. The study demonstrated a need for adding phosphate when remediating some natural waters and shows that nitrite may accumulate in large amounts when phosphate is limiting.

MATERIALS AND METHODS

Groundwater. The well used as a source of groundwater was located in northeastern Colorado and was adjacent to a cattle pen. Depth to groundwater in this well was ~ 14 meters. Soils overlying the area are a relatively flat (<3% slope), unstructured deep, Vona sandy loam a well to excessively drained, mixed, mesic Ustollic Haplargids (USDA Soil Conservation Service, 1982). Core samples of surface soils (to a depth of 140 cm) and well water were collected and analyzed for phosphate and nitrate content.

Laboratory Scale Denitrifying Walls. Denitrifying columns, 2.5 x 30 cm sand filled glass chromatography columns containing 0.93 g (1 mL) of vegetable oil as a carbon substrate, were used as laboratory scale models of denitrifying walls. Control columns received groundwater that contained no supplemental phosphate, and treatment columns received groundwater supplemented with sufficient potassium phosphate to yield a final concentration of 0.040, 0.080 or 0.160 mg/L-P. Groundwater was pumped through these columns at a rate of 110-120 mL/day for 7 weeks and was supplied to the columns fully oxygenated. Samples of the column effluent water were collected at regular intervals and analyzed for nitrate and nitrite.

In a second study denitrifying columns were supplemented with rock phosphate or Biofos™ to evaluate the ability of these phosphate sources to serve as sources of phosphate for denitrification walls. Five-gram amounts of Biofos™ (Ranch-Way Feed Mills, Fort Collins, CO) or rock phosphate were blended into the denitrifying walls in the sand columns, 1.86 g (2 mL) of vegetable oil added, and groundwater pumped through the columns for 10 weeks. Samples of the column effluent water were collected at regular intervals and analyzed for nitrate and nitrite content.

Analysis. Nitrate and nitrite were analyzed with an HPLC equipped with a UV detector (Hunter et al., 1997). Phosphate was determined by the Ascorbic Acid Method (Greenberg et al., 1992) modified by the addition of H_2SO_4 to a final concentration of 10mM to dissolve precipitates and by an HPLC equipped with a conductivity detector.

RESULTS AND DISCUSSION

Manure is a good source of both nitrate and phosphate and as would be expected surface soils beneath the cattle pen were high in both. Nitrate was present throughout the soil cores. In the upper 80 cm nitrate-N ranged from 4 to 60 mg/L. In deeper soils the concentration of nitrate was higher and the nitrate-N concentration was highest, > 300 mg/L, in the 100 to 118 cm sample. Phosphate was highest in the upper 0 to 5 cm of soil, reaching a concentration of 110 mg/L. Phosphate generally declined with depth and was present at 8.8 mg/L in the deepest sample at a depth of 132-140 cm.

A well ~ 60 meters south of the cattle pen was used as a source of water

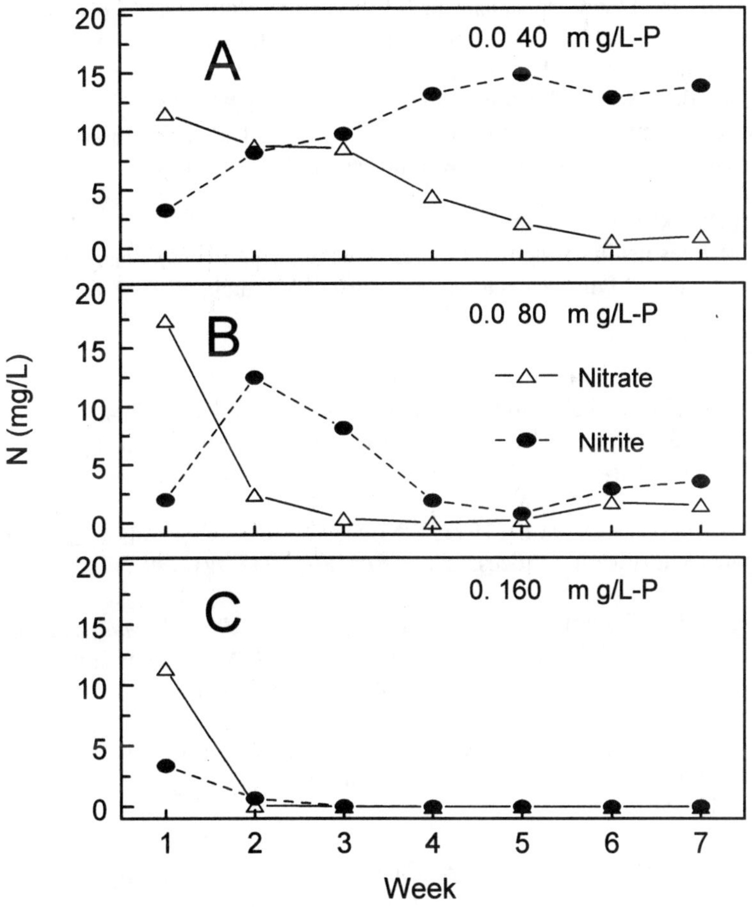

FIGURE 1. Nitrate and nitrite concentrations in effluents from columns supplemented with different concentrations of phosphate in the influent water.

for column denitrification studies. The well water was contaminated with 16 to 18 mg/L nitrate that originated in the adjacent cattle pens (Hunter, 1998). Although the surface soils above this well contained considerable phosphate, the well water contained ~ 0.010 mg/L-P. When water from this well was pumped through the denitrifying columns no denitrification was observed during the first three weeks. In the fourth week nitrate levels in the effluent water decreased and significant amounts of nitrite accumulated. When supplemental phosphate was added to the well water to give a final concentration of 0.040 mg/L-P (N/P = 400) and the water pumped through the denitrifying columns, the nitrate levels in the water declined slowly over a 5 week period, but again nitrite accumulated in large, almost stoichiometric, amounts (Figure 1A). Increasing phosphate to 0.080 mg/L-P (N/P = 200) resulted in a more rapid decrease in nitrate and transient accumulations of large amounts (> 5 mg/L) nitrite-N that lasted several weeks (Figure 1B). Increasing phosphate to 0.160 mg/L-P (N/P = 100) resulted in a rapid decrease in nitrate and only a transient accumulation of nitrite, lasting only a few days, in the column effluents (Figure 1C).

The addition of solid rock phosphate or Biofos™ (5 g of either) to the denitrifying columns at the time of packing provided adequate phosphate for denitrification for a 10 week period with no accumulation of nitrite.

CONCLUSIONS

These results illustrate the importance of assuring that adequate phosphate is available in denitrifying walls to prevent nitrite accumulation.

REFERENCES

Blowes, D. W., W. D. Robertson, C. L. Ptacek, and C. Merkley. 1994. "Removal of Agricultural Nitrate from Tile-Drainage Effluent Water using In-Line Bioreactors". *J. Contaminant Hydrology 15*:207-221.

Flere, J. M., and T. C. Zhang. 1999. "Nitrate Removal with Sulfur-Limestone Autotrophic Denitrification Processes". *J. Env. Eng. 125*:721-729.

Greenberg, A. E., Clesceri, L. S., and Eaton, A.D. (Ed.) 1992. *Standard methods for the Examination of Water and Wastewater, 18th Ed.* Am. Pub. Health Assoc., Am. Water Works Assoc. and Water Environ. Fed: Washington, D.C.

Hunter, W. J., R. F. Follett, and J. W. Cary. 1997. "Use of Vegetable Oil to Stimulate Denitrification and Remove Nitrate from Flowing Water". *Trans. ASAE 40*:345-353.

Hunter, W. J. 1998. "Nitrate Levels and Denitrification Activity Beneath a Cattle-Holding Pen". In *Animal Production Systems and the Environment: Volume II*, pp 593-598. Iowa State University Press: Ames, IA.

Hunter, W. J. 1999. "Use of Oil in a Pilot Scale Denitrifying Biobarrier". In A. Leeson and B. C. Alleman (Ed.), *Bioremediation of Metals and Inorganic Compounds,* pp. 47-51. Battelle Press: Columbus, OH.

Lampe, D. G., and T. C. Zhang. 1997. "Sulfur-Based Autotrophic Denitrification for Remediation of Nitrate-Contaminated Drinking Water". *In* B. C. Alleman and A. Leeson (eds), *In Situ and On-Site Bioremediation: Volume* 3 pp. 423 - 428. Battelle Press: Columbus, OH.

Robertson, W. D., and M. R. Anderson. 1999. "Nitrogen Removal from Landfill Leachate using an Infiltration Bed Coupled with a Denitrification Barrier". *Ground Wat. Monit. Remed. 19*:73-80.

Robertson, W.D., and J. A. Cherry. 1995. "*In Situ* Denitrification of Septic-System Nitrate using Reactive Porous Media Barriers: Field Trials". *Ground Wat. 33*:99-111.

Schipper, L., and M. Vojvodic-Vukovic. 1998. "Nitrate Removal from Groundwater using a Denitrification Wall Amended with Sawdust: Field Trial". *J. Env. Qual. 27*: 664-668.

USDA Soil Conservation Service. 1982. *Soil Survey of Weld County, Colorado Northern Part*. U.S. Government Printing Office, Washington, D. C

Volokita, M., A. Abeliovich, and M. I. M. Soares. 1996a. "Denitrification of Groundwater using Cotton as Energy Source". *Wat. Sci. Tech. 34*:379-385.

Volokita, M., S. Belkin, A. Abeliovich, and M. I. M. Soares. 1996b. "Biological Denitrification of Drinking Water using Newspaper". *Wat. Res. 30*:965-971.

COST EFFECTIVE BIOLOGICAL/CHEMICAL TREATMENT OF AMMONIA/AMMONIUM IMPACTED MEDIA

Warren D. Brady (IT Corporation, Baton Rouge, LA)
Duane Graves, Ph.D. and Julia Klens (IT Corporation, Knoxville, TN)
Dan Strybel (IT Corporation, Farmington, MI)

Abstract: Treatability studies were performed to investigate inexpensive treatment methods for ammonia/ammonium-impacted soil and groundwater at two sites. The first site contained two areas of varying ammonia/ammonium levels in soil with low pH (<4). At this site, high ammonia levels varied from 140 mg/kg to 3,500 mg/kg. The effects of pH control and microbial stimulation by amendment with a carbon source and an inoculum (activated sludge) were evaluated to determine methods for effectively reducing ammonia levels. The second site contained ammonia-impacted groundwater. Site soil and groundwater samples were subjected to testing to evaluate the effectiveness of amendment (carbon substrates) addition and oxygenation on the reduction of groundwater ammonia concentrations in situ.

Nitrogen pathway analyses were performed during both studies to verify that nitrification or other nitrogen transformations were not taking place and simply converting ammonium to a more mobile form. The results of these treatability studies identified easily implemented, cost-effective methods to reach environmentally acceptable end points and are presently being incorporated into full-scale remediation systems.

INTRODUCTION

Groundwater and surface water impact from both agricultural and industrial nitrogen sources is a global problem that requires cost-effective management and land use practices. From an agricultural perspective, intensive management practices have been implemented to increase yields and have in many cases increased the potential for non-point source pollution of soil, surface water, and groundwater. In an industrial setting nitrogen compounds like nitrate and ammonia are often evaluated like more traditional contaminants (i.e., metals, volatile organic compounds, fuel hydrocarbons) that require expensive remedial design. As a result, many of the fundamental properties and benefits of nitrogen compounds are ignored and cost-effective remedial solutions are problematic.

Though toxic at high concentrations, nitrogen compounds are limiting nutrients for biological activity at the macro and microscale. As such, they should not be addressed in the remedial design in a manner similar to contaminants that provide no direct role in metabolic functions. By defining the usefulness of nitrogen compounds under natural conditions up front and working backward to man-made problems, cost-effective remedial solutions for impacted soil and water can be developed. This paper provides the methodology for evaluating remedial options and developing cost-effective treatment technologies for two sites impacted by ammonia.

AMMONIA REMEDIAL TECHNOLOGY EVALUATION

Nitrogen can be present in soils or groundwater as ammonia (NH_3) and/or ammonium (NH_4^+). These two forms are often discussed interchangeably; however, their distribution and behavior in soil and groundwater environments are dissimilar. Ammonia is a gas in neutral to high pH soils and groundwater and is quickly volatilized. Ammonia disperses easily, since it is lighter than air (NH_3 density is 60% that of air). The ammonium form (NH_4^+) is present as a cation in acid to neutral pH soil and groundwater. Analytical data may be misleading since ammonia/ammonium analysis does not differentiate between ammonia (NH_3) or ammonium (NH_4^+) mass, but rather combines the two for a total ammonia number (EPA, 1983).

Ammonia can be transformed to other nitrogen compounds or removed from soil and groundwater systems by a number of the biotic and abiotic mechanisms. The possible fate(s) of ammonia and ammonium are summarized:

- Sorption - the dominant form of ammonia in soils and groundwater at nominal pH is ammonium (NH_4^+), a cation subject to sorption processes.
- Nitrification and Denitrification - transformation to NO_2^- by *Nitrosomonas* and *Nitrobacter* bacteria and further oxidized by *Nitrobacter* bacteria to NO_3^- (Payne, 1981).
- Incorporation Into Stable Organic Matter or Biomass –conversion of inorganic nitrogen to organic forms.
- Volatilization – loss of NH_3 to the atmosphere
- Plant Uptake –assimilation by plants.

When evaluating potential remedial technologies for ammonia, the potential fate(s) of ammonia listed above should be considered in the remedial design. Conceptually, the remedial design should, to the extent possible, rely on one or more of these processes as possible remedial scenarios. For instance, remediation of ammonia-impacted soil and/or shallow groundwater may be possible by simply providing a plant species that has a high nitrogen requirement. Alternatively, facilitating ammonia volatilization and/or sorption may be an applicable technology for large quantities of soil in a landfarm type application. Regardless of the application, potentially adverse transformations (i.e., nitrification) that may result from treatment should be evaluated prior to pilot or full-scale implementation.

SITE #1- AMMONIA IMPACTED SOIL

Soils from two surface impoundments contained elevated ammonia levels resulting from previous manufacturing process. The first impoundment (Impoundment I) contained NH_3-N concentration ranging from 5 to 165 mg/kg. The second impoundment (Impoundment II) contained higher NH_3-N concentrations ranging from 10 mg/kg to 3,500 mg/kg. Both impoundments had low pH (<4). The regulatory requirement at the site was to remediate ammonia to a soluble concentration of (30 mg/L) as determined by a deionized water extraction (DIWET). To evaluate ammonia treatment options, a laboratory evaluation was performed using impacted soil from the site.

Impoundment I. The treatment process for Impoundment I evaluated pH adjustment as a treatment to enhance the volatilization of ammonia. Soil pH was adjusted to 7, 7.5, and 8 in triplicated samples to evaluate biogeochemical changes that would be expected by soil treatment and to determine lime requirements. Samples were collected weekly from these treatments and a control for eight weeks and analyzed for ammonia, nitrate, nitrite, and total kjeldahl nitrogen (TKN).

As shown on Figure 1a, the pH 7.5 and 8.0 treatments effectively met the remedial action goal of 30 mg/L total ammonia (mg/kg) after 4 weeks and TKN results indicated that ammonia was being lost from the soil system (volatilization). Nitrate and nitrite concentrations were all below the lower detection limit indicating that nitrification was not taking place at a significant rate.

Impoundment II. The same testing process with pH adjustment was performed for soil in Impoundment II. The results of this test indicated that raising the pH was effective in volatilizing ammonia. However, pH adjustment alone was not sufficient to remediate ammonia in a timely and cost effective manner. To complement the pH adjustment, organic amendments were added to evaluate the conversion of ammonia to organic nitrogen. The proof of concept experiment included the following active treatments:

- pH adjusted - soil with a pH adjusted to approximately 7.
- Molasses amended and pH adjusted - This treatment was identical to the pH adjusted treatment except that molasses was added to the soil to give an organic carbon concentration concentration between 10,000 and 20,000 mg/kg.
- Activated sludge, molasses amended, and pH adjusted - This treatment was identical to the molasses amended and pH adjusted however activated sludge was added to the soil to provide an active microbial population.

As with the lab study for Impoundment I, all testing was performed in triplicate with controls. Samples were collected at 5, 7, 9, 12, 14, 16, 19, and 21 days and analyzed for ammonia, nitrate, nitrite, and TKN.

Figure 1b provides the ammonia treatment results for Impoundment II. The treatment process for Impoundment II took advantage of microbial growth. Very little microbial growth occurred naturally in the soil because the pH was low and there was an inadequate available food source (organic carbon). By supplying a constant source of food (molasses) to the soil and adjusting the pH to neutral using calcium oxide (CaO), microorganisms became active and began to assimilate the ammonia. Molasses was added regularly to maintain a constant organic carbon concentration so that the availability of a food source did not limit microbial growth. Oxygen was available from the atmosphere and water was added to maintain a constant moisture content. The only limiting factor for microbial activity was the availability of ammonia. Thus, as the microbial population increased, the available ammonia was consumed. Nitrate and nitrite concentrations remained below the lower detection limit indicating that nitrification was not taking place at a significant rate.

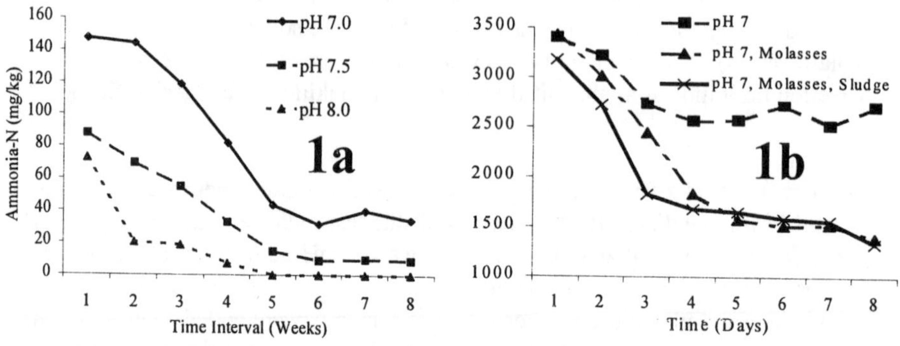

FIGURE 1 – Ammonia Treatment Results for Impoundment I (1a) and Impoundment II (1b).

SITE #2 – AMMONIA IMPACTED GROUNDWATER

Groundwater impacted with ammonia at this site discharged into a surface water body. Groundwater analytical data indicate that ammonia was present at concentrations greater than the groundwater/surface water interface criterion established by the state. Air sparging and conversion of ammonia to organic nitrogen through the addition of a carbon source was identified as a potential approach that could be used to remediate the ammonia and ensure surface water protection.

An ammonia laboratory study was performed using impacted groundwater and soil from the site. Three treatments and a control were established in triplicate to evaluate each of these potential technologies. The treatments were as follows:

- Air Sparge Treatment– Soil water slurries were shaken to aerate the samples and evaluate the effects of air sparging on the ammonia impacted groundwater.
- Hydrogen Release Compound (HRC®) (Regenesis Inc.) Treatment – HRC was added to each soil/water slurry to characterize the effect of this carbon source on ammonia treatment.
- Molasses treatment - molasses was added to each soil/water slurry to characterize the effect of this carbon source to ammonia treatment.

Samples were collected weekly for four weeks and analyzed for TKN, nitrate, nitrite, ammonia, and pH to characterize ammonia loss and transformation.

The air sparge treatment was performed to evaluate ammonia response to aeration of groundwater. One of the concerns with aerating these samples was the potential for ammonia to be converted to nitrate, which is more toxic and soluble in groundwater than ammonia. Figure 2 provides a graphical representation of the shaking control results and shows that ammonia levels decreased over time. Nitrification appears to occur as indicated by the increase in nitrate and nitrite after one week of treatment. The observed nitrification was greater than the reported nitrate and nitrate values in the control treatments, as shown on Figure 2, indicating that nitrate may be produced at a higher rate as a result of aeration.

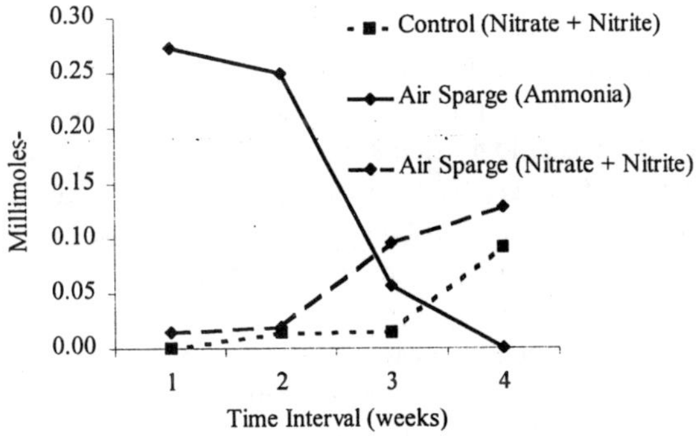

FIGURE 2 –Air Sparge Groundwater Treatment Results

Both the HRC and molasses treatments were performed to determine if the addition of a carbon source to the groundwater system would result in ammonia concentration reduction by conversion to stable organic nitrogen. Figures 3a and

3b provide a graphical representation of the HRC and molasses treatments, respectively. Nitrate and nitrite were only detected after the third week of treatment at concentrations below the levels reported in the control, indicating that nitrification was negligible with the addition of a carbon source. Based on this testing, the addition of a carbon source was viewed as a viable and easily implemented groundwater restoration technique for in situ remediation of ammonia.

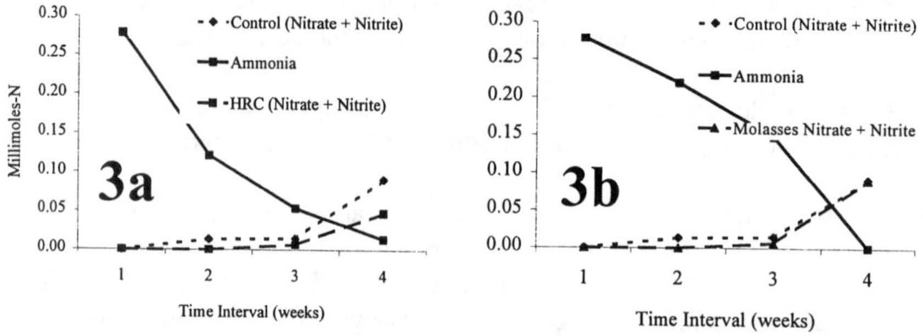

FIGURE 3 – HRC (3a) and Molasses (3b) Groundwater Treatment Results

DISCUSSION

These applications provide a simple methodology by which ammonia impacted soil and groundwater can be remediated in a cost-effective manner. In the soil treatment example from Impoundment I, manipulation of the direct relationship between pH and the ammonia form (NH_3 or NH_4^+) provided a basis for treatment that was verified in bench testing.

Soil treatment from Impoundment II demonstrated that microbial growth can be stimulated by adding a carbon source. Microorganisms require water, a food source (organic carbon), inorganic nutrients (ammonia and other minerals available in soil), and a respiratory substrate (oxygen in this case) to survive. When these requirements are met, microbial populations increase and so does nutrient assimilation, as demonstrated in the treatability testing.

In the carbon amended treatments, the presence of an organic carbon source (HRC or molasses) provided a competitive advantage to ammonia assimilating bacteria that converted ammonia to organic nitrogen rather than nitrate. The addition of HRC or molasses simply provided a carbon source, which increased the carbon to nitrogen (C/N) ratio and facilitated ammonia conversion to an organic form through biological processes.

Nitrification was not observed in the soil treatments for Impoundment I or II, indicating there was a low potential for adverse ammonia transformation. The laboratory results for groundwater also indicated that the addition of an organic substrate would facilitate the removal of ammonia without causing a significant accumulation of nitrate or nitrite. If nitrification took place in response to carbon addition, the nitrate would likely be quickly denitrified since the addition of a carbon source encourages reduction, which favors denitrification. The potential for nitrification should be examined when considering remediation of ammonia/ammonium impacted sites to avoid treatment technologies favoring nitrification, as was the case for air sparging in this investigation.

The technologies described above are inexpensive and can be easily implemented using readily available amendments (lime, HRC, molasses). The field scale soil treatment applications described for Impoundment I and II are being performed with standard agricultural equipment such as tractors, plows and tillers, spray rigs, and fertilizer spreaders. For the groundwater ammonia treatment, carbon can be introduced into an aquifer by pressure injection using direct push technologies or through existing wells. Application strategies include injecting the carbon source throughout the entire ammonia impacted area, application only to hot spots, or application along the down gradient edge of plumes to form a barrier to off-site migration.

REFERENCES

Jordan, J.L., S.L. Morgan, and A.H. Elnagheeb. 1994. "An economic analysis of cover use in Georgia to protect groundwater quality. Georgia Agricultural Experiement Station. University of Georgia. *Research Bulletin No 419.*

Payne, W.J. 1981. *Denitrification.* Wiley, New York.

USEPA. 1983. Methods for chemical analysis of water and wastes. EPA 600/4-79-020, Revised March, 1983. NTIS number PB84-128677.

ENHANCED IN SITU BIOLOGICAL DENITRIFICATION: COMPARING TWO AMENDMENT DELIVERY SYSTEMS

Clay Jones, ***H. Eric Nuttall*** (University of New Mexico, Albuquerque, NM)
Bart Faris (New Mexico Environmental Department, Albuquerque, NM)

ABSTRACT: Nitrate contamination in groundwater is of growing health and regulatory concern. Conventional nitrate removal processes, which include pump and treat, ion exchange, and reverse osmosis are costly and only concentrate the nitrate whereas bioremediation chemically transforms nitrate into harmless nitrogen gas. A method of enhanced *in situ* biological denitrification (EISBD) is being developed by the University of New Mexico, which promises to offer considerable advantages over conventional methods of nitrate treatment. To advance the EISBD process, two distinctly different amendment delivery systems are being tested at a site located near Mountain View in Albuquerque's South Valley. At this location, an over-fertilized vegetable farm operated in the fifties created a one square mile nitrate plume with nitrate concentrations over 300 mg/L nitrate-nitrogen. This paper presents and discusses the results from the passive direct push amendment delivery system and mentions briefly the ongoing inverted five spot system.

INTRODUCTION

The scientific principles underlying biological denitrification have been well established in the literature and through six years of research at the University of New Mexico (UNM). The work at UNM is documented in several studies (Chen, 2000; Deng, 1998; Jones, 2001 Lu, 1998; Lu, et al., 1999; Nuttall, 1997; Nuttall, et al., 1999). The experimental amendment delivery system discussed herein is designed to demonstrate *in situ* bio-denitrification on a large scale suitable for commercial application and expose challenges to such applications.

Nitrate is a national as well as worldwide water contaminant. The two primary sources of nitrate contamination in groundwater are over-fertilization and animal farms such as dairies (Spaulding and Exner, 1993). The US EPA estimates that 2.4% of domestic wells in the United States exceed the Maximum Contaminant Level (MCL) of 10 mg/L nitrate as nitrogen (Technology Overview, 2000). Nitrate is a groundwater contaminant that is of growing regulatory concern. High levels of nitrate in water can have adverse effects on the environment and on humans and animals.

Principles of Biological Denitrification. The denitrification reaction is an energy yielding reaction carried out by indigenous bacteria, which transform nitrate in groundwater into harmless nitrogen gas. Every metabolic reaction involves an electron donor and an electron acceptor. In the denitrification reaction, nitrate is the electron acceptor. An electron donor must be supplied to allow the reaction to proceed. Lu (1998) outlines many aspects of using several different electron donors (glucose, ethanol, methanol, acetate, and lactate). In the field demonstrations we have elected to use acetate as the electron donor because laboratory tests have shown that it produces least biomass and therefore helps to prevent biofouling of the aquifer (Chen, 2000, Deng, 1998,).

The bacteria use nitrate both in the energy yielding denitrification reaction and also for synthesis of cell components. When acetate is used as the electron donor, the stoichiometry of these reactions is as follows (Lu, 1998):

Metabolic denitrification reaction:

$$0.625Ac + NO_3 \text{ Æ } 1.25HCO_3 + 0.25N_2 \qquad (1)$$

Cell component synthesis reaction:

$$3.5Ac + NO_3 \text{ Æ } C_5H_7O_2N + 2HCO_3 \qquad (2)$$

Combined metabolic and cell synthesis reactions
97% metabolic + 3% cell synthesis:

$$0.712Ac + NO_3 \text{ Æ } 0.485N_2 + 0.03C_5H_7O_2N + 1.273HCO_3 \quad (3)$$

Where $C_5H_7O_2N$ represents the chemical composition of cellular material and Ac represents acetate. Therefore, the design of the field experiments must deliver acetate at a 0.712 to 1.000 molar ratio of acetate to nitrate.

It has been found that the addition of phosphate is necessary to promote the denitrification reaction in water from the South Valley. A phosphate nutrient, in the form of sodium trimetaphosphate (TMP), is also supplied to give a concentration of 21mg/L of phosphate in the aquifer. 3 mg/L is the theoretical minimum TMP concentration, but higher concentrations have been seen to not adversely affect the rate of denitrification (Nuttall et al., 1999).

AMENDMENT INJECTION SYSTEMS

Two uniquely different amendment delivery systems (one passive and the other active) that were tested in the field side by side and so that their performance and costs can be directly compared.

The first field scale demonstration was a direct push amendment delivery system. The direct push demonstration was deployed in March, 2000. A direct push delivery is comprised of a one-time injection of amendment directly into the nitrate-laden aquifer. Direct push is therefore a passive amendment delivery, which relies on natural mixing effects (i.e., diffusion, dispersion, advection) to bring the amendment into contact with the nitrate-laden water. Amendment was injected through nine direct push points over a rectangular grid pattern of 9. 14 m x 9.14 m (30' x 30'). Approximately 3.79 m^3 (1000 gallons) of amended water was injected per point at a rate of 0.03 – 0.0379 m^3/min (8-10 gpm) and using an injection pump pressure of 1.72 mPa (250 psig).

The second delivery system is an inverted five spot pattern covering an area of 30.5 m x 30.5 m (100' x 100'). The inverted five spot is an active amendment delivery system. Five permanent wells were installed at the corners of the square pattern. In a recirculating mode, water was pumped from each of the four corners and re-injected with amendment addition into the center well. This design ultimately remediates a circular region emanating from the center injection well.

This paper discusses the design and performance of the direct push system. The results of the inverted five spot pattern is still in operating and the results will be presented in a future article.

Site Layout. The field demonstration of EISBD discussed in this paper is located in the South Valley of Albuquerque, NM. The South Valley is ideal because it is near the University of New Mexico and there is a large nitrate plume there. The nitrate plume covers an area of about 2.2 square kilometers (0.85 square miles). The nitrate plume was caused by over-fertilization of a vegetable farm in the 1950s (6). Though the average nitrate-nitrogen concentration in the groundwater of the South Valley is 90-110 mg/L, the field demonstration is centered over what is thought to be the heart of the plume where the nitrate-nitrogen concentration is 215-250 mg/L. This plume of highest nitrate concentration is dangerously close to the community of Mountain View, which until recently has drawn its drinking water from the aquifer.

The water table is at 14.3 m(47 ft) bgs and 12.8 m (42 ft) below the ground is a 1.5 m (5 ft) thick clay zone relatively impermeable to water. Below this upper clay aquitard is a 4.6 m (15 ft) thick sandy/loam soil aquifer. Another clay aquitard exists directly beneath the aquifer, at a depth of approximately 19.9m (62 ft). The nitrate contamination is confined within the aquifer, which lies between the two aquitards. The sand in the aquifer is homogenous with a saturated hydraulic conductivity of 10^{-3} cm/s (2.8 ft/day) and a flow gradient of 0.005m/m (0.005ft/ft). The groundwater flow direction is in the East-Southeast direction.

The direct push amendment delivery is comprised of nine successive injections, one at each of nine separate injection points. The nine injection points are laid out in a three by three square grid pattern with 4.6 m (15 ft) between each point. Reaction progress is monitored using three permanent monitoring wells installed at the site. See Figure 2 for the layout of the direct push.

The direct push delivery relies on natural effects such as diffusion and advection to mix the amendment-laden injectate with the nitrate-laden groundwater.

Experimental Procedures. 3800 L (1000 gal) of amended nitrate-free water were injected into the aquifer at each of the nine injection points. The amended water contained sodium acetate, sodium trimetaphosphate (TMP), and sodium bromide.

Amounts of chemical amendments correspond to the amount of water in the injection area and level of nitrate contamination in the water. Since the injection area was 9.1 m by 9.1 m (30 ft by 30 ft) and the aquifer is approximately drawn by hand using a bailer. 1.2 well volumes of water were discarded from each well before a sample was collected.

Bromide concentrations were determined using an Orion Ion Specific Electrode for bromide measurement. Acetate, nitrate, nitrite, sulfate, and chloride concentrations were obtained from a Dionex DX-500 HPLC (Ion Chromatography) system.

The results graph showing nitrate and bromide concentrations versus time is given in Figures 1. Since considerably less denitrification was observed in comparison to that expected, an analysis of amendment leakage into the vadose zone was performed to better estimated the amount of lost amendment.

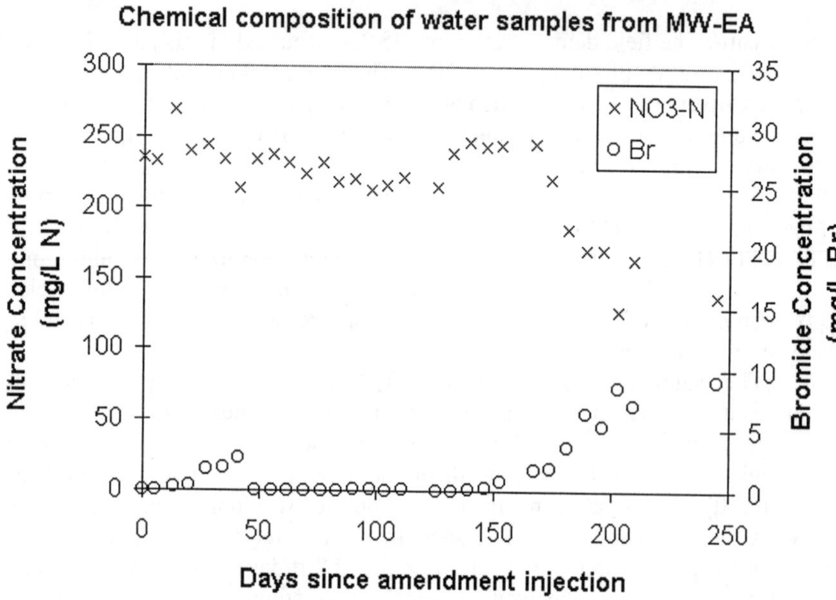

FIGURE 1. Nitrate/Bromide Concentration Versus Time for MW-EA

Lost Amendment versus Amendment Injected into Aquifer. Direct push access to the aquifer was accomplished by hammering a series of pipe segments from the surface into the aquifer. It is likely that hammering a pipe into the soil may create an annular space around the injection pipe where soil has been pushed aside by the vibrations of hammering. This small but signification annular space would then have relatively low resistance to flow compared to the soil matrix of the surrounding aquifer. Injecting at high pressures in such a situation may lead to preferential flow up this annular space instead of into the aquifer. This situation would lead to a large percentage of the injected amendment escaping the aquifer and entering the vadose zone. In order to evaluate the likelihood of significant amounts of amendment being lost this way, a mathematical model was constructed and evaluated.

The approach of such a mathematical comparison is to obtain mathematical expressions for flow and pressure drop both in an annulus and in the aquifer. The height of the injection rod can then be divided into differential

4.6 m (15 ft) deep, the volume of soil involved is 380 cubic meters (13,500 cubic feet). The soil of the aquifer has an estimated void fraction of 0.3, so the volume of water to be treated by the direct injection experiment is approximately 114,000L (30,200 gallons).

Acetate requirements were determined by assuming a nitrate concentration of 215 mg/L of nitrate-nitrogen. The experimental work reported by Nuttall, et al., 1999, shows denitrifying bacteria use nitrate and acetate at a constant molar ratio. The nitrate to acetate ratio she found is 1.000 moles of nitrate to 0.712 moles of acetate. TMP requirements were based on the value of 21 mg/L, and bromide requirements were based on approximately 100mg/L in the injection liquid for ease of detection.

Chemical amendment was mixed with water in two 1900 L (500 gal) tanks for a total of 3800 L (1000 gal) of water injected at each injection point. The water used for injection was clean and nitrate-free. The first of the two tanks was equipped with a mixer for dissolving the amendments in the water prior to injection. Sodium acetate, sodium trimetaphosphate, and sodium bromide, all in powder form, were added to the water in the tank with the mixer. Since only one of the two tanks was equipped with a mixer, water was always injected from this tank to assure that the injection liquid was well mixed.

The water was injected through 3.18cm (1.25 in) diameter pipe that was hammered into the ground to the desired depth. The drilling rig was mounted on the back of a truck, and the mixing tanks were brought in on a trailer

At each of the nine injection points, the 1900 L (1000 gal) of amended water was injected in three parts. The 3.18 cm (1.25 in) diameter pipe was hammered into the earth to a depth of 19 m (62 ft), which roughly corresponds to the bottom of the aquifer. At this point, one third of the amended water (633 L or 333 gal) was injected. Then the pipe was withdrawn 1.2 m (4 ft) so that the end of the pipe was approximately in the center of the aquifer, at a depth of 17 m (57 ft), where the second third of the amended water (633 L or 333 gal) was injected. Finally the pipe was withdrawn another 1.2 m (4 ft) so that the end of the pipe was at the top of the aquifer, at a depth of 16 m (53 ft), where the final third of the amended water (633 L or 333 gal) was injected.

BioManagement Services, Inc. was the company responsible for injecting the amendment. Arts Manufacturing and Supply, Inc. was the company responsible for conducting the direct-push drilling.

The injection procedure was performed at each of the nine injection points over a period of three days, from 14 March 2000 to 16 March 2000. It took the drillers approximately 10 minutes to hammer the 3.18 cm (1.25 in) diameter pipe to the desired depth and approximately 2 hours to inject the 1900 L (1000 gal) of water. Therefore the average injection rate was 31 L/min (8.3 gpm). Injection pressure was about 1.7 MPa (250 psi).

Experimental Results. To evaluate the effectiveness of the direct push experiment, water samples from the three monitoring wells were analyzed weekly. The extent of mixing and denitrification can be determined from the changes in nitrate, acetate, and bromide concentrations over time. Samples were elements. At each element the amount of amendment flowing up the annulus versus into the aquifer can be calculated by finding the relative flow rates that will cause equal pressure drops in each direction.

An approximate quantitative comparison of flow into the aquifer versus flow up the annulus around the injection rod can be obtained from pressure drop calculations. The pressure drop equation for laminar flow in an annulus was from Bird, et al. (1960). An equation describing pressure drop of flow into a confined aquifer was from Kresic (1997).

Assumptions involved in the annulus pressure drop calculation:

- Smooth, impermeable, pipe-like walls. The outer ring of compressed soil only roughly fits this description.
- Reynold's number less than 2,000. This assumption is satisfied only for flow in the annulus of less than 2 gpm. However, the Reynold's number is always less than 10,000, which is less than an order of magnitude too high.

Assumptions involved in the pressure drop calculation for flow into a confined aquifer:

- Injection pipe is screened across the entire thickness of the aquifer. This assumption is not met. The injection rod used for the direct push procedure was an open ended pipe which emitted water from its tip only.
- Saturated hydraulic conductivity of the aquifer is constant at 10^{-3} cm/s.
- Boundary condition: The depth to water (piezometric head) did not change at MW-NW during injection. This was field verified during injection. Therefore the radius of influence of the injection is less than the distance from any injection point to MW-NW.

Since the thickness of the annulus around the injection rod is unknown, it was left as a variable in the calculations. The annulus pressure drop equation and the aquifer pressure drop equation can be solved to give pressure drop as a function of flow rate. At any point up the annulus, the pressure drop of the liquid traveling up the annulus must be equal to the pressure drop of the liquid traveling into the aquifer. The calculations showed how much injection liquid will escape the aquifer through the annulus as a function of annulus thickness. If the annulus is thicker than 2.5mm (0.1 in) then more than 85% of the injection liquid will escape into the vadose zone. We estimate that only about 15% or less of the amendment entered the aquifer and the reminder escaped into the vadose zone.

SUMMARY AND CONCLUSIONS

The direct push amendment delivery system is a passive amendment delivery system, which relies on natural processes such as advection, dispersion, and diffusion to mix the injected amendment with the nitrate-laden groundwater. The objectives of the demonstration were

- Perform a direct push injection of amendment into the aquifer.
- Investigate the effects that injection pressure/flow rate have on amendment introduction to/mixing in the aquifer.
- Monitor reaction progress and mixing of amendment with groundwater via sampling from permanent monitoring wells.

The direct push amendment delivery system was deployed over a three-day period from 14-Mar-2000 to 16-Mar-2000. The site is located near the heart of a nitrate plume located in Albuquerque's South Valley. 3800L (1000 gal) of amendment were injected into each of nine injection points. The amendment consisted of acetate (carbon source), trimetaphosphate (phosphorous nutrient), and bromide (inert tracer).

Amendment injection was carried out under high pressure (around 1.7 MPa or 250 psig). The high pressure was used to obtain high amendment flow rate (about 31 L/min or 8.3 gpm). High amendment flow was desired to keep the injection time to a minimum (~2 hrs) and thereby reduce labor and equipment costs.

Calculations and experimental results show that injection at high pressures may lead to amendment escaping into the vadose zone via an annular space of relatively low resistance to flow. This annular space can be formed around the direct push rod by the act of hammering the rod into the ground. Therefore injection at lower pressures is necessary to allow the amendment to enter the aquifer and not escape into the vadose zone.

REFERENCES

Bird et al (1960) Bird, Stewart, and Lightfoot. 1960. *Transport Phenomena*, John Wiley & Sons, New York.

Jones (2001) Jones, C. 2001. Enhanced In Situ Bioremediation. M.S. Thesis, Chemical and Nuclear Engineering Department, University of New Mexico.

Deng (1998) Deng, L. 1998. In situ biological denitrification of groundwater. M.S. Thesis. Chemical and Nuclear Engineering Department, University of New Mexico.

Chen (2000) Chen, J. 2000. Continuous Biodenitrification of Groundwater: In Situ Field Test, M.S. Thesis in progress, Chemical and Nuclear Engineering Department, University of New Mexico.

Kresic (1997)	Kresic, Neven. 1997. *Quantitative Solutions in Hydrogeology and Groundwater Modeling*, CRC Lewis, Publishers, New York,
Lu (1998)	Lu, Y. 1998. Sequential bioremediation of nitrate and uranium in contaminated groundwater Ph.D. dissertation, Chemical and Nuclear Engineering Department, University of New Mexico.
Lu et al. (1999a)	Lu, Y., H. E. Nuttall, and W. Lutze 1999. "Denitrification Kinetics: Model and Comparison to Laboratory and Field Data". Proceedings from the In Situ On-site Bioremediation; 5th International Symposium, April 19-22, 1999, *San Diego, California, Bioremediation of Metals and Inorganic Compounds*, 5(4), pgs.65-72. Battelle Press; Columbus, OH.
Nuttall (1997)	Nuttall, H. E. 1997. "Restoring New Mexico's groundwater using Biodenitrification," *New Mexico Journal of Science*, 54-73
Nuttall et al (1999b)	H. E., L. Deng, A. Abdelouas, Y. M. Lu, and W. Lutze, "In situ Denitrification: A field demonstration" Proceedings from the In Situ and On-site Bioremediation, 5th International Symposium, April 19-22, 1999, *San Diego, California, Bioremediation of Metals and Inorganic Compounds*, 5(4), Battelle Press, pgs.59-64. Battelle Press; Columbus, OH,

"Technology Overview: 2000. Emerging Technologies for Enhanced In Situ Biodenitrification of Nitrate Contaminate Groundwater," *ITRC (Interstate Technology Regulatory Cooperation) Publication*,

HYDROCARBON BIODEGRADATION RATES AND WATER POTENTIAL IN NITROGEN AUGMENTED DESERT SOILS

Claudia Walecka-Hutchison (University of Arizona, Tucson, Arizona)
James L. Walworth, (University of Arizona, Tucson, Arizona)

ABSTRACT: The effects of flux in water potential resulting from N augmentation on hydrocarbon degradation rates were evaluated in a bench scale study using 4 arid region soils. Degradation was determined respirometrically via oxygen consumption, and water potential (matric plus osmotic) was analyzed using a thermocouple psychrometer. Increasing N concentrations resulted in a decrease in total water potential and correlated with decreased respiration rates. Highest respiration was observed in the 250 mg N/kg soil treatments and application of 1000 mg N/kg soil reduced respiration between ½ to ⅓ the maximum amount. Soils with greater levels of native salinity showed reduced respiration rates at all levels of N fertilization. Microbial preference for NH_4^+ over NO_3^- was observed and NO_3^- amended soils consumed only 65-70% as much O_2 as those treated with NH_4^+. The water potential depression resulting in a 50% reduction in respiration was much greater than that observed in humid region soils, suggesting a greater salt tolerance by microbial populations of arid region soils.

INTRODUCTION

The widespread use of petroleum products has inadvertently led to extensive soil and groundwater contamination. Restoration of hydrocarbon contaminated sites not only entails great expense, but often, through application of physical-chemical treatment methods such as absorbents or technologies such as pump-and-treat, merely transfers the contaminant to another medium. *In situ* bioremediation is considered a potentially favorable alternative because it transforms petroleum hydrocarbons to stable non-toxic end-products (carbon dioxide, water, and biomass under optimal conditions), and has the potential to reduce costs.

Significant parameters of *in situ* hydrocarbon biodegradation include available soil water, oxygen, redox potential, nutrients, pH, and temperature (Sims et al., 1993). Frankenberger (1991) recommends 50% to 70% of soil water holding capacity as the optimal range of soil moisture for bioremediation. Both nitrogen (N) and phosphorous (P) are found in short supply in hydrocarbon contaminated ecosystems (Alexander, 1999). However, the higher bacterial consumption rate of N relative to P during the catabolic breakdown of hydrocarbons often results in N becoming the primary limiting inorganic nutrient for oil bioremediation (Alexander, 1999). Therefore, to sustain optimum microbial growth and biotransformations required for bioremediation, N and often P must be supplemented.

Estimations of exogenously supplied N requirements have generally been based on empirical C:N:P ratios (Lewandowski and DeFilippi, 1998). Dibble and Bartha (1979) found optimal hydrocarbon degradation at a C:N ratio of 60:1, and a

C:P ratio of 800:1. Paul and Clark (1989) recommended a 30:5:1 C:N:P ratio (C:N ratio of 6:1) to allow unrestricted growth of soil bacteria. Litchfield (1993) recommended a ratio of 100:10:2 (10:1 C:N ratio) to compensate for decreased elemental bioavailability. Sims et al. (1989) recommended a C:N:P ratio of 120:10:1 (12:1 C:N ratio), while the US EPA (1993) proposed a C:N:P range of 100:10:1 - 100:1:0.5.

Alternatively, the amount of N has been determined from synthesis and cell growth oxidation/reduction reactions (McCarty, 1972). Alexander (1999) described a 10:1 microbial C:N ratio and a substrate assimilation efficiency of 30%. Therefore, during bioremediation of 1000 g of organic C, 300 g of C will be converted to microbial biomass. This would require 30 g of N (because the biomass has a C:N ratio of 10:1), resulting in an the overall soil C:N ratio (the ratio of contaminant C to required N) of 1000:30 or 33:1. Alexander stressed that the efficiency of biomass production is dependent on many factors including the species carrying out the transformation, the composition of the substrate as well as its concentration, and temperature. Graham et al. (1995) concluded that biodegradation of various subclasses of hydrocarbons requires different nutritional amendments. Cleland et al. (1997) reported that varying nutrient conditions has an effect on both bioactivity and the microbial species distributed. Therefore, the validity of applying a single C:N ratio in all situations becomes questionable.

Numerous nutrient augmentation studies have demonstrated positive effects via accelerated hydrocarbon biodegradation rates (Atlas and Bartha, 1972; Dibble and Bartha, 1979; Bragg et al., 1994; Zhou and Crawford, 1995). However, under certain conditions N supplementation can have negative effects on hydrocarbon degradation. Cason (1994) suggested that a C:N:P of 300:15:1 (C:N = 20:1) limited biodegradation. Zhou and Crawford (1995) found that whereas C:N ratios of 50:1, 18:1, and 15:1 favored microbial growth, excessive addition of N (C:N = 1.8:1) almost stopped hydrocarbon biodegradation. Graham et al. (1995) found low hydrocarbon biodegradation rates at both the expected low N supply levels (~500 mg/kg N, due to N-limited growth conditions) as well as the unexpected high N supply levels (~1500 mg/kg N). Braddock et al. (1997) observed highest hydrocarbon degradation rates as well as numbers of hydrocarbon degraders under lowest (100 mg/kg N) rather than highest (300 mg/kg N) nutrient addition and concluded that at 200 to 300 mg/kg N microbial activity and hydrocarbon degradation were inhibited. They hypothesized that the decline in microbial activity resulted from the increased salinity attributable to the partitioning of the fertilizer salts into the pore water (total soil water potential ranged from -0.2 MPa in the control to -1.5 MPa at the highest level of fertilizer added). Walworth et al. (1997a) correlated soil moisture with the *in situ* effective nutrient concentration and concluded that, depending on the soil texture, even low fertilization could lead to toxic effects on microbial populations due to osmotic stress. Walworth et al. (1997a,b) noted that hydrocarbon respiration rates are restricted by both low bioavailable N in unfertilized soils, as well as the osmotic soil water potential attributable to the dissolution of fertilizer salts under high N conditions. They observed highest respiration in a humid-region soil at total water potential of -0.3 to

-0.4 MPa or greater. As the water potential became more negative, respiration declined and was reduced by 50% at -0.8 MPa. Haines et al. (1994) found that increasing salinity (from 0 to 30 mg/kg NaCl) increased the lag phase and inhibited oil biodegradation rates in slurry reactions.

Given the 10-fold variation in published C:N ratios, as well as variability in microbial species, contaminant type, contaminant concentration, and soil texture, optimal N augmentation remains elusive. Arid region soils with naturally high levels of soluble salts and salt tolerant microbial populations might be expected to be more tolerant to large doses of N fertilizers and the resulting drop in soil water potential than less saline soils. Previous works (Braddock et al., 1997; Walworth et al., 1997 a,b) have studied over fertilization in non-saline soils. The purpose of this study is to evaluate the role of N augmentation on arid soil water potential and determine the effect on hydrocarbon biodegradation rates. The resulting data are compared to that of similar studies performed in more humid soils.

MATERIALS AND METHODS

Four arid soils of the southern Arizona region, Gilman silt loam (hyperthermic,Typic Torrifluvent), Casa Grande sandy clay loam (hyperthermic,Typic Natrargid), Grabe sandy clay loam (Grabe 1) and Grabe clay (Grabe 2)(thermic, Typic Torrifluvents), were analyzed in this study. Soils were refrigerated in their naturally moist state prior to use. All soils were sieved using a 2 mm sieve, wetted to 50% - 70 % of field capacity, and re-sieved to achieve uniformity. The specific soil moisture content of each soil after the moist-sieving is given in Table 1.

On dry weight basis the soils were amended with 5000 mg/kg diesel fuel as carbon source and 150 mg/kg P ($Ca(H_2PO_4)\cdot H_2O$). After a thorough mixing, 200g (dry weight) of each soil were placed into 500 ml bottles, supplemented with N (NH_4NO_3) at 0 (control), 250, 500, and 1000 mg/kg and mixed. The different levels of N addition were run in triplicate for each soil. Additionally, the Case Grande soil was used to determine the effects of alternate N sources (NH_4Cl and $NaNO_3$) at the same N concentrations.

Contaminant degradation was determined using a computerized respirometer (N-CON Comput-OX System 12). KOH pellets placed in a container inside each bottle were used to trap CO_2 generated during respiration. As CO_2 was produced and O_2 consumed, O_2 was replaced to retain ambient conditions. The quantity of replacement O_2 (or microbial O_2 uptake rate) was recorded every ½ hour. Each run lasted a total of 500 hours at a constant temperature of 25 °C.

Water potential (ψ, matric plus osmotic) was determined in triplicate for all soils and N concentrations using a thermocouple psychrometer (Decagon Tru Psi). At the end of each respirometric run, soil from each reactor was used to determine the recoverable inorganic N (NH_4^+ and NO_3^-) via the steam-distillation method (Mulvaney, 1996).

RESULTS AND DISCUSSION

Increasing N concentrations corresponded with a decrease in total water

TABLE 1. Selected Soil Properties and Soil Water Potential (matric plus osmotic)

Soil	Texture	% Field Capacity	Native Inorganic N (mg/kg)	N source	Water potential (MPa)			
					0 mg N/kg soil	250 mg N/kg soil	500 mg N/kg soil	1000 mg N/kg soil
Gilman	silt loam	66	120.05	NH_4NO_3	-8.43	-9.01	-9.37	-9.68
Grabe 1	sandy clay loam	60	NA	NH_4NO_3	-7.37	-7.87	-8.17	-8.80
Casa Grande	sandy clay loam	67	2.78	NH_4NO_3	-0.84	-1.06	-1.18	-1.77
Grabe 2	clay	46	147.83	NH_4NO_3	-2.26	-2.29	-2.76	-3.42
Casa Grande	sandy clay loam	70	3.20	NH_4Cl	-0.73	-1.37	-2.06	-3.29
Casa Grande	sandy clay loam	70	2.2	$NaNO_3$	-0.70	-1.40	-2.17	-3.48

Table 2. Microbial Respiration Rates at Varying Additions of N

Soil	N Source	Respiration Rates (mg O_2/kg soil per day)			
		0 mg N/kg soil	250 mg N/kg soil	500 mg N/kg soil	1000 mg N/kg soil
Gilman	NH_4NO_3	13.3	20.4	24.9	9.5
Grabe 1	NH_4NO_3	4.7	0.14	0	0
Casa Grande	NH_4NO_3	32.7	90.0	61.2	37.9
Grabe 2	NH_4NO_3	63.6	90.0	81.6	45.6
Casa Grande	NH_4Cl	50.4	110.4	86.4	38.4
Casa Grande	$NaNO_3$	52.8	76.8	57.6	25.2

potential (matric plus osmotic) in all soils (Table 1). Depending on soil characteristics, the depression in ψ between the control and the 1000 mg/kg N treatments when NH_4NO_3 was used as the N source ranged from -0. 93 MPa in the Casa Grande soil, -1.16 MPa in the Grabe 2 soil, -1.25 MPa in the Gilman soil, and -1.43 MPa in the Grabe 1 soil (Table 1). When NH_4Cl and $NaNO_3$ were used, the reduction in ψ between the control and the 1000 mg/kg N treatments of the Casa Grande soil were -2.56 and -2.78 MPa respectively. Although the same soil texture with similar moisture content was used in the latter study, the greater decrease in water potential upon addition of NH_4Cl and $NaNO_3$ may be attributed to the presence of additional counter ions (Na^+ or Cl^-).

Table 2 shows the respiration rates of each soil with varying additions of N. In general, soils with greater levels of native salinity showed reduced respiration rates in all treatments. Respiration observed in the control treatments ranged from 5, 13, 33, 50 and 53, and 64 mg O_2/kg soil per day with respective ψ values of -7.37, -8.43, -0.84, -0.73 and -0.70, and -2.26 MPa (Grabe 1, Gilman, Casa Grande -NH_4NO_3, Casa Grande-NH_4Cl and $NaNO_3$, and Grabe 2 soil respectively, Table 1). Although the Grabe 1 control treatment had a higher ψ relative to that of the Gilman soil, its lower respiration may be related to the soil's very high sodium level that, in conjunction with the osmotic effect induced by any level of N augmentation appeared to completely inhibit this soil's already low microbial activity (Table 2). The higher respiration rate at the lower ψ of the Grabe 2 control treatment relative to that of the Casa Grande soil may be attributed to its high level of inorganic N (148 mg/kg N) relative to the low N (2.7 mg/kg N) naturally present in the Casa Grande soil (Table 1).

Greatest respiration was almost always seen in the 250 mg/kg N treatments of all N sources (Table 2), with the highest rates correlating to highest water potentials and to the lowest recoverable inorganic N of the N fertilized soils(data not show). In the 250 mg/kg N-NH_4NO_3 treatments of the Casa Grande, Grabe 2, and Gilman soils, respiration rates were 90, 90 and 20 mg O_2/kg soil per day respectively, with ψ of -1.06, -2.29 and -9.01 MPa (Table 1) respectively. Although the Grabe 2 soil had less than the recommended soil moisture content (46% of field capacity, Table1) as well as a lower ψ relative to the Casa Grande soil, its clay texture allowed for greater water retention resulting in less osmotic stress induced via N fertilization (slight decrease in ψ of -0.03 MPa from the control to the 250 mg/kg N treatment relative to -0.22 MPa in the Casa Grande soil, Table 1) resulting in the favorable degradation rate observed. The highest respiration in the Gilman soil was found in the 500 mg/kg N treatment at 25 mg O_2/kg soil per day. However, the 250 mg/kg N treatment was nearly identical for most of the duration of the experiment.

Amendment of variable N sources (NH_4Cl and $NaNO_3$) demonstrated microbial preference for the reduced NH_4^+ over that of the oxidized NO_3^- form. At all levels of application, NO_3^- amended soils consumed 65-70% as much O_2 as those treated with NH_4^+ (Table 2). Again, maximum respiration occurred in the 250 mg/kg treatments with rates of 110 and 77 mg O_2/kg soil per day correlating with the NH_4Cl and $NaNO_3$ amendments respectively (Table 2).

Regardless of the N source, the 1000 mg/kg N treatments demonstrated lower respiration rates relative to the control treatments and therefore appeared to inhibit aerobic petroleum degradation (Table 2). More specifically, addition of 1000 mg/kg N reduced the respiration rates to between ½ to ⅓ the maximum amount (50%, 42%, 38%, 35 %, and 33% of the maximum respiration in the Grabe 2, Casa Grande-NH_4NO_3,Gilman,Casa Grande-NH_4Cl,and Casa Grande-$NaNO_3$ soils respectively). Greatest reduction in respiration was observed in soils which demonstrated greater reductions in ψ (Casa Grande-NH_4Cl and Casa Grande-$NaNO_3$ soils respectively, Table 1 and 2). Respiration was least inhibited in the Grabe 2 soil which also had a lower reduction in ψ, most likely due to the soil's fine texture and increased ability to retain water.

CONCLUSION

Regardless of initial salinity and ψ, control treatments of all soils under study demonstrated some microbial activity. Generally, greatest respiration rates correlated with highest ψ and natural abundance of inorganic N. Nitrogen fertilization increased respiration in all but one soil where the osmotic stress induced by N augmentation in conjunction with the soil's sodic condition resulted in nearly complete microbial inhibition. Maximum respiration was observed in the 250 mg/kg N treatments regardless of the N source used. Again, highest respiration rates at this level of fertilization correlated with highest ψ among fertilized soils. Microbial preference for NH_4^+ over NO_3^- was observed at all levels of application, with NO_3^- amended soils consuming only 65-70% as much O_2 as those treated with NH_4^+. Addition of any source of N at 1000 mg/kg reduced respiration rates between ½ to ⅓ of the maximum rate observed. This rate reduction was related to the lower ψ attributed to higher N amendments. Fertilizer chemistry played a role as the dissolution of counter ions in addition to N salts contributed to lower ψ, resulting in lower observed respiration at higher N treatments. N induced reduction in respiration was not related to initial soil salinity indicating that native microbial populations were equally as sensitive to the decrease in ψ attributable to fertilization, regardless of initial soil conditions.

Recoverable inorganic nitrogen (NH_4^+ and NO_3^-, data not shown) confirmed that increased microbial respiration was indicative of diesel degradation and not nitrification. For example, in the 250 mg/kg N (NH_4NO_3) treatments of the Grabe 2, Casa Grande, and Gilman soils, only 6.4, 6.6, and 15.3% of the maximum O_2 consumed, respectively, could be from nitrification, assuming *all* ammonium loss during the experiments was due to nitrification. In the 250 mg/kg N (NH_4Cl) treatment of the Casa Grande soil, only 8.9% of the O_2 consumed could be attributed to nitrification. Lastly, the Casa Grande soil that received $NaNO_3$ treatments (in which no nitrification could have occurred) demonstrated identical trends in O_2 consumption, i.e. maximum and inhibitory respiration rates were observed in the 250 and 1000 mg/kg N soil treatments respectively.

Previous work done in humid soils (Braddock et al., 1997; Walworth et al., 1997 a,b) demonstrated highest respiration rates at total water potential of -0.3 to -0.4 MPa, with 50% reduction in respiration at a -0.4 MPa decrease in ψ. This study showed greatest respiration at -1.06 to -9.01 MPa, with a 50 % reduction in respiration attributable to a -1.13 MPa decrease in ψ. Therefore, the results of this study suggest greater tolerance by microbial populations of arid soils to osmotic stress resulting from N fertilization.

The dissolution of N salts, as well as counter ions present in N fertilizers, into the soil solution results in a decrease in the osmotic component of the total water potential. Although this component is not usually considered in the assessment of N augmentation, it has a significant effect on microbial populations, potentially affecting their locomotion, removal of metabolic by-products, as well as solute and substrate diffusion and supply. Soil texture is of key importance in N augmentation as the higher the water retention of the soil, the less osmotic stress induced by N supplementation because fertilizer N dissolves into a greater volume of water.

REFERENCES

Alexander, M. 1999. *Biodegradation and Bioremediation*, 2nd edition. Academic Press, Inc., Sand Diego, CA.

Atlas, R.M., and R. Bartha. 1972. "Degradation and Mineralization of Petroleum in Sea Water: Limitation by Nitrogen and Phosphorous." *Biotechnology and Bioengineering. 14*:309-318.

Braddock, J.F., M.L. Ruth, P.H. Catterall, J.L.Walworth, and K.A. McCarthy. 1997. "Enhancement and Inhibition of Microbial Activity in Hydrocarbon-Contaminated Arctic Soils: Implications for Nutrient-Amendd Bioremediation." *Environmental Science and Technology. 31*(7): 2078-2084.

Bragg, J.R., R.C. Prince, E.J. Harner, and R.M. Atlas. 1994. "Effectiveness of

Bioremediation for the Exxon Valdez Oil Spill." *Nature*. *368*: 413-418.

Cason, T. 1994. "Nitrogen Supply Key to Degradation." *Soils*. May: 8-13.

Cleland, D.D., V.H. Smith, and D.W. Graham. 1997. "Microbial Population Dynamics During Hydrocarbon Biodegradation Under Variable Nutrient Conditions."In B. Alleman and A Leeson (Eds.), *In Situ and On-Site Bioremediation*, pp. 105-110. Vol. 4. Battelle Press, Columbus, Ohio.

Dibble, J.T., and R. Bartha. 1979. "Effect of Environmental Parameters on the Biodegradation of Oil Sludge." *Applied Environmental Microbiology*. *37*: 729-739.

Frankenberger, W.T. 1991. "The Need for a Laboratory Feasibility Study in Bioremediation of Petroleum Hydrocarbons." In P.T. Calabrese and Kostecki (Eds.), *Hydrocarbon Contaminated Soil and Groundwater, Volume II*. Lewis Publishers, Boca Raton, FL.

Graham, D.W., V.H. Smith, and K.P. Law. 1995. "Application of Variable Nutrient Supplies to Optimize Hydrocarbon Biodegradation." In R.E. Hinchee, D.B. Anderson, and R.E. Hoeppel (Eds.), *Bioremediation of Recalcitrant Organics*, pp.331-340. Battelle Press, Columbus, Ohio.

Haines, J.R., M. Kadkhodayan, D.J. Mocsny, C.A. Jones, M. Islam, and A.D. Venosa. 1994. "Effect of Salinity, Oil Type, and Incubation Temperature on Oil Degradation." In R.E. Hinchee, D.B. Anderson, Jr, F. B. Metting, and G.D. Sayles, (Eds.), *Applied Biotechnology for Site Remediation*, pp. 75-83. Lewis Publishers, Boca Raton.

Lewandowski, G.A., and L.J. DeFilippi. 1998. *Biological Treatment of Hazardous Wastes*, pp. 87, 248-250, 366-367. John Wiley & Sons, Inc. New York.

Litchfield, C.D., 1993. *In Situ Bioremediation: Basis and Practices*. Levin, M.A., Gealt, M.A.,(Eds.), pp. 167-196. McGraw-Hill, New York.

McCarty, P.L., 1972. "Stoichiometry of Biological Reactions", International Conference: *Toward a Unified Concept of Biological Waste Treatments Design*, Atlanta, GA.

Paul, E.A., and F.G. Clark. 1989. *Soil Microbiology and Biochemistry*. Academic Press, San Diego.

Sims, J.L., R.C. Sims, and J.E. Mathews. 1989. *Bioremediation of Contaminated Surface Soils*. EPA 600/9-89/073. United States Environmental Protection Agency.

Sims, J.L., Sims, R.C., DuPont, R.R., Matthews, J.E., & Russell, H.H. 1993. *In Situ*

Bioremediation of Contaminated Unsaturated Subsurface Soils. U.S. Environmental Protection Agency, Office of Solid Wast and Emergency Response, Washington D.C.

USEPA 1993. *Selecting Remediation Techniques for Contaminated Sediment*, Office of Water, EPA 823-B93-001, June, 1993.

Walworth, J.L., C.R. Woolard, J.F. Braddock, and C.M. Reynolds, 1997a. "Enhancement and Inhibition of Soil Petroleum Biodegradation through the use of Fertilizer Nitrogen: An Approach to Determining Optimal Levels." *Journal of Soil Contamination. 6*(5): 465-480.

Walworth, J.L., C.R. Woolard, J.F. Braddock, and C.M. Reynolds. 1997b. "The Role of Soil Nitrogen Concentration in Bioremediation." *Bioremediation 4*(4): 283-288.

Zhou, E., and R.L. Crawford. 1995. "Effects of Oxygen, Nitrogen, and Temperature on Gasoline Biodegradation in Soil." *Biodegradation. 6*: 127-140.

ELECTROKINETIC MOVEMENT OF BIOLOGICAL AMENDMENTS THROUGH NATURAL SOILS TO ENHANCE IN-SITU BIOREMEDIATION

David Gent, R. Mark Bricka,
US Army Engineer Research and Development Center, Vicksburg, MS
Dr. Dennis D. Truax and Dr. Mark E. Zappi
Mississippi State University, Starkville, MS

ABSTRACT: Electrokinetics is an in-situ electrochemical process capable of transporting ions through sand as well as low permeable soils. Ionic nutrients can be amended near each electrode and transported into the soil by ion migration and electroosmosis. This study investigated and demonstrated the feasibility of nutrient and carbon source transport in soils by electrokinetic for potential *in-situ* bioremediation.

A series of bench scale column studies were conducted to assess the migration of acetate and nitrate amendments in a simulated *in-situ* environment. Electrolysis reactions at the electrodes result in changes in hydrogen and hydroxide ion concentrations. Strategies for managing shifts in pH using acid and base solutions at the cathode and anode in the subsurface were investigated. Experimental variables included two current densities, three soil types, and two amendments at varying concentrations.

The hydraulic conductivity of the sand, silt, and clay were 43.2 m/day ($5x10^{-2}$ cm/s), $8.2x10^{-5}$ m/day ($9.5x10^{-8}$ cm/s), and $8.4x10^{-7}$ m/day ($1x10^{-9}$ cm/s) respectively. The linear transport rates of amendments in the sand were between 60-80 m/day. The linear transport rates of amendments in the silt and clay were between 1.1-2.5 m/day and 0.9-2.5 m/day respectively.

The results indicate that electrokinetic induced ion migration can be used to enhance amendment flux relative to hydraulic flow characteristics though various subsurface soils in substantial concentrations. This study demonstrated that electrokinetics has the potential to increase the availability of amendments in soil pore water and groundwater systems where traditional injection techniques have been incomplete or ineffective.

INTRODUCTION

Technologies used for cleanup and contaminant removal belong in two primary categories, physical/chemical technologies and biological treatment technologies. Most physical cleanup technologies involve *ex-situ* processes such as physical separation, incineration, air stripping, activated carbon adsorption, and advanced oxidation. Many of these technologies can be generally categorized as pump-and-treat, solidification and landfill disposal.

The problem with most of these treatment schemes is their high cost and the physical removal of contaminated soils. Pump and treat injection systems can become clogged or fouled with biomass and by the precipitation of minerals. For *in-situ* biological treatment to be sustainable the hydraulic conductivity of the groundwater system must be approximately 10^{-4} cm/s (100 ft/yr) or greater (Norris et al., 1993). Currently it is difficult to pump and treat fine-grained soils. Contaminated silts and clays with low hydraulic conductivity pose a significant problem to any type of *in-situ* remediation. Hydrofracturing the soil creates macropores where the soil in the immediate vicinity to the macropores can be treated but the soil that is not fractured cannot be treated. Many organic chemicals become bound to the soil and are hydrophobic or have a low affinity to the liquid phase. The concentration of these chemicals may be in such low concentrations, that they may not serve as an adequate carbon source by themselves.

Electrokinetics (EK) may be an effective and efficient way to move nutrients into aquifers, contaminated fine grain soils, loose-grained soils, as well as unsaturated soils to facilitate the biodegradation of the contaminant *in-situ*. EK offers a different approach to *in-situ* bioremediation since a hydraulic head is not needed to move contaminants. Electrokinetics may also be capable of transporting essential elements for microbial activity into contaminated layers. Ionic nutrients can be amended to the soil near each electrode and transported into the soil pore water by ion migration and electro-osmosis.

MATERIAL AND METHODS

A bench-scale system was developed to facilitate the transport of anion amendments to determine the feasibility of electrokinetic transport with no influence of hydraulic head. The experimental concept is shown in Figure 1.

Electrokinetics Bench System. The EK system consisted of a data acquisition system, 300V-3.5A DC power supply, multi-channel peristaltic pump for recirculation with Tygon® pump tubing, optical prism flow sensors, 2 mixers, 2 digital ph controllers with electrodes, 2 pH control tanks, 2 digital pH transmitters, 1 acrylic EK cells with 2 resin impregnated carbon electrodes (Figure 2).

EK cells were made from clear acrylic sheets with interior dimensions approximately 2 inches (5.3 cm) wide, 6 inches (15.24 cm) in height and 15 ¼ inches (37.8 cm) in length. A solid cap was attached to one end of the cell using ¼"-20 stainless steel screws with ⅛ inch (0.3175 cm) neoprene gaskets and 30-45 micron filter paper placed on each end. The nutrient analysis method selected was USEPA SW846 Method 300A using a Dionex DX500 Ion Chromatograph.

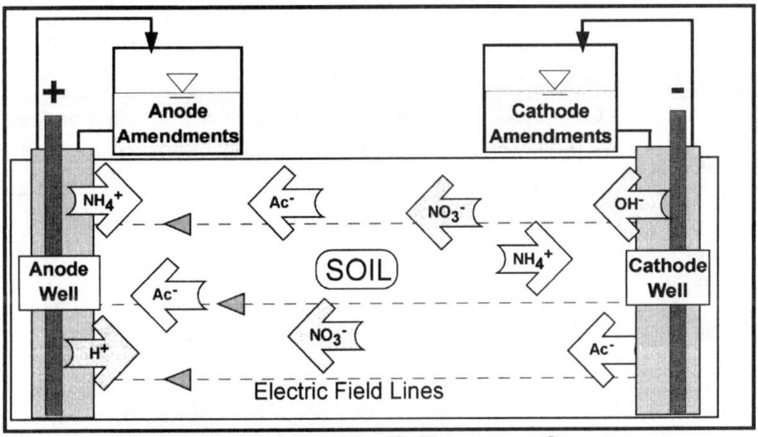

FIGURE 1. Electrokinetic Transport Concept.

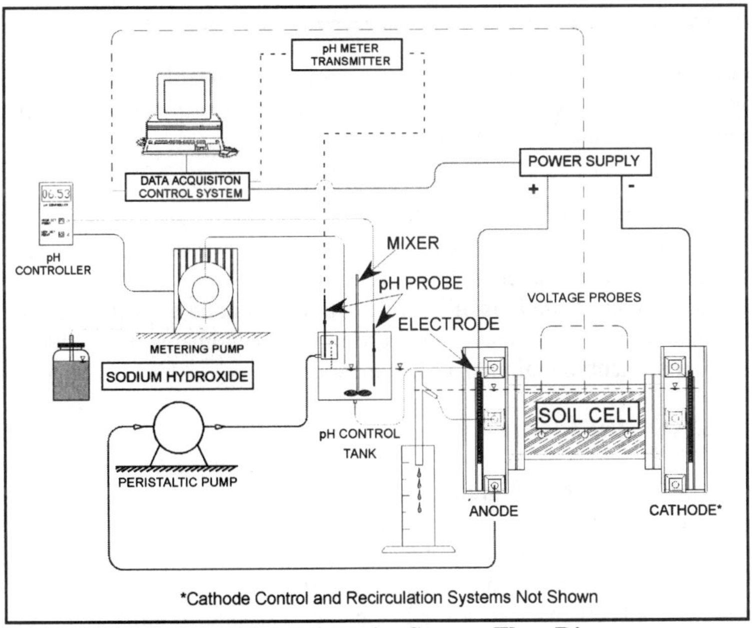

FIGURE 2. Electrokinetics System Flow Diagram.

Soils. This study used natural soils to better simulate actual field conditions. The soils used for these series of tests, sand, silt, and clay. They were: Alligator Clay, WES Reference (silt), and Ottawa Sand. Test Parameters for all tests are listed in Table 1.

TABLE 1. Summary of EK Test Parameters

	Ottawa Sand			WES Reference				Alligator Clay			
Parameter	*TEST A*	*TEST B*	*TEST C*	*TEST A*	*TEST B*	*TEST C*	*TEST D*	*TEST A*	*TEST B*	*TEST C*	*TEST D*
Current applied (mA)	4.1	4.1	9	4.6	5.0	10.0	4.0	4.6	4.9	10.0	4.0
Duration of Test (days)	7.8	7.8	7.8	36.1	23.2	27.7	16.0	42.5	23.2	28.4	19.0
Soil Classification (UCS)	SP	SP	SP	CH	CH	CH	CH	CL	CL	CL	CL
%Sand	97.6	97.6	97.6	0.5	0.5	0.5	0.5	2.8	2.8	2.8	2.8
% Fines	2.4	2.4	2.4	99.5	99.5	99.5	99.5	97.2	97.2	97.2	97.2
Hydraulic conductivity (cm/day)	175	175	175	8.2E-03	8.2E-03	8.2E-03	8.2E-03	8.6E-05	8.6E-05	8.6E-05	8.6E-05

Amendments. Amendments to the system must depolarize the electrode reactions in the anode and cathode electrode compartments. When inert electrodes are used, the following electrolysis reaction occurs.

$$2H_2O - 4e^- \rightarrow O_2(g) + 4H^+ \qquad E^o = -1.229 \quad \text{(anode)} \qquad (1)$$

$$2H_2O + 2e^- \rightarrow H_2(g) + 2OH^- \qquad E^o = -0.8277 \quad \text{(cathode).} \qquad (2)$$

The cathode amendments selected for this study were nitric acid and acetic acid while the anode amendment was sodium hydroxide. These cathode amendments contained the ions of interest, while the anode amendment was used just to control the pH of the anode compartment. Amendments were added to the cathode compartment to maintain a constant source.

RESULTS AND DICUSSION

Ottawa Sand. The nitrate migrated from the cathode to the anode in less than 24 hours and reached 100% of the cathode concentration in 120 hours (5 days) (Figure 3). The nitrate amendment was added daily to the cathode electrode compartment to maintain a constant concetration of 80 mg/L. The linear EK movement rate was 60 cm/day for the low power tests and 80 cm/day for the high power test.

Silt and Clay. The clay and silt tests were ran under varied conditions to determine if EK transport was feasible (Table 1). The WES Reference soil (silt) and the Alligator Clay anode and cathode fluid results were similar. Representative results are shown in Figure 4. Since the duration of these low hydraulic permeable soils were much longer than the Ottawa sand test, the pH was automatically adjusted with the acetate amendment by means of a digital pH controller. The breakthrough of acetate or nitrate was not achieved after 520 hours (22 days) of testing. Figure 4 illustrates the continuing increase in acetate concentration in the cathode compartment while the nitrate concentration was maintained relatively steady. The increase in acetate concentration was due to its use as a pH control solution. The presence of nitrite and ammonium in the cathode indicates electrolytic reduction of nitrate. This reduction must be taken into account, when determine power and nutrient requirements.

Silt and Clay Pore Water Results. Nutrient analysis of the pore water along the soil core was required to determine nutrient migration. The pore water was extracted from each soil at the end of each test then centrifuged to separate the pore water from the soil.

The supernate was decanted into sample vials and immediately analyzed for anions by Ion Chromatography (IC). Figure 5 is representative of the overall soil pore water results. Figure 6 illustrates the acetate results for all tests. Acetate migrated throughout all soil sections from cathode to anode. However there was little evidence of this migration in the anode electrode compartment fluid. The nitrate migration was a different matter entirely. In the tests with 100 mg/L of nitrate maintained in the cathode electrode compartment, there was little evidence of nitrate migration into the soil. However, there was evidence of nitrate reduction. The linear EK movement rate for the silt and clay tests ranged from 1.1 to 4.1 cm/day and 0.9 to 2.1 cm/day respectively.

CONCLUSIONS

Ottawa Sand. Nitrate under an electric potential moved though the Ottawa Sand with relative ease and with minimal reduction of the nitrate to other nitrogen species in the cathode. The nitrate breakthrough time to the anode electrode compartment was less than 16 hours for all three Ottawa sand test. The nitrate anode compartment concentration reached 50% of the cathode compartment nitrate concentration, 100 mg/L NO_3^-, in less than 90 hours for all three tests. The linear rate of ion migration for the sand 60-80 cm/day was lower than the hydraulic conductivity at the power applied.

Silt and Clay. The total amount of pore water nitrate decreased in every test except the one with 1000 mg/L nitrate was added to the cathode. The amount of nitrate reduction, to nitrogen gases is unknown. The test with 1000 mg/L nitrate added at the cathode demonstrates that significant amounts of nitrate can migrate through this soil under an electric field (Figure 5). Increased acetate concentrations were found in all soil sections for the silt and clay tests. Only trace amounts of acetate were found in the anode compartment for all tests. It was assumed that the acetate ion decarboxylates into methyl radicals (CH3•) and carbon dioxide at the anode electrode. There was no significant nitrate or nitrogen ions concentration in the anode compartment during any silt or clay EK test. The test with 1000 mg/L nitrate added at the cathode supports the hypothesis that nitrate can be moved into a clay soil by electromigration (Figure 5).

The linear rate of acetate ion migration for the silt and clay was 3 to 4 orders of magnitude higher than the hydraulic conductivity for silt and clay respectively. This electrokinetic feasibility study demonstrates that ion migration can serve to transport amendments for bioremediation where hydraulic flow is impractical.

These results demonstrate that electrochemical movement of organic acid anions can be engineered under single-phase dc fields to stimulate *in-situ* remediation of environmental contaminates in heterogeneous aquifers and leaky aquitards.

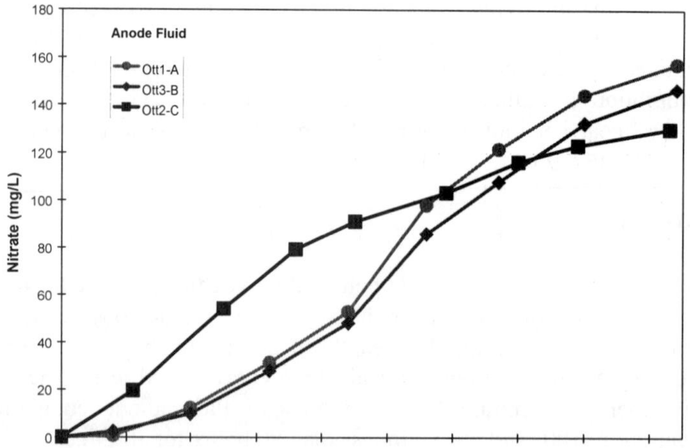

FIGURE 3. Anode Fluid Nitrate Concentration (Ottawa Sand Tests).

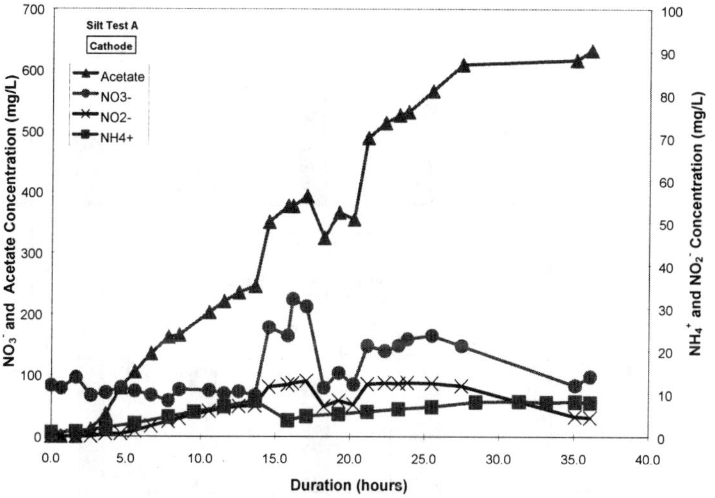

FIGURE 4. Cathode Acetate and Nitrogen Ion Concentrations

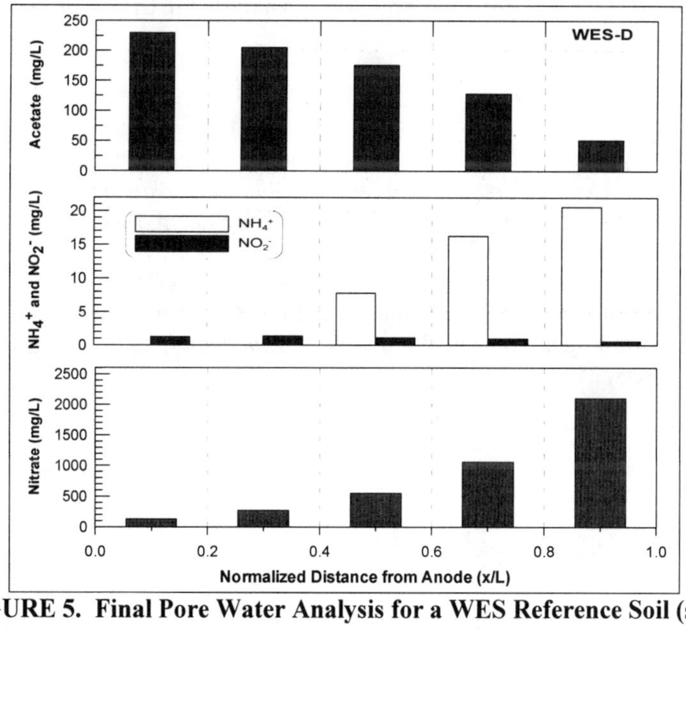

FIGURE 5. Final Pore Water Analysis for a WES Reference Soil (silt).

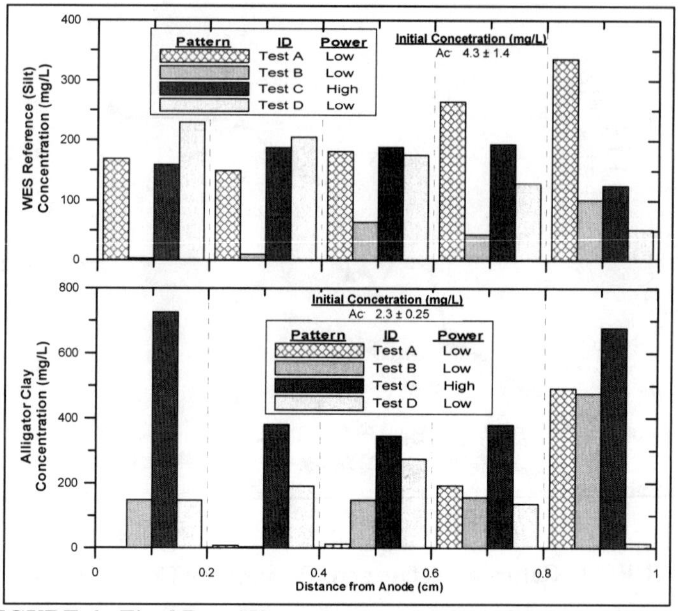

FIGURE 6. Final Pore Water Acetate Analysis for Clay and Silt.

REFERENCES

Acar, Y. B., S. Puppala, R. Marks, R. J. Gale, and M. Bricka, (1993). *An Investigation of Selected Enhancement Techniques in Electrokinetic Remediation*, Report presented to US Army Waterways Experiment Station, Electrokinetics Inc., Baton Rouge, Louisiana, 230p

Norris, Hinche, Brown, McCarty, Simprini, Wilson, Kampbell, Reinhard, Bouwer, Borden, Vogel, Thomas, Ward (1993). *Handbook of Bioremediation*, Lewis Publishers, 257p.

BIOREMEDIATION OF PERCHLORATE IN GROUNDWATER AND REVERSE OSMOSIS REJECTATES

M. E. Losi, Vitthal Hosangadi, and David Tietje
(Foster Wheeler Environmental Corporation, Santa Ana, CA)
Tara Giblin and William T. Frankenberger, Jr.
(University of Calif., Riverside, CA)

ABSTRACT: Cost-effective techniques are being investigated for removal of perchlorate (ClO_4^-) from groundwater and various process wastes. In this study, we assessed the efficacy of a packed bed biological reactor (PBR) to remove ClO_4^- from groundwater and from reverse osmosis (RO) process wastes (rejectates). The data showed that the PBR system was capable of treating groundwater containing ClO_4^- concentrations of approximately 800 μg L^{-1} to non-detectable concentrations (<4 μg L^{-1}), with a residence time as low as 0.3 hr. ClO_4^- concentrations were also reduced in RO rejectate from approximately 5.0 mg L^{-1} to <4.0 μg L^{-1} (residence time of 0.8 hr), and in secondary RO rejectate from approximately 10 mg L^{-1} to 0.2 mg L^{-1} (residence time of 2.1 hr). Nitrate was also removed from all three feeds by this system, while sulfate reduction was not observed. This study demonstrated the potential feasibility of the PBR system for removing ClO_4^- from groundwater and from RO process waste.

INTRODUCTION

Perchlorate (ClO_4^-), a component of rocket fuel, explosives and other pyrotechnics, has recently been identified as a groundwater contaminant in the United States, and is currently impacting a number of drinking water aquifers. An interim action level (IAL) of 18 μg L^{-1} has been established by the California Department of Health Services (CDHS), which is currently observed by most states. Biotreatment and reverse osmosis (RO) are recognized as potentially applicable technologies for removing ClO_4^- from contaminated groundwater (Urbansky, 1998). Biotreatment destroys ClO_4^- via biological reduction in the presence of a suitable electron donor (carbon substrate), thus converting it to chloride and oxygen, which are relatively innocuous. Biotreatment is potentially the most economical technique for treating ClO_4^- contaminated groundwater, however questions exist regarding potability of biotreated water (Urbansky, 1998). In contrast, RO is capable of producing potable effluent, but does not destroy ClO_4^-. Rather, it is collected in a concentrated waste stream (rejectate), which must be further treated prior to disposal.

A packed bed reactor (PBR) is a type of fixed film bioreactor, which is potentially applicable for treating low concentrations of ClO_4^- in groundwater (Logan, 1999). This technology requires relatively minimal equipment, and has the potential to achieve reasonably short residence times, thus minimizing the footprint needed for a full-scale system. We conducted studies to assess the efficacy of a bench-scale, acetate-fed, PBR to treat low concentrations of ClO_4^- (approximately 800 μg L^{-1}) in groundwater collected from an impacted site, and higher ClO_4^- concentrations in simulated RO rejectates.

Objective. The objective of this work was to evaluate achievable effluent ClO_4^- levels, and to estimate residence times necessary for optimal ClO_4^- removal from three treatment streams: 1) actual groundwater collected from an impacted site (containing <1 mg ClO_4^- L^{-1}); 2) simulated primary RO rejectate resulting from treatment of the groundwater; and 3) simulated secondary RO rejectate, resulting from RO treatment of the primary rejectate. Secondary goals included determining the fate of nitrate (NO_3^-) and sulfate (SO_4^{2-}), which were also present in the influent feeds.

MATERIALS AND METHODS

Packed Bed Reactor Specifications. The PBR consisted of a cylindrical Plexiglas column containing Celite (World Minerals Corp., Lompoc, CA), a diatomaceous earth product, as a solid support medium. Column specifications are given in Table 1. A Master Flex 4S (Cole Parmer, Vernon Hills, IL) peristaltic pump was used to deliver the start-up medium and influent feeds to the reactor.

TABLE 1. Packed bed bioreactor column specifications.

Parameter	Specification
Inside Diameter	13.5 cm
Total Height	21.4 cm
Bed height	12.5 cm
Total volume	3062 mL
Bed volume	1789 mL
Pore volume	1236 mL
Packing material	Celite (R-635)

Inoculation/Start-up. The inoculum consisted of a ClO_4^--reducing bacterial isolate, perc1ace, previously characterized by Herman and Frankenberger (1999). Batches of this isolate were grown in flasks containing Celite and autoclaved FTW (minimal salts) medium (Herman and Frankenberger, 1999) supplemented with ClO_4^- (500 mg L^{-1}) and acetate (1 g L^{-1}) to allow biomass to attach to the Celite. The Celite/perc1ace mixture was poured into the column, and FTW medium containing ClO_4^- and acetate was circulated through the column for two weeks to stabilize the culture on the support medium.

Operation. Following start-up, influent feed was introduced in an up-flow mode. Three influent feeds were tested, including groundwater, simulated primary RO rejectate, and simulated secondary RO rejectate (RO2), from RO treatment of the primary rejectate. Compositions of the groundwater and simulated reverse osmosis rejectates are given in Table 2. The rejectates were formulated based on preliminary RO testing data. These data indicated that initial concentration of the groundwater by a factor of 5, and secondary concentration of the rejectate from the initial process by a factor of 2 yielded permeates with ClO_4^- levels near or below the IAL (18 μg L^{-1}). Simulated rejectate feeds were therefore formulated to contain (as closely as possible) 5 (RO) and 10 (RO2) times the ion concentrations present in the original groundwater (based on TDS [total dissolved solids] and ClO_4^- analytical data from the RO tests). Note that in formulating the rejectates, additions of Ca^{2+} and SO_4^{2-} additions were minimized in an effort to avoid precipitation, and Na^+ and Cl^- concentrations were increased accordingly to compensate for the corresponding loss of TDS.

TABLE 2. Compositions of the groundwater and simulated reverse osmosis rejectates used as influents in the PBR study.

Analyte	Average concentrations, groundwater[1] (mg L^{-1})	Simulated RO rejectate (mg L^{-1})	Simulated RO2 rejectate (mg L^{-1})
Cl^-	24	336	672
ClO_4^-	0.8	5	10
NO_3^--N	14	71	142
Mg^{2+}	19	95	190
SO_4^{2-}	27	27	54
HCO_3^-	155	775	1550
CO_3^{2-}	0.3	5	10
Ca^{2+}	50	25	50
K^+	2.5	15	30
Na^+	23	427	853
Fe^{2+}	0.1	0.5	1.0
TDS	318.3 (measured)	2025 (calculated)	4050 (calculated)

[1]Averages of concentrations measured over several years of monitoring.

Each of the three tests lasted approximately 4 weeks. Acetate (as sodium acetate) was added as the energy source via an in-line feed pump at a relatively high target concentration of 0.5 g L^{-1} (stoichiometric calculations for reduction of O_2, NO_3^-, and ClO_4^- by acetate at representative levels indicate that less acetate is needed). The rationale for this was that actual acetate utilization rates were not known, and in order to estimate residence times for optimal ClO_4^- removal, it was necessary that acetate not be limiting in the system. Minor amounts of ammonium and phosphate (as NH_4Cl and KH_2PO_4) were also added as nutrient supplements. The general approach used was to begin with a residence time that yielded non-detectable effluent ClO_4^- concentrations, and incrementally increase the flow rate, thereby decreasing the residence time, until ClO_4^- breakthrough occurred. Flow rates of 5, 10, 25, 50, and 75 mL min^{-1}, corresponding with residence times of 4.2, 2.1, 0.8, 0.4, and 0.3 hr, were evaluated during various portions of the three tests. NO_3^-, SO_4^{2-} and acetate were also tracked analytically in various tests.

Analytical. Influent and effluent ClO_4^- concentrations were analyzed by ion chromatography (IC) in accordance with EPA Method 300.0, modified (detection limit of 4 μg ClO_4^- L^{-1}). NO_3^-, SO_4^{2-}, and acetate were also analyzed by IC (detection limit for NO_3^-, SO_4^{2-} and acetate of 0.2 mg L^{-1}).

RESULTS AND DISCUSSION

Treatment of Groundwater. Influent and effluent ClO_4^- concentrations for the various residence times over the duration of the groundwater experiment are shown in Fig. 1. Influent ClO_4^- concentrations were generally measured between 700 and 800 μg L^{-1}. Data from acetate analysis (not shown) indicated that acetate

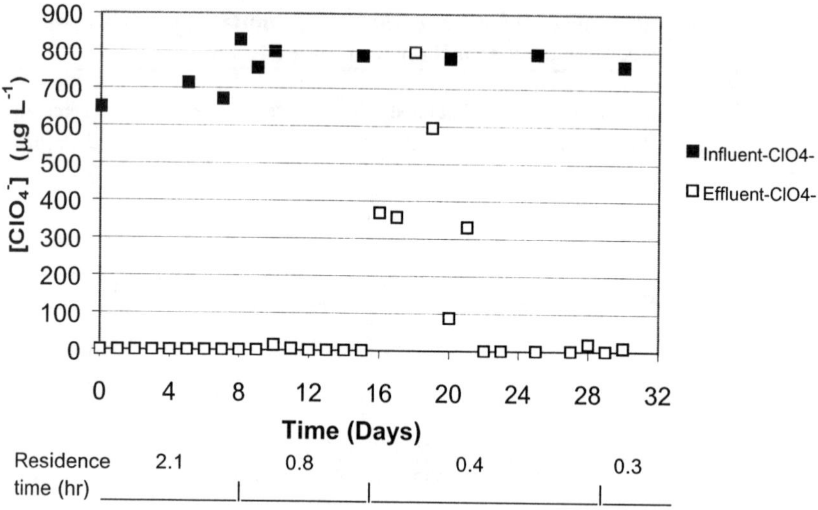

FIGURE 1. Influent and effluent perchlorate concentration measured over time during groundwater PBR study.

was present in the system in excess (present in the effluent) throughout the experiment, except in one instance (noted below).

The first 15 days of the experiment evaluated residence times of 2.1 and 0.8 hours. Effluent ClO_4^- concentrations were non-detect over this period with the exception of days 10 and 11, where the ClO_4^- levels were 13.6 and 4.3 μg L^{-1}, respectively. Based on these results, the flow rate was increased incrementally, to evaluate residence times of 0.4 and 0.3 hours. For the remainder of the experiment, effluent ClO_4^- concentrations were below, or slightly above, the detection limit, with the exception of days 16-21, during which the supply of acetate was interrupted due to a pump failure. When acetate was re-supplied to the system (day 18), ClO_4^- concentrations dropped rapidly and by day 22, were consistently below the detection limit. Although inadvertent, this verified that the observed ClO_4^--removal was coupled with utilization of acetate.

On day 27, the flow rate was again increased, which decreased the residence time to 0.3 hr. This flow rate was maintained for only 3 days, due to imminent exhaustion of the groundwater. Over this period, effluent ClO_4^-

concentrations remained very low (non-detect or slightly above). At this point, the experiment was terminated due to exhaustion of the influent water supply.

Much of the previously published work on ClO_4^- PBR systems has focused on concentrations well above the 1 mg L^{-1} range, and relatively little is known regarding treatment of water containing ClO_4^- at levels below 1 mg L^{-1}. For example, other PBR studies have demonstrated effective removal of ClO_4^- from relatively high level feeds: approximately 35 mg ClO_4^- L^{-1} in synthetic feed (Logan et al., 1999); and 500-1500 mg ClO_4^- L^{-1} in process waste (Wallace et al., 1998). Data presented here supports results of previous and concurrent experimentation in other laboratories, and supplements these other studies by demonstrating that non-detectable ClO_4^- concentrations can also be achieved with comparable residence times, but much lower influent ClO_4^- levels.

Influent and effluent NO_3^- and SO_4^{2-} levels measured over the duration of the groundwater experiment are presented in Figure 2.

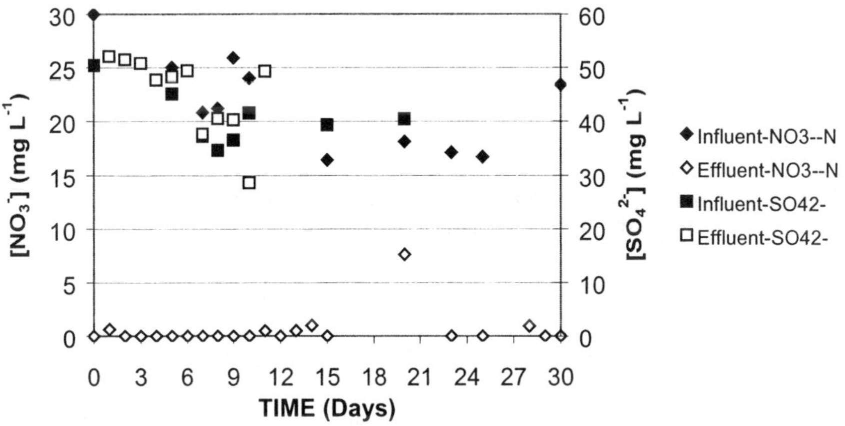

FIGURE 2. Influent and effluent nitrate and sulfate concentrations measured over time during groundwater PBR study.

As shown here, significant NO_3^- removal was observed with the exception of day 20, during which (as noted above) the system was recovering from "upset" caused by the absence of acetate. Data in Figure 2 also indicate that reduction of SO_4^{2-} did not occur in the reactor. Reduction of SO_4^{2-} would be undesirable with regard to ClO_4^- treatment systems because it generates hydrogen sulfide (H_2S), which is an irritant and this would trigger various air-quality and effluent discharge regulations. It is also noted that the characteristic smell of H_2S was not encountered during any of the experiments.

Treatment of Primary RO Rejectate (RO). Influent and effluent ClO_4^- concentrations for the RO rejectate experiment are shown in Figure 3.

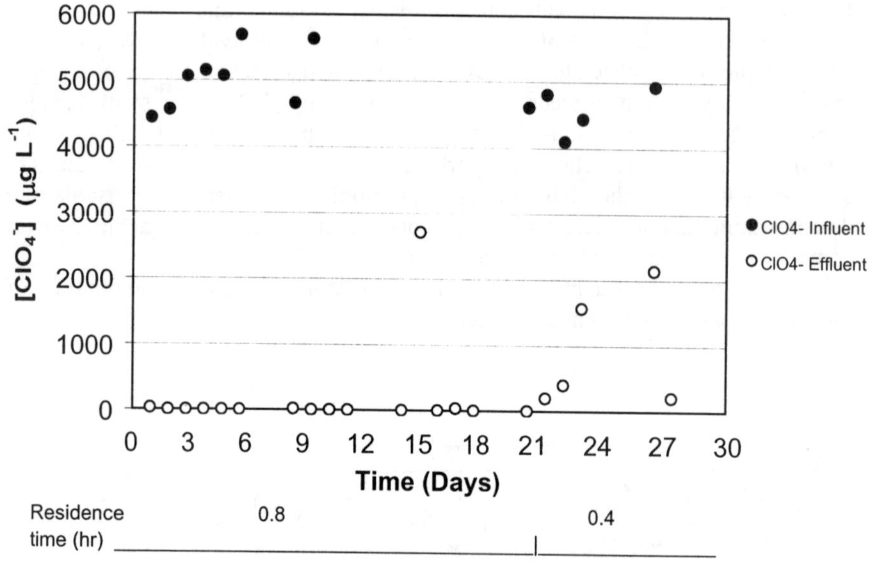

FIGURE 3. Influent and effluent perchlorate concentrations measured over time during primary RO rejectate PBR study.

As with the groundwater experiment, acetate was supplied in excess, and was not limiting in the reactor. Data presented in Figure 3 indicate that removal of ClO_4^- from simulated RO rejectate was almost complete with initial concentrations of approximately 5000 μg L^{-1} at a residence time of 0.8 hr. When the residence time was reduced to 0.4 hr, on day 22 of the operation, ClO_4^- breakthrough was observed. NO_3^- was not detected in the effluent at a residence time of 0.8 hr, but breakthrough was observed when the residence time was reduced to 0.4 hr (data not shown).

This residence time is greater than that required to bring about complete removal of ClO_4^- and NO_3^- from groundwater. The decreased efficiency compared with the groundwater experiment was likely due to the high level of concentrated salts, which generally increases the metabolic demands of the bacteria (leading to slower growth), in conjunction with increased levels of NO_3^- and ClO_4^- requiring reduction.

Treatment of Secondary RO Rejectate (RO2). The bioreactor efficiency was tested for removal of ClO_4^- from a secondary rejectate stream (RO2) resulting from further concentration of the primary RO rejectate (to minimize waste). The same initial residence time used in the RO study (0.8 hr) was initially tested, on days 0-11. Data from the RO2 experiment are presented in Figure 4. Data in Figure 4 suggest that the column required approximately five days to acclimate to

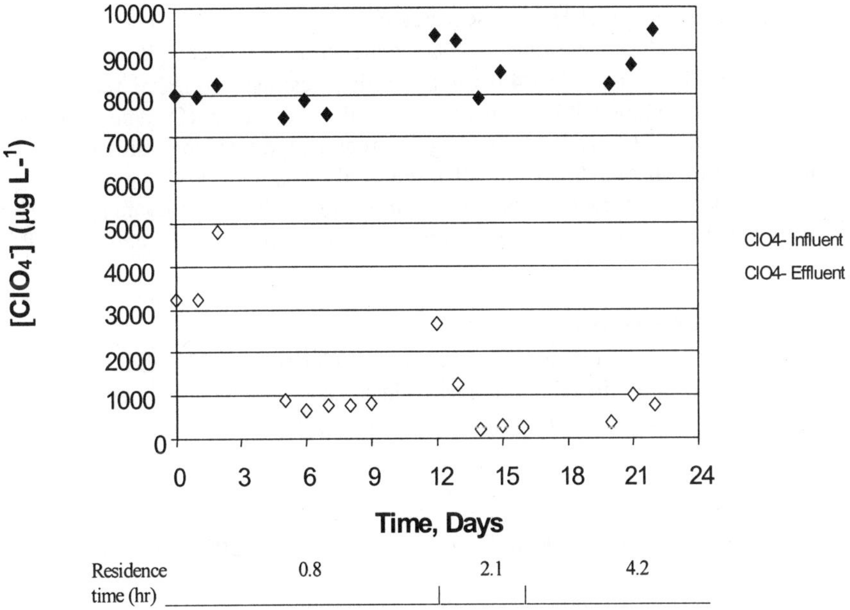

FIGURE 4. Influent and effluent perchlorate concentrations measured over time during secondary RO rejectate (RO2) PBR study.

the RO2 feed, probably due to the higher TDS content of the water.

On day 5, 87% of ClO_4^- was removed with a residence time of 0.8 hr. After 9 days, it appeared that the system had stabilized and that this residence time was inadequate for complete ClO_4^- removal. The residence time was then increased to 2.1 hr from day 12-16. At this residence time, approximately 98% of ClO_4^- was removed so that less than 200 μg L^{-1} remained in the effluent (days 14-16). For the final 6 days of the study (beginning on day 17), the residence time was increased further to 4.2 hr. Breakthrough of ClO_4^- was still evident and problems with clogging of the bioreactor were noted. Acetate was present in the effluent, and was therefore not limiting, at all times after day 3. Interestingly, NO_3^- was completely removed from the influent at all flow rates (data not shown).

As with the prior experiment, it is likely that increased NO_3^- and TDS levels contributed to the decreased reactor performance. In addition, visual inspection of the bioreactor showed that precipitate, or "scale," was forming due to the high levels of salts in the simulated RO2, which may have contributed to a decrease in microbial activity.

CONCLUSIONS

These experiments showed that the PBR/perc1ace system was capable of treating groundwater containing ClO_4^- concentrations of approximately 800 µg L^{-1} to non-detectable concentrations (<4 µg L^{-1}), with a residence time as low as 0.3 hr. This has important ramifications with regard to applicability, as many ClO_4^--impacted sites in California contain ClO_4^- at concentrations well below 1 mg L^{-1} (California Department of Health Services [CDHS] website, 2000). With regard to the primary RO rejectate, ClO_4^- concentrations were lowered from approximately 5.0 mg L^{-1} to <4.0 µg L^{-1} with a residence time of 0.8 hr. The reactor also reduced the ClO_4^- concentration in the secondary RO rejectate (RO2) from approximately 10 mg L^{-1} to 0.2 mg L^{-1}, requiring a residence time of 2.1 hr. Significant removal of NO_3^- from all three feeds was also noted, while sulfate reduction was not observed. This study demonstrated the potential feasibility of the PBR/perc1ace system for removing ClO_4^- from groundwater and from RO process waste.

ACKNOWLEDGEMENT

The authors recognize and thank Ed Coppola of Applied Research Associates for his work in calculating salt additions to match target concentrations for the RO rejectates.

REFERENCES

California Department of Health Services (CDHS). 2000. Perchlorate in California drinking water. www.dhs.cahwnet.gov/ps/ddwem/chemicals/perchl/perchlindex.htm.

Herman, D.C., and W. T. Frankenberger, Jr. 1999. Bacterial Reduction of Perchlorate and Nitrate in Water. *J. Environ. Qual.*, *28*: 1018-1024.

Logan, B. E. 2000. Evaluation of Biological Reactors to Degrade Perchlorate to Levels Suitable for Drinking Water. In: E. T. Urbansky (*ed*) *Perchlorate in the Environment*. Environmental Science Research, Vol. 57. Plenum Press: New York, NY.

Logan, B. E., K. Kim, J. Miller, P. Mulvaney, and R. Unz. 1999. Biological Treatment of Perchlorate Contaminated Waters. In: A. Leeson and B. C. Alleman (*eds*) *Bioremediation of Metals and Inorganic Compounds*. pp 147-151, 5(4), Battelle Press: Columbus, OH.

Urbansky, E. T. 1998. Perchlorate Chemistry: Implications for Analysis and Remediation. *Bioremediation J. 2*:81-95.

Wallace, W., S. Beschear, D. Williams, S. Hospadar, and M. Owens. 1998. Perchlorate Reduction by a Mixed Culture in an Up-Flow Anaerobic Fixed Bed Reactor. *J. Ind. Microbiol. Biotech. 20*:126-131.

EVALUATION OF IN SITU BIODEGRADATION OF PERCHLORATE IN A CONTAMINATED SITE

Zhong Zhang, Terri Else, Penny Amy, and ***Jacimaria Batista***
University of Nevada Las Vegas, Las Vegas, NV, USA

ABSTRACT: Parameters affecting perchlorate biodegradation in the Las Vegas Wash area in Las Vegas, NV were investigated at field and laboratory. Two types of tests were performed in water samples from the Wash and the contaminated site, enumeration of perchlorate degrading bacteria and anaerobic microcosm testing. Microbial enumeration was performed by colony counts using standard plateing techniques. Microcosm testing was conducted in serum bottles using standard anaerobic procedures. A mineral/nutrient and buffer medium was supplemented to the testing bottles and lactate was used as the electron donor. The microbial count results show that perchlorate reducers are present along the Wash and in the contaminated area and numbers range from 3.0×10^3 to 7.7×10^4 counts/mL. The microbial counts were higher in the zone where the Wash meets Lake Mead. No correlation was found between the concentration of perchlorate in the Wash and the number of microorganisms present. The results of the microcosm testing show that natural biodegradation of perchlorate in the Las Vegas site is limited by the lack of an electron donor, the presence of nitrate, and the salinity levels in the area.

INTRODUCTION

Perchlorate (ClO_4^-), an important component of rocket fuel and explosives, has been detected in several water supplies throughout the United States (AWWARF, 1997). The contaminant is known to interfere with the functioning of the thyroid gland (Chiovato et al., 1997). A site in Las Vegas, Nevada, is highly affected and contaminated groundwater has seeped into Lake Mead and the Colorado River. In Las Vegas, it is presumed that perchlorate-containing wastes, that were discharged into unlined ponds by two perchlorate manufacturers, migrated to the groundwater and from there to the Las Vegas Wash and Lake Mead, the water source for the 1.2 million inhabitants of the Las Vegas Valley.

Very little is known about the specific interactions of perchlorate with various environmental components (e.g., soils and microorganisms). It has been demonstrated that perchlorate is easily biodegradable by several bacterial strains (Bliven, 1996; Catts, 1999; Coats et al., 1999; Herman and Frankenberger, 1998; Korenkov et al., 1976; Logan, 1998; Van Ginkel et al., 1998). In the biodegradation process, perchlorate is used as a terminal electron acceptor and it is reduced to innocuous chloride. Perchlorate degrading microorganisms have been found to be ubiquitous to a large spectrum of environments. Despite the ubiquity of perchlorate degrading microorganisms, the reduction of perchlorate concentrations in contaminated sites, such as the Las Vegas Wash area, has not been observed.

This research, performed on the highly contaminated site in Las Vegas, examines the extent of perchlorate interaction with various environmental components and their effects on the fate of perchlorate in the environment. The objectives of the work reported here were: (a) to identify whether indigenous microorganisms would be able to promote in-situ biodegradation of perchlorate, and (b) to determine the most relevant parameters and limitations to perchlorate biodegradation in the contaminated site.

MATERIALS AND METHODS

The distribution of perchlorate concentrations was determined by measuring perchlorate and other anion concentrations in aqueous samples at about 0.4-mile intervals along the Las Vegas Wash (Figure 1). The measurements started at the Las Vegas Wastewater Treatment Plant (LVWTP) and extended to the entrance of Lake Mead covering a distance of 6 miles. Water samples from the contaminated seepage (Basic Management Area in Figure 1) and the Wash were collected into pre-sterilized plastic bottles and cooled with Ecogel™ ice packs for transport to the laboratory. Water samples were analyzed for perchlorate, nitrate, sulfate using a Dionex DX –120 Ion Chromatograph, and total organic carbon (TOC) using a Schimadzu 5000A TOC.

Two types of tests, microcosm testing in serum bottles and colony counts by standard plateing techniques, were performed to investigate the role of indigenous microorganisms on perchlorate biodegradation in the perchlorate-contaminated Las Vegas Wash area. It was hypothesized that under favorable conditions perchlorate biodegradation may occur naturally in the contaminated site.

For the microcosm studies, water samples were collected along the Wash and incubated within 24 hours after being transferred to the laboratory. To stimulate microbial growth, minerals/nutrients, phosphate buffer and a carbon source (lactate) were added into sterilized serum bottles. The composition of the minerals/nutrient broth used has been reported elsewhere (Mulvaney, 1999). Control bottles were additionally prepared for monitoring purposes. A lactate concentration of 300 mg/L was added to all bottles. This addition was based on previous research that showed a perchlorate to lactate ratio of 1:3 (w/w) is required for perchlorate biodegradation to take place at a reasonable rate (Liu and Batista, 1999). Control bottles contained only raw water sample with no added lactate and nutrient. All the serum bottles were incubated at 20°C. Concentrations of perchlorate, nitrate, sulfate, TOC, and chloride were recorded with time.

To enumerate the possible perchlorate-respiring colonies, spread-plates were prepared by pouring about 25 mL sterilized growth media into a Petri dish. Finished spread-plates were inoculated and placed into an anaerobic hood for microbial proliferation. The growth media used contains 41mg/L sulfate and no

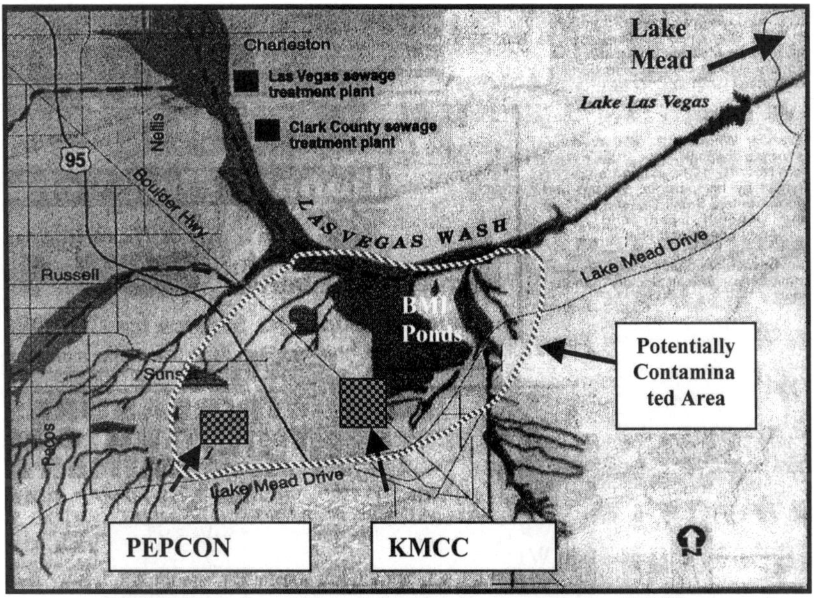

FIGURE 1. Location of the Contaminated site (BMI) along the Las Vegas Wash. (Source: Modified from the LVRJ)

Nitrate. The contaminated water used contains both nitrate and sulfate, but the volume of water used for inoculation was minimal. Thus, sulfate and perchlorate were the only electron acceptors, with perchlorate being preferred to sulfate. Viable perchlorate-respiring cells were counted on a Quebec colony counter.

RESULTS AND CONCLUSIONS

Perchlorate profiles along the Las Vegas Wash show that perchlorate concentrations along the Wash range from zero to 1000 ppb. Furthermore, this distribution can be separated into two sections. The first section originates from the Las Vegas Wastewater Treatment Plant and ends at the seepage area with very low perchlorate concentrations. The second section starts at the seepage and goes down the Wash to Lake Mead with concentrations ranging from 500-1000 ppb. The contaminated seepage contains about 100 ppm perchlorate, 2000 ppm sulfate, 50 ppm nitrate, and 100 ppm chlorate.

Table 1 shows microbial counts for 15 water samples from the Las Vegas Wash and from the contaminated site. Microbial counts for the Las Vegas Wash and the contaminated site varied from 3×10^3 to 7.7×10^4 counts/ml. Contrary to what was expected, the smallest counts of indigenous perchlorate degrading microorganism were found within the contaminated area. The highest counts were found in the Las Vegas Bay (entrance of Lake Mead). Along the Wash the number of perchlorate reducing microorganisms does not correlate with the perchlorate concentrations.

TABLE 1. Microbial counts for perchlorate reducing microorganisms in the Las Vegas Wash area.

Sample ID	Count (#/ml of water)
Oct2299-W3	3.2 E+04
Oct 2299-W5	3.3 E+03
Oct2299-W1	1.5 E+04
Oct2299-W2	8.5 E+03
Oct2999-W4	2.9 E+04
Nov1999-W1	3.0 E+04
Nov1999-W2	2.5 E +04
Nov1999-W3	3.0 E+04
Jan2100-W1	3.6 E+04
Jan2100-W2	2.3 E+04
Jan2100-W3	1.8 E+04
Jan2100-W4	2.0E+04
Feb1100-W1	7.7 E+04
Feb1100-W2	1.4 E+04
Feb1100-W3	1.6 E+0.4

Figure 2 shows representative data obtained from the microcosm studies. The data clearly show that indigenous perchlorate reducing bacteria are present at reasonable numbers in the perchlorate-contaminated site in Las Vegas. Further, the data indicate that when an electron donor (lactate) was not present, perchlorate biodegradation was not observed (Figures 2a, 2b, 2e, and 2f). The TOC of the Wash water is about 5 mg/L. This concentration is considerably low and this carbon is apparently not available for microbial utilization. Thus, one of the factors limiting biological perchlorate reduction in the Las Vegas Wash and in the contaminated area is the lack of an electron donor. The data also reveal (Figures 2a and 2b) that for most of the highly contaminated samples (100 ppm range), a perchlorate biodegradation rate of 148 μg day^{-1} – 187 μg day^{-1} during the initial 30 incubation days was observed. In the succeeding 30-60 days, a reduced rate in the order of 33μg day^{-1} – 34 μg day^{-1} was found. Figures 2e and 2f show that for samples with lower initial perchlorate concentrations (200-600 ppb), the rate of biodegradation is much lower (1.4 μg day^{-1} – 2 μg day^{-1}). The reduction in perchlorate biodegradation rate in all tests is the result of the decrease of lactate (electron donor) concentration with time. For the samples collected in the Las Vegas Bay, the microbial counts are the highest and perchlorate biodegradation was the fastest. This implies that indigenous perchlorate-respiring microorganisms are ubiquitous in the Las Vegas site and that they rapidly acclimate to the new environments. Water samples collected from trenches inside the contaminated area containing very high salinity levels showed very slow biodegradation rates (Figures 2c and 2d). Even though these two water sources are highly contaminated (perchlorate content between 60-70 ppm), perchlorate

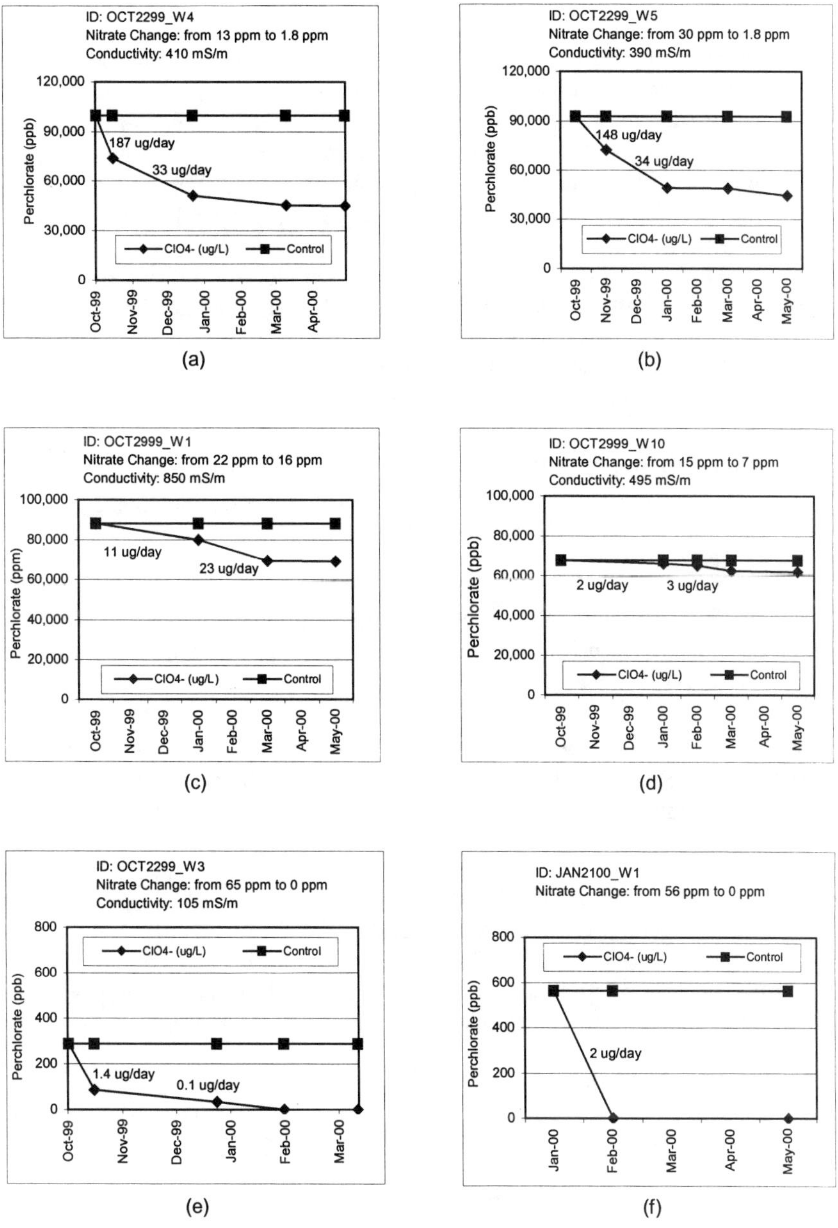

FIGURE 2. Represenative data from microcosm tests performed with water samples from the perchlorate contaminated sites in Las Vegas

biodegradation rates were very low (2 μg day^{-1} – 23 μg day^{-1}). Slow biodegradation rates in these samples results from the lower microbial population and the high salinity levels of the water samples from these trenches.

In samples where both nitrate and perchlorate concentrations were high, nitrate was found to be a preferred electron acceptor to perchlorate (Figure 2a, 2b, 2e, and 2f). However, when the perchlorate levels were low (<1 ppm) as compared to those of nitrate (50 ppm range), complete perchlorate biodegradation was observed.

The results indicate that indigenous perchlorate-respiring microorganisms are ubiquitous in the Las Vegas site and that they acclimate to the new environment promptly. Natural perchlorate biodegradation in the studied area seems to be limited by three factors, namely, lack of an electron donor (carbon), presence of high nitrate levels, and high salinity levels.

ACKNOWLEDGEMENT
This research is funded by the U.S. Environmental Protection Agency.

REFERENCES

AWWARF (American Water Works Association Research Foundation). 1997. *Final report of the perchlorate research issue group workshop*, September 30 to October 2. AWWARF, Denver, CO.

Bliven, A.R. 1996. "Chlorate respiring bacteria: isolation, identification, and effects on environmentally significant substrates". M.S. Thesis, Department of Chemical and Environmental Engineering, University of Arizona, Tucson, AZ.

Catts, J.G. 1999. "The biochemical removal of perchlorate from San Gabriel Basin Groundwater and potable uses of the treated water". *Extended Abstracts, 218th ACS Meeting,* New Orleans, August 22-24, 39, 107-109.

Chiovato, L., F. Santini, and A. Pinchera. 1997. "Treatment of hyperthyroidism. Pisa, Italy". http://www.thyrolink.com/thyint/2-95int.htm#intro

.

Coates, J.R., U. Michaelidou, R.A. Bruce, S.M.. O'Connor, J.N. Crespi, and L.A. Achenbach. 1999. "Ubiquity and diversity of dissimilatory (Per)chlorate reducing bacteria". *Appl. Environ. Microbiol.* 65(12):5234-5241.

Herman, D.C., and W.T. Frankenberger Jr. 1998. "Microbial-Mediated Reduction of Perchlorate in Groundwater". *J. Environ. Qual.* 27:750-754.

Korenkov, V.N., V. Ivanovich, S.I. Kuznetsov, and J.V. Vorenov. 1976. "Process for purification of industrial waste waters from perchlorates and chlorates". U.S. Patent No. 3,943,055, March 9.

Liu, J., and J. Batista. "Perchlorate Removal from Waters by a Membrane-immobilized Biofilm". Proc. *AWWA Inorganics Conference.,* Albuquerque, New Mexico, Feb 28, 2000.

Logan, B.E. 1998. "A review of chlorate and perchlorate respiring microorganisms". *Bioremediation J.* 2(2):69-79.

Mulvaney, Peter 1999. "Perchlorate and chlorate reduction by axenic cultures". M.S. Thesis, Environmental Pollution Control, Pennsylvania State University.

Van Ginkel, G.C., Kroon, A.G.M., Rikken, G.B., and S.W.M. Kengen. 1998. "Microbial conversion of perchlorate, chlorate, and chlorite." *In: Proceedings of the Southwest Focused Groundwater Conference: Discussing the Issue of MTBE and Perchlorate in the Ground Water*, pp. 92-95. National Ground Water Association.

BIOLOGICAL PERCHLORATE REMOVAL FROM DRINKING WATERS INCORPORATING MICROPOROUS MEMBRANES

Jacimaria R. Batista and Jian Liu
University of Nevada Las Vegas, Las Vegas, NV, USA

ABSTRACT: This research investigated a bench-scale biological treatment system to remove perchlorate from waters, in which the perchlorate-degrading microorganisms are physically separated from the water being treated by a microporous membrane, thus minimizing post treatment requirements. The system consists of a diffusion reactor (DR) that holds the perchlorate containing water and a biological reactor (BR), where microorganisms are grown as a biofilm on the membrane. Because of concentration gradient, the perchlorate in the DR tank diffuses across the membrane into the BR tank, and is metabolized in the biofilm. The results show that the diffusion coefficient of perchlorate through the membrane is significantly smaller than that of perchlorate in water. Perchlorate biodegradation testing using immobilized biofilms has shown excellent perchlorate biodegradation. Perchlorate to chloride ratios, calculated for several reactor runs, were found to be about one, indicating good reliability of the reactor design. The interference of nitrate, sulfate, and total dissolved solids (salinity as NaCl) has been investigated and the results show that high nitrate and salinity levels negatively impact perchlorate biodegradation. Sulfate was found not to interfere with perchlorate biodegradation.

INTRODUCTION

Recently, perchlorate has been found in surface and ground waters in the western United States (Renner, 1999). Perchlorate is a health concern because it competes with iodine affecting the production of hormones by the thyroid gland (Cooper, 1991). Conventional water treatment technologies, such as air stripping, and adsorption, have no effect on perchlorate removal from waters. However, perchlorate is very biodegradable. The prompt biodegradation of perchlorate has been reported by several researchers (Attaway and Smith, 1994; Korenkov et al., 1976; Rikken et al., 1996) and it provides the prospect for biological perchlorate removal treatment systems. Mixed and pure bacterial cultures have been demonstrated to reduce perchlorate biologically. In the reduction process, which occurs under anaerobic conditions, perchlorate is used as an electron acceptor, and is reduced to innocuous chloride when an electron donor (organic carbon or hydrogen), nutrients/minerals and buffer are provided.

In the last three years, there have been several reports on biological reactors to remove perchlorate from waters (Catts, 1999; Coppola, 1998; Greene and Pitre, 1999; Kim, 1999). These treatment systems include anoxic fluidized bed methanol-fed reactor, sand filter bed, and hydrogen gas-phase reactor. A disadvantage of the above systems is that bacteria used to degrade perchlorate will come directly into contact with the water being treated. For the drinking water industry that translates to the presence of a larger number of microbes, some uncommon in drinking waters, that has to be inactivated by disinfection. In addition, if a carbon source is used as the electron donor, residual organic carbon and microbial debris may promote biofilm growth in the distribution system, thereby decreasing the quality of the drinking water. The membrane-immobilized biofilm system, investigated here, can separate the perchlorate-contaminated water from the microbes, greatly minimizing the presence of microbes and by-products in the finished water.

MATERIALS AND METHODS

The diffusivity of perchlorate through the membrane, prior to biofilm growth, was determined experimentally to evaluate the migration of perchlorate from the diffusion reactor (DR; left side) to the biofilm reactor (BR; right side) (Figure 1). A Memcor polyvinylidene fluoride (PVDF) membrane with 0.45 μm pore size was used in this investigation. The characteristics of the membrane are shown in Table 1. The diffusion coefficient was determined by placing a circular piece of membrane (10 cm diameter) between the two tanks. The diffusion chamber was then filled with 5 liters of deionized (DI) water containing 1000 mg/L perchlorate. The perchlorate concentration in the biofilm reactor was monitored every 15 minutes. The diffusivity test was performed in duplicate. Diffusivity was calculated by using Fick's law. The diffusivity was determined by least-square regression of the transformed experimental data.

FIGURE 1. Membrane-immobilized biofilm reactor set-up.

A perchlorate degrading enrichment master culture was developed from the return activated sludge (RAS) from the Clark County Sanitation District (CCSD) wastewater treatment plant in Las Vegas. This enrichment culture was used in all experiments reported here.

To investigate the biodegradation of perchlorate by an immobilized biofilm, it is necessary to first establish a biofilm on the surface of the membrane.

TABLE 1. Characteristics of the membrane used to immobilize the biofilm.

Membrane Type	**Memcor PVDF**
Pore Size, μm	0.45
Thickness, μm	99
Pore Fraction, %	70
Diffusion Coefficient, cm^2/sec (testing)	**3.75×10^{-6}**
Diffusion Coefficient, cm^2/sec (Wilke-Chang Method)	**1.53×10^{-5}**

A membrane was placed between the two reactors. The biofilm was established on the BR reactor over a week period prior to the experiments. Five liters of DI water containing 1000 mg/L lactate, 200 mg/L perchlorate (lactate/perchlorate ratio of 5:1), nutrients/minerals and buffer solutions and seed microbes (0.1 g/L) from the master reactor were added to the BR reactor. Five liters of DI water were also added to the DR reactor to keep the hydraulic pressure on both sides of the membrane the same. A YSI 54A oxygen meter and a Corning pH/Ion meter 450 was placed in the BR to continuously control the oxygen concentration and the pH. Deoxygenation in the BR reactor was obtained and kept by stripping with nitrogen gas using fine-bubble ceramic diffusing stones. Oxygen levels were kept undetectable at all times. Daily, samples were taken from both reactors and analyzed for perchlorate, lactate, and chloride.

In the batch testing of nitrate interference, about 100 mg/L of perchlorate, 500 mg/L of lactate, buffer, nutrient/minerals and nitrate (0 ppm, 10ppm, 30ppm and 60 ppm) were added to four 125-ml serum bottles. In the batch testing of sulfate interference, about 10 mg/L of perchlorate, 150 mg/L of lactate, buffer, nutrient/minerals, and sulfate 90 ppm, 20 ppm, 100 ppm and 500 ppm) were added to four 125-mL serum bottles. The bottles were sealed with a butyl rubber cap and crimped with an aluminum ring. The tests were performed in duplicate.

The influence of salinity on perchlorate biodegradation was investigated using culture tube. In the culture tube testing, 8-mL glass tubes, caps, nutrient/minerals, buffer, lactate, and perchlorate solutions were autoclaved prior to their use in the experiments. The tubes were filled with the solutions and inoculated with the master culture. Microbial activity was assessed by measuring, with a Spectrophotometer (Spectronic 20), the changes in optical density, at 600 nm, due to microbial growth.

RESULTS AND DISCUSSION

Figure 2 shows the results of duplicate tests for perchlorate diffusion through the PVDF membrane. The duffisivity was calculated from the experimental data to be 3.75×10^{-6} cm^2/sec. For comparison, the diffusivity of perchlorate in water, without a membrane, was calculated by using the Wilke-Chang Method (Table 1), in which the molar volume of perchlorate was calculated using the Method of LeBas (Reid et al., 1987). The diffusion coefficient of perchlorate calculated by the Wilke-Chang method (Reid et al.,

1987) was found to be 1.53 x 10^{-5} cm^2/sec. Therefore, the diffusivity of perchlorate through the microporous membrane tested is significantly smaller than that in water. This result confirms that perchlorate will migrate through the semi-permeable membrane by diffusion to the biodegradation reactor.

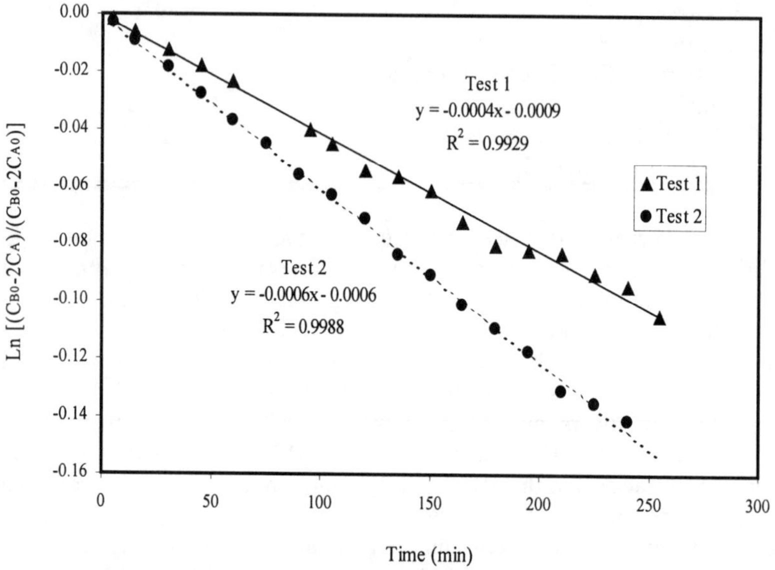

FIGURE 2. Perchlorate diffusion through Memcor PVDF membrane

Figure 3 shows the biodegradation of perchlorate by the biofilm immobilized on the PVDF membrane. Notice that perchlorate biodegradation rate was very fast in the first 6 days of the test (0.16 moles ClO_4^- biodegraded/m^2-day). The perchlorate concentration in the DR reactor decreased from 207 mg/L to about 83 mg/L during this period. After the sixth day, the perchlorate biodegradation rate decreased significantly. This decrease in biodegradation rate was caused by the limited availability of lactate (carbon source). The lactate concentration in the BR reactor decreased sharply and it reached very small concentrations after the sixth day. The concentration of chloride increased with time as perchlorate was biodegraded. The ratio of moles of perchlorate biodegraded to those of chloride generated was found to be 0.99. This is an excellent mass balance, considering the large size of the reactor used.

The final concentrations of chloride in both reactors were approximately the same. Contrary to perchlorate and lactate, chloride concentrations change significantly from one reactor to another. This is due to two factors: (a) diffusion of chloride through the membrane due to its small size, (b) movement of water from the DR reactor (smaller concentration of ions) to the BR reactor (higher concentration of ions) by osmotic pressure. In all the tests performed with this membrane, migration of water from the DR to the BR reactor was observed. That means, with time, the volume of water in BR increased. As can be observed in

Figure 3, chloride concentration stabilizes when both reactors contained about the same concentration of chloride.

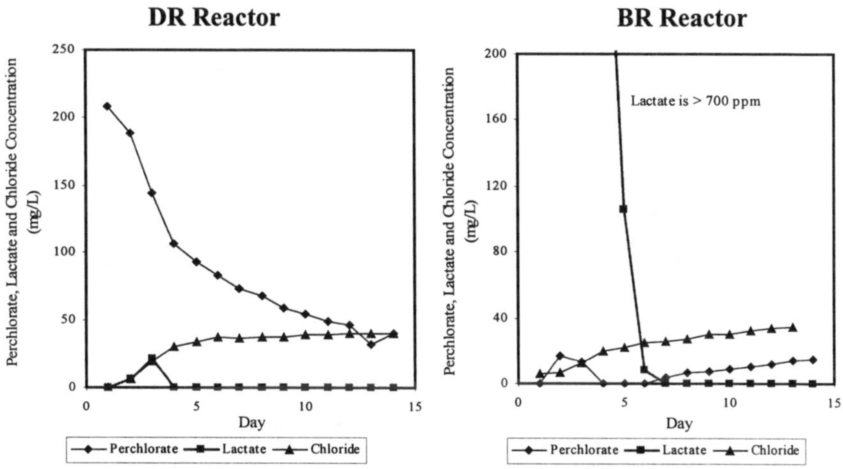

FIGURE 3. ClO_4^-, lactate, and Cl^- concentration during biodegradation of by ClO_4^-a biofilm immobilized on a PVDF membrane.

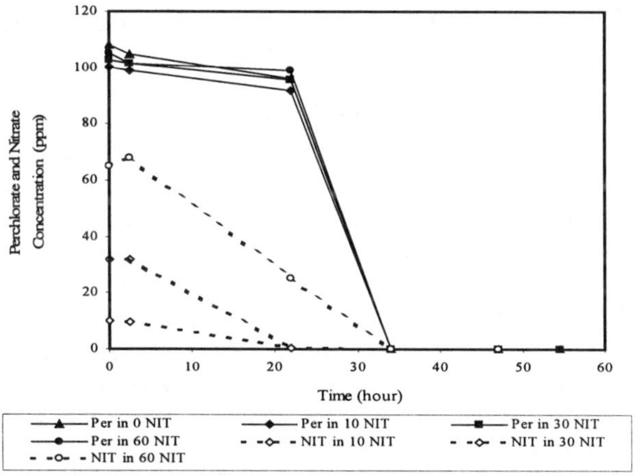

FIGURE 4. Batch testing of nitrate interference on perchlorate biodegradation using "BALI" culture.

Figure 4 shows the results for the batch testing on the interference of nitrate on perchlorate biodegradation. The lower three dashed lines show the decrease in the concentration of nitrate with time in the presence of approximately 100 mg/L perchlorate. The top four solid lines show the decrease in perchlorate concentration with time in the presence of about 10 mg/L, 30mg/L and 60 mg/L of nitrate. As shown in Figure 4, the rate of nitrate biodegradation is much greater than that of perchlorate. In addition, significant perchlorate biodegradation only started after the nitrate had been almost completely biodegraded. These results show that the enrichment mixed culture prefers nitrate to perchlorate as an electron acceptor. Thus, nitrate negatively affects perchlorate biodegradation.

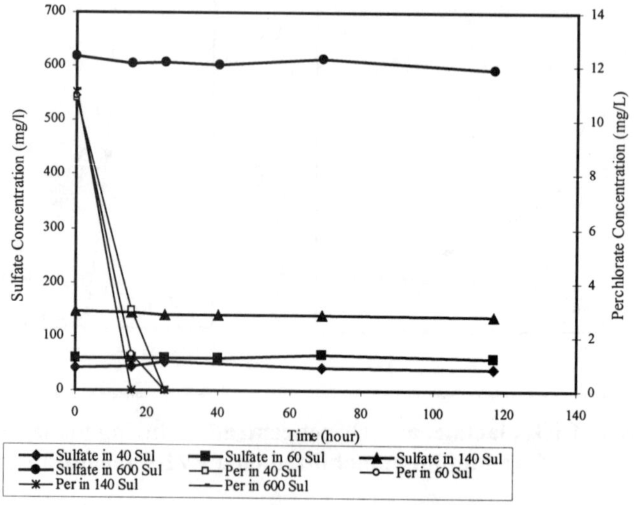

FIGURE 5. Interference of sulfate on perchlorate biodegradation (batch testing).

Figure 5 shows the result of sulfate interference on perchlorate biodegradation in batch tests. For this test, the initial concentrations of perchlorate were about 10 mg/L, while the initial concentrations of sulfate varied from 40 mg/L to 600 mg/L. The nutrient/minerals broth used for microbial growth contained a background sulfate concentration of approximately 41 mg/L. The results showed that perchlorate biodegraded from about 10 mg/L to undetectable levels within 16 hours, while the sulfate concentration remained practically constant during the same period. Thus, sulfate concentrations ranging from 40 to 600 mg/L seem not to interfere with perchlorate biodegradation. This is confirmed by the fact that sulfate biodegradation started after all the perchlorate had been biodegraded.

Figure 6 shows the results of the interference of salinity on perchlorate biodegradation. The transmittance through the culture tubes with time for different salinity levels was measured. Notice that the higher the salinity level, the higher the transmittance, indicating that the microbes grew much slower at higher salinity levels. Also notice that for each individual salinity level, the transmittance first decreased with time until it reached a minimum, and then it increased slowly with time and reached a constant level. In the test tubes, flocculation of the microbes was observed after they reached maximum growth. That explains the increase in transmittance observed in the experiment. No microbial growth was observed at salinity levels above 4%.

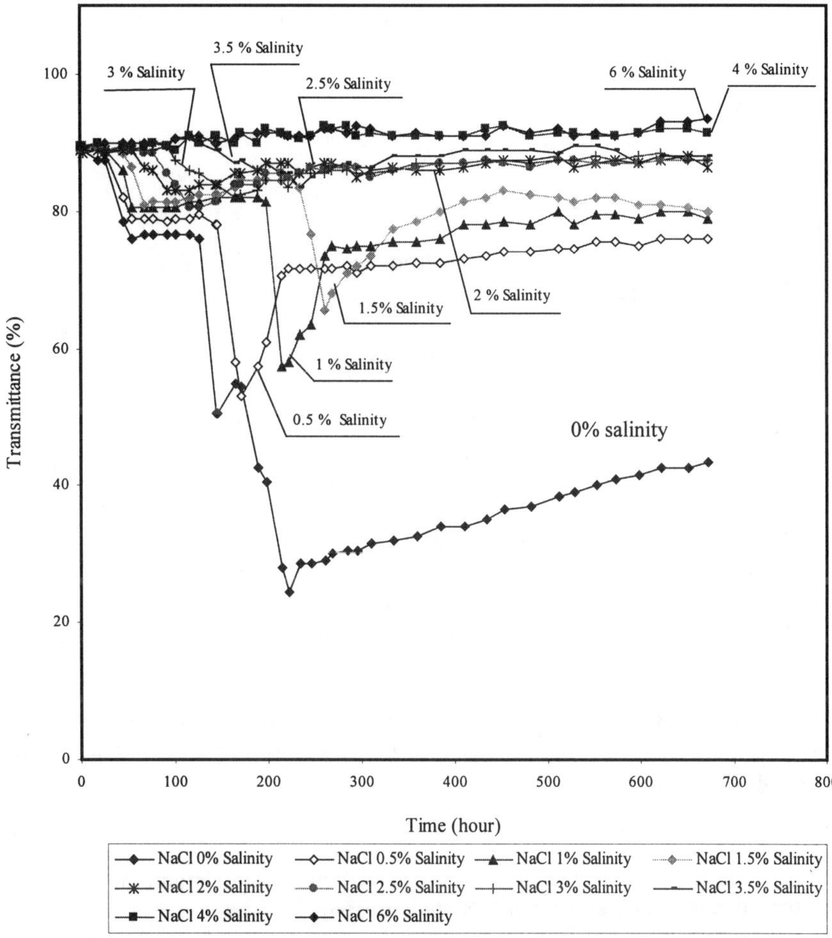

FIGURE 6. Transmittance at different salinity concentration.

In conclusion, the results of the experiments have shown that perchlorate easily migrates through semipermeable membranes by diffusion, eliminating the need of energy input. The diffusivity of perchlorate through the microporous membrane tested is significantly smaller than that of perchlorate in water. Perchlorate biodegradation by the membrane-immobilized biofilm was found to be fast and steady. When a carbon source (lactate) is added to the BR reactor, perchlorate quickly degrades to chloride. Calculated ratios of perchlorate to chloride are very close to the theoretical ratios. Nitrate and salinity were found to negatively affect perchlorate biodegradation. Sulfate does not impact perchlorate biodegradation.

The presence of the membrane allows for controlled diffusion of perchlorate to the BR reactor, so that the concentration of perchlorate in the finished water can be kept at a desired level without fluctuations or sporadic spikes. In addition, the amount of microbes in the product water is potentially very small, given the contaminated water is isolated from the microbes by the membrane.

ACKNOWLEDGMENTS
This research was funded by the American Water Works Association Research Foundation.

REFERENCES
Attaway, H. and M.D. Smith. 1994. "Propellant Wastewater Treatment Process". U.S. Patent 5,302,285.

Catts J. G. 1999. "The Biochemical Removal of Perchlorate From San Gabriel Basin Groundwater and Potable Use of the Treated Water". *Division of Environmental Chemistry Preprints of Extended Abstracts*. 39(2):107-109.

Cooper, D.S., 1991, Treatment of Thyrotoxicosis, In. L. E. Braverman and R. D. Utiger (Eds.), *In -The Thyroid: A Fundamental and Clinical Text*, 6th ed. J.B. Lippincott, Philadelphia, PA. pp 887-916.

Coppola, Edward N. 1998. "Perchlorate Biodegradation Technology: Multiple Applications", *NGWA Southwest Focused Ground Water Conference, Anaheim, CA.*

Greene, Mark and Michael P. Pitre. 1999. "Treatment of Groundwater Containing Perchlorate Using Biological Fluidized Bed Reactors With GAC Or Sand Media". *Division of Environmental Chemistry Preprints of Extended Abstracts*. 39(2):105-107.

Kim, Ki-jung. 1999. "Microbial Treatment of Perchlorate-Contaminated Water". Master's thesis. The Pennsylvania State University.

Korenkov, V.N., V. I. Romanenko, S.I. Kuznetsov, and J.V. Voronov. 1976. "Process for Purification of Industrial Waste Waters from Perchlorates and Chlorates." U.S. Patent No. 3,943,055.

Reid, R.C., J. M. Prausnitz, and B. E. Poling. 1987. *The Properties of Gases and liquids*. Fourth Edition. McGraw-Hill, Inc.

Renner, Rebecca. 1999. "EPA Draft Almost Doubles Safe Dose of Perchlorate in Water". *Environmental Science & Technology. News*. *33*:110-111.

Rikken, G.B., A.G.M. Kroon, and C.G. van Ginkel. . 1996. "Transformation of (Per)chlorate into Chloride by a Newly Isolated Bacterium: Reduction and Dismutation." *Applied Microbiolology Biotechnology*. *45*:420 -426.

ENHANCED NATURAL ATTENUATION OF PERCHLORATE IN SOILS USING ELECTROKINETIC INJECTION

W. Andrew Jackson (Texas Tech University, Lubbock, Texas 79409 USA)
Mi-Ae Jeon (Texas Tech University Lubbock, Texas, 79409 USA)
John H. Pardue (Louisiana State University, Baton Rouge LA USA)
Todd Anderson (Texas Tech University, Lubbock, Texas 79409 USA)

Abstract: Perchlorate (ClO_4^- or PC) contamination of groundwater has recently become a major concern across the nation. Electrokinetic injection (EKI) has been used successfully to remove contaminants (charged and uncharged) from low permeability media without the limitations of preferential flow. The overall objective of this study was to demonstrate that EKI can be economically used to promote the simultaneous removal of PC and stimulation of PC degradation by injected substrates in soils from contaminated sites. Experiments were conducted in horizontal packed columns. Kaolin, sand, and a silty clay subsurface soil were examined. Sodium glycine and lactic acid were used to control the pH and provide substrates for injection. PC was effectively removed from both clay and sand mainly by ionic flux. In addition lactic acid was successfully injected through the column. The ability of lactic acid and glycine amendments to stimulate PC degradation was confirmed in bottle studies. Injection of substrates appears to be able to promote near complete degradation of PC before transport out of the column in contaminated soils. Electrokinetic removal of PC with the simultaneous electrokinetic injection of organic substrates offers a potentially efficient means of PC removal followed by long-term enhanced natural attenuation of PC in field sites.

INTRODUCTION

Perchlorate (ClO_4^- or PC) contamination of groundwater has recently become a major concern across the nation. The Environmental Protection Agency (EPA) placed PC on the Contaminant Candidate List in 1998 and may develop a Health Advisory in the short term. The State of California has established a current advisory action level of 18 ppb in water. Manufacturing or processing of compounds such as ammonium perchlorate, a component of solid rocket fuel, has taken place in 44 states, and groundwater and surface water contamination by PC has been reported in 14 states (Damian and Pontius 1999). Observation of many contaminated sites over the past twenty years has demonstrated the potential value of in situ degradation for many organic chemicals by indigenous microorganisms with and without added enhancements. These successes include the biological remediation of petroleum contaminated water and sediment, and the now widespread use of anaerobic biological degradation (natural attenuation) of chlorinated solvents. PC degradation appears to be similar to chlorinated solvents and can act as an electron acceptor as the indigenous microorganisms degrade present or added organic species. In this sequence, the PC can be reduced to Cl^-.

One successful strategy that has been employed to remediate chlorinated solvents is injection of substrates (organic acids) to promote degradation. Some limitations to this technique have included low permeability media and problems with preferential flow. Electrokinetic injection (EKI) has been used successfully to remove contaminants (charged and uncharged) from low permeability media and without the limitations of preferential flow (Loo and Chilingar, 1997; Bruell et al., 1992). Combining these two technologies (removal by EKI with the simultaneous EKI of organic material to promote degradation) could allow for the efficient removal of PC while simultaneously promoting degradation of PC. The overall objective of this study was to demonstrate that EKI could be economically used to promote the simultaneous removal of PC and stimulate PC degradation by injection of substrates in soils from contaminated sites. Laboratory experiments were carried out to achieve three specific objectives: (1) to determine the rate of removal of PC from contaminated clay, sand, and mixed media columns (10 cm diameter by 66 cm length) by EKI; (2) to determine the ability to promote degradation by injection of lactic acid and glycine; and (3) to demonstrate the ability to simultaneously remove PC and promote PC degradation.

MATERIALS AND METHODS

Microcosm Studies. Degradation potential of the unamended and amended soil was determined in microcosm studies. Contaminated sediments from the Longhorn Ammunition Site were used to construct microcosms. Microcosm studies followed the EPA's recommended procedure (Wilson et al., 1996). Details of treatments examined are listed in Table 1. An autoclaved control was also included. Samples were analyzed for PC every 3-5 days. First order degradation rates were determined for each treatment.

Table 1. Experimental conditions for PC degradation studies.

Treatment	Soil	Water	PC	Lactic	Glycine
Autoclaved	50g	50ml (DI)	10mg/kg	-	-
Control	50g	50ml (DI)	10mg/kg	-	-
Uncontaminated	50g	50ml (DI)		-	-
Lactic	50g	50ml (DI)	10mg/kg	3mM	-
Glycine	50g	50ml (DI)	10mg/kg	-	3 mM
Lactic/Glycine	50g	50ml (DI)	10mg/kg	3 mM	3 mM

Column Studies All EKI experiments were conducted in 60 cm long 10 cm diameter plexiglass columns (Figure 1). These apparatuses allow the sampling of electrode reservoirs (end caps) and the column along its length. Solution level in the electrode reservoirs was kept constant by the use of a combination of overflow lines and level-controlled pumps in order to quantify the osmotic flux and prevent it from altering the water level. The pH control was achieved by using a dilute lactic acid solution pumped through the cathode reservoir and dilute

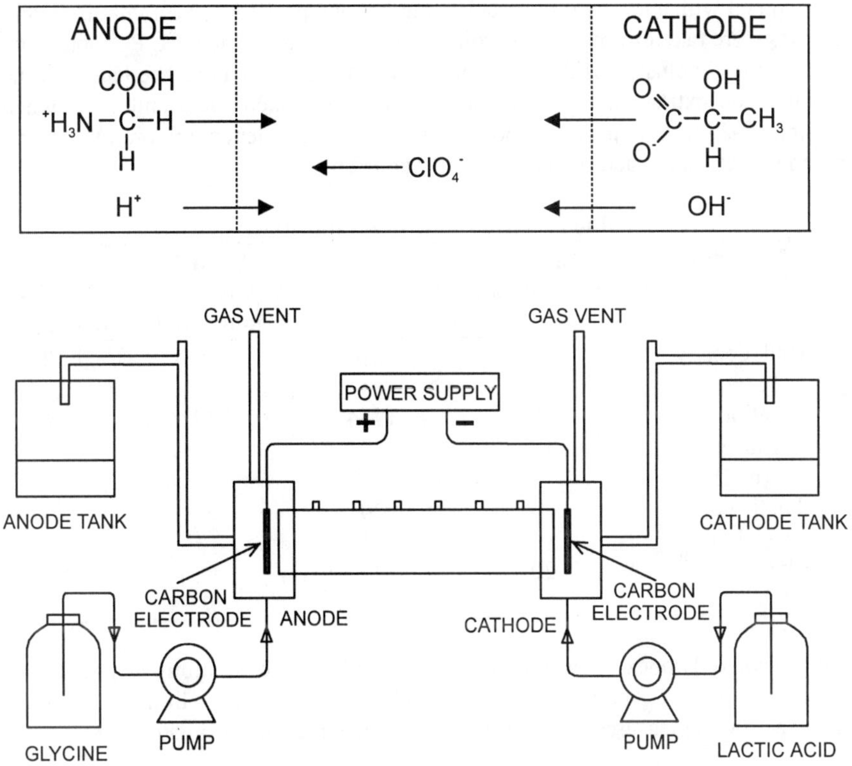

FIGURE 1. Schematic of electrokinetic set-up

glycine ($NaOOCCH_2NH_3$) solution pumped through the anode reservoir. Pumping rates were adjusted daily to compensate for H^+ and OH^- generated in the anode and cathode respectively. All experiments used graphite disc electrodes (10 cm diameter). Three types of material were examined: kaolin, sand, subsurface soil (silty-clay). The sand was washed of any fines or trace contaminants. Soil (5 cm-30 cm BGS) was obtained from the Longhorn Ammunition site near building 25C and contained approximately 200 ppb of PC. All materials (sand, kaolin, and soil) were spiked with additional PC (10-100 mg/kg) and packed in the columns. Daily samples were taken from the reservoir overflows and tested for pH, PC, and lactic acid. At the conclusion of the experiment cores were taken along the length of the columns and extracted. Extracts were tested for PC, lactic acid and COD when possible. 3-25 milliamp current potentials were applied depending on the packing material. Details of each run are listed in Table 2.

Table 2. Details of EK column studies.

Material	Kaolin 1 100 % kaolin	Kaolin 2 100 % kaolin	Sand 1 No.3 BlastSand	Sand 2 Grade 4 BlastSand
Bulk density (g/cm^3)	1.0	1.0	1.64	1.52
Duration (hr)	694	1134	852	547
Current (mA)	17	24	3	7
Lactic conc. (mM)	20	20	5	5
Glycine conc. (mM)	20	20	5	5

Analytical. A Dionex IC was used to measure PC concentrations following the general method of Anderson and Wu, (2001). The detection limit was 5 ppb. Lactic acid was measured using High Performance Liquid Chromatography.

RESULTS

Perchlorate Removal. PC from both the contaminated sand and clay was efficiently removed. Table 3 lists the total % PC removed at the cathode and anode as well as the total energy expended, for each replicate. Total PC removed differed between experiments due to the total time for which the experiments were conducted. Removal rates were nearly identical (Figure 2) for replicates of kaolin but differed between sand replicates. Differences in removal rates for the sand replicates is probably due to differences in the osmotic flux which were caused by pH (discussed later). PC was primarily removed by migration towards the anode caused by the ionic flux. Total removal of PC was generally similar between the sand and clay replicates although the rate of removal was higher in sand columns.

Table 3. PC removal from sand and kaolin by electrokinetics.

	Duration (hours)	% Removal (Anode)	% Removal (Cathode)	% Removal (Total)
Kaolin 1	694	86.69	0.23	86.92
Kaolin 2	1134	96.71	0.19	96.90
Sand 1	852	112.48	1.64	114.12
Sand 2	547	91.28	1.06	92.34

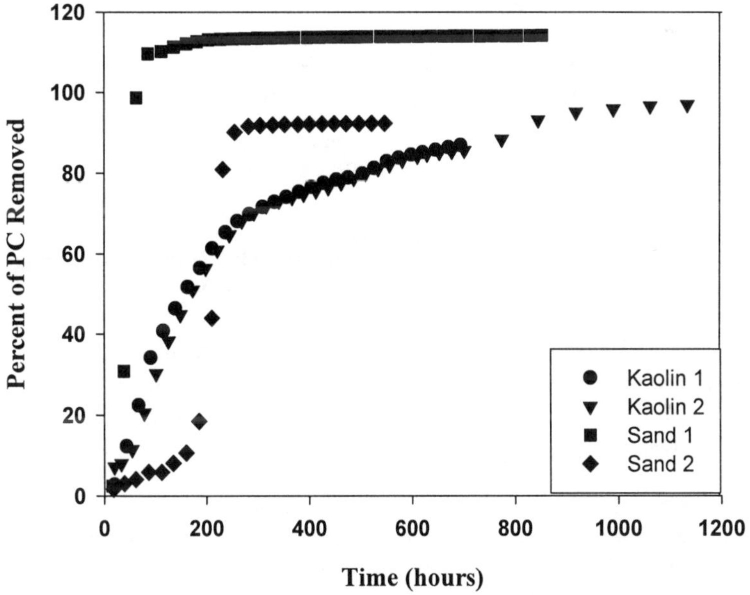

FIGURE 2. PC removal from sand and clay columns using electrokinetics.

The osmotic flux in columns packed with kaolin was constant after an initial equilibration period and did not vary in direction, anode to cathode (Figure 3). Osmotic flux in the sand replicates was more varied initially and generally lower in magnitude. The large initial osmotic flux to the anode in the sand 2 replicate may be responsible for the initial lag in PC removal in this replicate. Direction of flow switched during the experiment probably due to reversals in the pH at the cathode and anode. Control of pH was normally maintained between 6-8

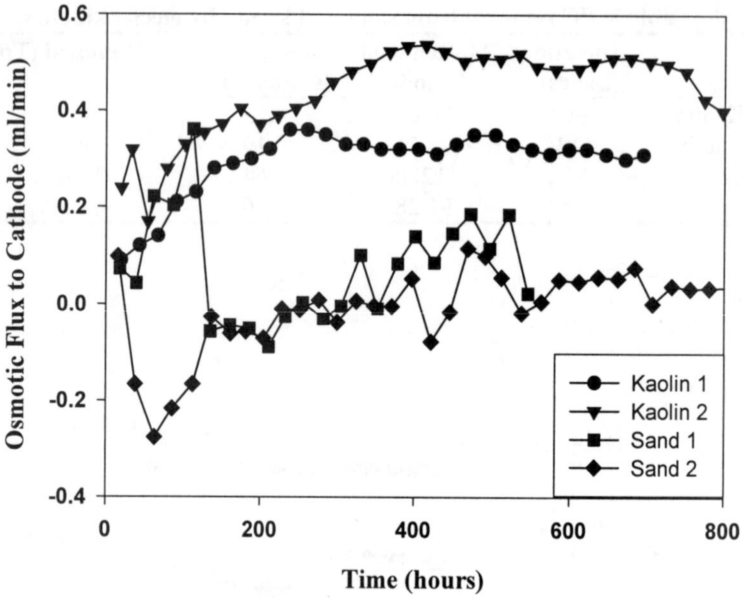

FIGURE. 3. Osmotic flux in sand and clay columns subjected to electrokinetic removal of PC.

at both anode and cathode. Neutrality is important to increase ionic mobility of PC (i.e. high OH^- concentrations reduce the percentage of ionic flux by PC). In addition, since the goal is to both remove PC and stimulate intrinsic degradation, strong acid or base fronts are undesirable.

Transport of Lactate. At the end of each experiment cores or pore water were removed from the columns and tested for lactate. Concentration profiles are presented in Figures 4. Lactate was transported into the columns at concentrations sufficient to promote degradation. In microcosm studies, lactate and glycine additions increased the degradation rate of PC (control = 0.185 d^{-1}; glycine =1.62 d^{-1}; lactate = 0.748d^{-1}). Lactic acid pumped into the cathode is quickly transformed into lactate due to the OH^- ions generated at the electrode. Lactate is transported into the column due to ion flux. Lactate concentrations were significantly less in the clay column than sand columns. This corresponds with the faster rate of PC removal from sand, probably due to the increased osmotic flux in the clay cores which impedes the ionic flux. No data is available on glycine transport due to our inability to specifically analyze for this compound however it could be transported into the column due to its positive charge and/or by the osmotic flux.

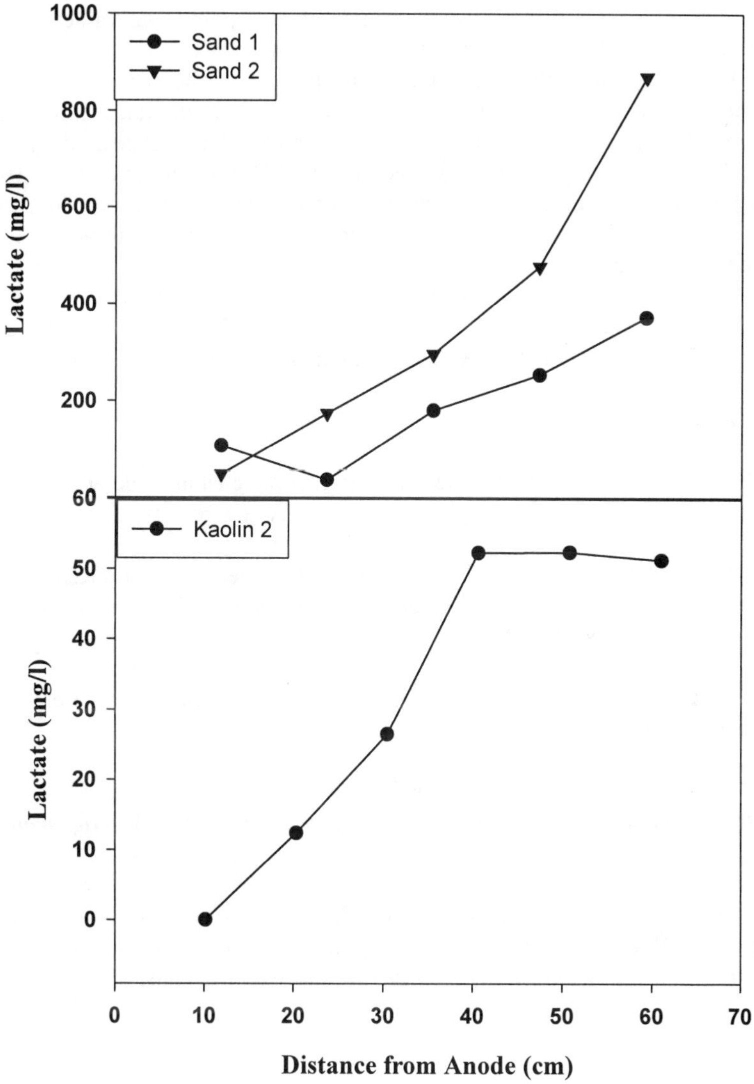

FIGURE 4. Lactate distribution in columns at the end of each experiment.

CONCLUSIONS

PC can be removed from both sand or clay material by electrokinetics. Spiked material was quantitatively removed of PC within 15-30 days using very low applied currents (3-25 milliamps). The mechanism appears to be by ionic flux rather than by osmotic flux. Osmotic flux is higher in clay columns than sand. PH control was effectively maintained at biologically appropriate values at both the cathode and anode by pumping lactic acid or sodium glycine at very low rates (0.5-1 ml/min). In addition to transporting PC out of the contaminated material lactate, added as lactic acid to control the pH, also migrated throughout the column. Separate microcosm studies have shown that lactate and glycine are capable of promoting intrinsic degradation of PC. Glycine is also suspected of transport into the material. Experiments (not reported) using contaminated soil from a PC contaminated site appear to show most PC degrading before transport out of the column.

Electrokinetic removal of PC with the simultaneous electrokinetic injection of organic substrates offers a potentially efficient means of PC removal followed by long-term enhanced natural attenuation of PC in field sites.

REFERENCES

Anderson, T. A., and T. H. Wu. 2001. "Extraction, cleanup, and analysis of the perchlorate anion in tissue samples". J. Chrom. A. (In review).

Bruell, C.J., A.S. Burton, and T.W. Walsh. 1992. "Electroosmotic Removal of Gasoline Hydrocarbons and TCE from Clay". *Journal of Environmental Engineering*, 118(1):68-101.

Damian, P., and W. Pontius. 1999 "From Rockets to Remediation". *Environmental Protection.* 10(6):24-32.

Loo, W.W., and G. V. Chilinger. 1997. "Advances I the Electrokinetic Treatment of Hazardous Wastes in Soil and Groundwater". *Remediation Management*. 2:38-41.

Wilson, B.H., J.T. Wilson, and D. Luce. 1996. "Design and Interpretation of Microcosm Studies for Chlorinated Compounds". *Symposium on Natural Attenuation of Chlorinated Organics in Ground Water*. Dallas, TX.

CASE STUDY OF EX-SITU BIOLOGICAL TREATMENT OF PERCHLORATE-CONTAMINATED GROUNDWATER

A. Paul Togna, William J. Guarini, Sam Frisch, and Michael Del Vecchio
(Envirogen, Inc., Lawrenceville, NJ USA)

Jonna Polk and Cliff Murray
(United States Army Corps of Engineers, Tulsa, OK USA)

David E. Tolbert
(Longhorn/Louisiana Army Ammunition Plant, Doyline, LA USA)

Abstract: Groundwater in certain areas at Longhorn Army Ammunition Plant (LHAAP) (Texas) contains volatile organic compounds (VOCs) and perchlorate from past operations at the site. Groundwater from a Burning Ground and Landfill is currently being remediated by pumping the water from an interceptor collection trench system to the surface, removing VOCs and metals in an *ex-situ* treatment process, and discharging the treated water to a nearby stream. In early 2000, the Army Corps of Engineers Tulsa District, which oversees the operation of the groundwater treatment plant, took steps to supplement the existing treatment process with a biological fluid bed reactor (FBR) to remove the perchlorate prior to surface water discharge. After preliminary FBR sizing and costing information was obtained, a laboratory treatability program was conducted to confirm the system design assumptions and confirm the effectiveness of the FBR process for treating the LHAAP groundwater. Approximately 650 gallons (2,460 liters) of LHAAP groundwater containing 11,000 to 23,000 (average 16,500) μg/L of perchlorate were used for the evaluation. Both acetic acid and ethanol were investigated as growth (i.e. electron donor) substrates. For the majority of the test, effluent perchlorate concentrations were below the quantitation limit of 5 μg/L, except when the laboratory FBR was operated at a low substrate load to determine the point of treatment failure. The target effluent perchlorate concentration was 350 μg/L. Based on the success of the laboratory test, a full-scale FBR system with the capacity to treat 50 gallons per minute (gpm) (190 liters per minute) of LHAAP water has been installed at the groundwater treatment plant. System start-up occurred in February 2001.

BACKGROUND

Groundwater in certain areas at Longhorn Army Ammunition Plant (LHAAP) contains volatile organic compounds (VOCs) and perchlorate from past operations at the site. Groundwater from a Burning Ground and Landfill is currently being remediated by pumping the water from an interceptor collection trench system to the surface, removing VOCs and metals in an *ex-situ* treatment process, and discharging the treated water to a nearby stream. In early 2000, the

Army Corps of Engineers Tulsa District, which oversees the operation of the groundwater treatment plant, took steps to supplement the existing treatment process to remove the perchlorate prior to surface water discharge. After a review of existing treatment options, the Army Corps of Engineers chose a biological fluid bed reactor (FBR) (supplied by the team of Envirogen and USFilter/Envirex Products) to remove the perchlorate based on the FBR's proven field effectiveness (Greene and Pitre, 2000; Hatzinger et al., 2000; Sutton and Greene, 1999). Prior to installing the FBR, a laboratory pilot study was conducted to confirm the system design assumptions and show the effectiveness of the FBR process for treating the LHAAP groundwater. Based on the success of the laboratory test, a full-scale FBR system with the capacity to treat 50 gallons per minute (gpm) (190 liters per minute) of LHAAP water was installed at the groundwater treatment plant. System start-up occurred in February 2001. This paper describes the FBR process for treatment of perchlorate, summarizes the results of the laboratory pilot study, and describes the full-scale FBR system. As of the preparation of this paper, no full-scale operating data were available.

THE FBR PROCESS

The effectiveness of a biological system is dependent on maintaining a highly active biomass concentration in the bioreactor. Fixed-film bioreactors cultivate organisms that prefer to grow attached to surfaces. FBRs are highly efficient fixed-film bioreactors that rely on the immobilization of microbes on a hydraulically fluidized bed of media particles (Figure 1). These particles provide an extremely large surface area for growth of biological films, thus producing a large amount of biomass in a small reactor volume. The bed of particles is fluidized to reduce back-pressure resistance to flow by directing reactor influent passes through a distribution system at the bottom of the bed that establishes a uniform upflow velocity distribution within the fluidized bed. The fluidization rate is constant and set at a rate that causes a 25 to 30 percent expansion of the bed. As biological films grow on the particles, making them less dense, the bed further expands. Feed flow and recycle flow are blended to provide the necessary up-flow velocity for fluidization. The choice of media depends on the nature and concentration of the target compounds to be treated. Sand is sometimes chosen for treatment of water containing high concentrations of organics, where the inventory of biomass in the FBR is expected to be large (i.e., high biofilm growth applications). Granular activated carbon (GAC) is often selected when the treatment criteria is very stringent [i.e., treatment down to µg/L levels]. If needed, oxygen is supplied by a bubbleless aeration device (shown in Figure 1).

For treatment of perchlorate (and nitrate), oxygen is not added to the reactor. Instead, perchlorate (and/or nitrate) functions as the terminal electron acceptor, and an appropriate electron donor (typically an organic substrate) is added to the reactor in an amount sufficient to consume all the perchlorate (and residual oxygen and nitrate) present. Several microbial strains have been identified with the ability to degrade perchlorate by using the molecule as a terminal electron acceptor during growth on either an inorganic or organic substrate. The enzymatic pathways involved in perchlorate reduction have yet to be fully elucidated. However, a perchlorate reductase enzyme appears to catalyze an initial two-step reduction of perchlorate (ClO_4^-) to chlorate (ClO_3^-) and then chlorite (ClO_2^-). The chlorite is further reduced by a chlorite dismutase enzyme to chloride (Cl^-) and oxygen (O_2). Thus, the microbial degradation of perchlorate produces two innocuous products, chloride and oxygen.

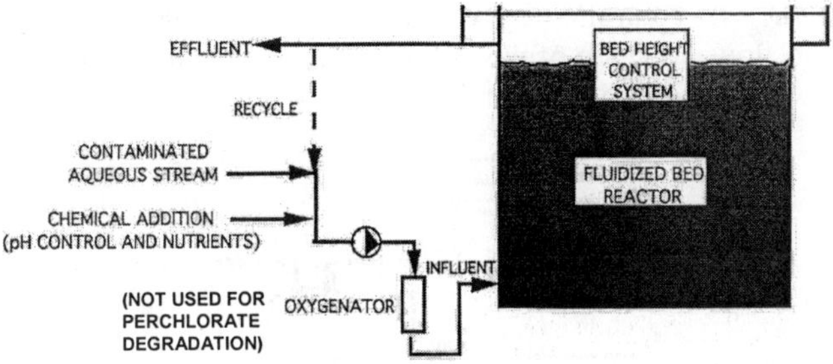

FIGURE 1. General FBR system process schematic.

FBR TREATABILITY STUDY

Overview. The laboratory FBR treatability program was designed to generate process data for use in designing and assessing the performance of a full-scale FBR to treat the LHAAP groundwater. Operating parameters were evaluated using two different electron donors (carbon sources). The program consisted of four phases:

Phase 1: Sample characterization;
Phase 2: Start-up and acclimation of biomass;
Phase 3: Steady operation of the laboratory FBR system;
Phase 4: Data interpretation and report generation.

Laboratory System Description. The laboratory FBR system was constructed of glass, stainless steel, and Teflon® materials to minimize abiotic chemical losses. The system had a total liquid volume hold-up of approximately 4 liters with an empty bed volume of approximately 1,000 mL. Figure 2 is a process flow schematic of the laboratory FBR system.

Groundwater samples from LHAAP were shipped to Envirogen in 55-gallon (200-liter) drums for processing through the laboratory FBR. The drums were kept in a cool environment prior to use. The groundwater feed to the FBR was introduced in the recycle line on the downstream side of the recycle pump. Based on the concentrations of perchlorate present, GAC was chosen as the biofilm support media for this application. Both acetic acid and ethanol were investigated as electron donor substrates for the LHAAP groundwater. These carbon sources were added with a syringe pump into the combined feed and recycle flow (i.e. influent flow) of the FBR at doses equal to or greater than the minimum theoretical amounts [i.e. minimum total organic carbon (TOC) requirements] needed to reduce all the perchlorate, chlorate, nitrate, and oxygen present in the feed, as well as support growth of biomass. The equations used to determine these amounts are similar to those used for nitrate reduction (i.e., denitrification) reactions (USEPA, 1993). The GAC bed in the reactor was initially charged to give a settled bed volume of approximately 800 mL, and was fluidized to approximately 1,000 mL using a peristaltic pump on the recycle line. The reactor pH was controlled by direct addition of caustic. A portion of the liquid that overflowed the top of the reactor was recycled to maintain fluidization; the balance exited as treated effluent.

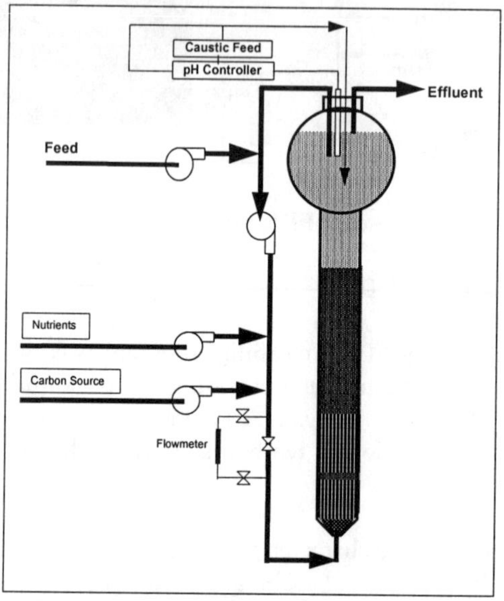

FIGURE 2. Laboratory FBR flow schematic.

The laboratory FBR system operated 24 hours a day, 7 days a week. The pre-acclimation of biomass (Phase 2) was accomplished by pumping a synthetic feed containing approximately 25,000 μg/L of perchlorate through the FBR. Pre-growth of biomass compresses the time required to attain steady-state conditions, thereby shortening the overall length of the test program. Groundwater was fed to the laboratory FBR during Phase 3 such that the hydraulic retention time (HRT) through the laboratory unit was the same as that which would occur if 50 gpm (190 L/min) were processed through a standard 5-ft (1.5-m) diameter full-scale FBR (i.e., standard EFB-5.0 design). Key process variables were routinely monitored and/or controlled, including pH, nutrient concentrations, perchlorate concentrations, and soluble TOC concentrations. Soluble nutrients [ammonia-nitrogen (NH_3-N) and ortho-phosphate-phosphorus (PO_4-P)] were continually added to the liquid in the recirculation line using a peristaltic pump to maintain residual NH_3-N and PO_4-P levels greater than 2 mg/L in the effluent. The pH of the reactor contents was automatically controlled through the addition of sodium hydroxide (NaOH). Influent and effluent grab samples were collected periodically for analysis of perchlorate and other compounds (see below). The quantitation limit for perchlorate was 5 μg/L. The temperature of the reactor was maintained at ambient conditions (70-75°F or 21-24°C).

Sampling and Analytical Methods. The initial wastewater characterization during Phase 1 was performed on a representative sample of the material received. The analytical methods used are outlined in Table 1. These same methods were used during the operation of the FBR in Phase 3. Additional analyses for chloride and sulfide were conducted in Phase 3 using EPA Methods 300.0 and 376.1, respectively. A standard toxicity screening assay was also conducted during Phase 1 to determine if anything inhibitory or toxic to microbial growth was present in the LHAAP groundwater; from the results of this assay, it was determined that nothing toxic was present.

During the operation of the laboratory FBR system, a log was maintained to record flowrates, temperatures, pH, dissolved oxygen (DO) readings, oxidation/reduction potential (ORP), and other pertinent operating data. Influent and effluent grab samples were collected as needed to assess the performance of the system.

Results and Conclusions. Twelve 55-gallon (200-liter) drums of treated groundwater from LHAAP arrived on 22 May 2000. These samples were collected after the LGAC unit at the LHAAP Groundwater Treatment Plant (following air stripping and metals precipitation also). The contents of three drums were analyzed for DO, perchlorate, chlorate, nitrate-nitrogen, and other components. The results are shown in Table 1. The groundwater samples contained an average of 3.8 mg/L of DO, 14.7 mg/L of perchlorate, 0.5 mg/L of chlorate, and 1.9 mg/L of nitrate-nitrogen. The samples contained minimal amounts of metals, suspended solids, organics, ortho-phosphate, and ammonia-nitrogen. Sulfate concentrations were approximately 300 mg/L.

The LHAAP groundwater began being processed through the laboratory FBR on 13 June 2000 following the Phase 2 start-up period. The flowrate of groundwater through the FBR was generally maintained at the flow needed to duplicate the hydraulic retention time (HRT) through a standard 5-ft diamater full-scale FBR (EFB-5.0) treating 50 gpm (190 L/min). For the first 3.5 weeks of testing, ethanol was used as the electron donor carbon source, and was added to the FBR at a rate that was higher than the theoretical requirement for complete reduction of all the perchlorate, chlorate, nitrate, and oxygen in the feed (including the amount of ethanol required for biomass growth). For the following 1.5 weeks of testing, the ethanol addition rate was reduced to the minimum theoretical requirement. For the final 3 weeks of testing, the electron donor carbon source was changed to acetic acid, and was fed at approximately the same excess TOC level as the initial feedrate of ethanol.

The treatability program ran smoothly, except for three periods when the groundwater feed supply to the FBR was interrupted due to problems with the feed pump. These events are shown in Figure 3, which summarizes the influent and effluent perchlorate data during the test. An effluent perchlorate concentration of less than 350 μg/L (the target perchlorate level for LHAAP) was maintained throughout the project using either ethanol or acetic acid as the carbon substrate, except when the laboratory FBR was operated at the minimum theoretical TOC requirement. For the majority of the experiment, effluent perchlorate concentrations were below the quantitation limit of 5 μg/L. In summary, the data showed that a 5-ft (1.5-m) diameter Envirogen/USFilter Envirex Products FBR designed with a standard expanded bed height or greater will achieve the desired perchlorate effluent quality when treating 50 gpm (190 L/min) of LHAAP groundwater (after air stripping and metals precipitation).

TABLE 1. Initial characterization of water from LHAAP groundwater treatment plant.

Parameter	Method	Units	Average	Std. Dev.
Oxygen	D.O. Probe	mg/L	3.8	0.4
Perchlorate	EPA 314.0	mg/L	14.7	0.4
Chlorate	EPA 300.0	mg/L	0.5	0.0
Nitrate-N	EPA 300.0	mg/L	1.9	0.1
Nitrite-N	HACH Method 8507	mg/L	0.013	0.003
Ortho-phosphate-P	EPA 300.0	mg/L	<0.2	Not Applicable
Ammonia-N	EPA 350.2	mg/L	<0.5	Not Applicable
Sulfate	EPA 300.0	mg/L	303	11.5
Chemical Oxygen Demand	EPA 410.4	mg/L	30	23.2
Total Organic Carbon	EPA 415.1	mg/L	<1	Not Applicable
Oil & Grease	EPA 413.1	mg/L	Less than 10 mg/L for a composite sample	Not Applicable
Total Suspended Solids	EPA 160.2	mg/L	10	5.3
Volatile Organic Contaminants	SW-846 8260	mg/L	Less than 0.10 to 0.05 mg/L for all on 8260 list except for acetone @ 0.18 mg/L (one of two samples)	Not Applicable
Priority Pollutant Metals	EPA 200.7 and EPA 245.1 (Hg)	ug/L	Less than PQL for all on 200.7 list (and Hg) except for Ni @ 1.8 ug/L and Zn @ 165 ug/L	0.1 for Ni and 47.4 for Zn
Broth Tube Toxicity/Inhibition Test	Internal SOP	N/A	Not Toxic or Inhibitory	Not Applicable

FULL-SCALE SYSTEM DESCRIPTION

Based on the success of the laboratory test, a full-scale FBR system with the capacity to treat 50 gpm (190 L/min) of LHAAP water has been installed at the groundwater treatment plant. System start-up occurred in February 2001. The electron donor carbon source for the reactor was chosen to be 50% acetic acid. As of the preparation of this paper, no operating data were available.

The FBR system is composed of a Fluidized Bed Reactor Vessel and an FBR Equipment Skid (see Figure 4). The reactor vessel is 5 ft (1.5 m) in diameter and 21 ft (6.4 m) tall. The influent to the FBR (i.e., the combined feed and recycle) is distributed through a proprietary distribution header at the bottom of the tank. This header is designed to distribute flow evenly and smoothly across the cross-sectional area of the reactor with a minimum amount of turbulence.

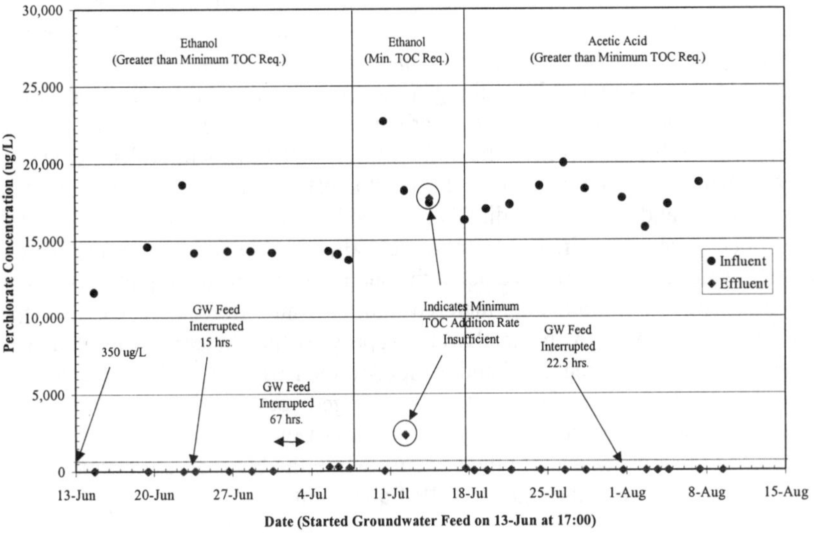

FIGURE 3. Performance of laboratory FBR.

FIGURE 4. Full-scale FBR system during construction.

To remove excess biomass, there are two patented biomass separation systems at the top of the bed and a third (patent pending) in-bed media cleaning system which can be positioned anywhere along the bed vertically. These devices function to remove excess biomass from the surface of the carbon particles, preventing them from being carried out of the reactor. When biological growth occurs on the fluid bed media, the diameter of the media increases and its effective density is reduced, resulting in an expansion of the media bed beyond that due to fluidization of the bare carbon media. The excess biomass that is separated from the media exits the system through the effluent collection system. Stray carbon particles that exit in the reactor effluent are collected in a Media Capture Tank, and can be periodically pumped back to the reactor.

Feed water is drawn from an equalization tank and combined with recycled water from the FBR vessel at the suction of the influent pump(s). Acetic acid and inorganic nutrients are metered into the water after the influent pump(s) at a rate proportional to the feed water supply to the system. The feed water leaves the FBR Equipment Skid and is piped to the FBR vessel. The treated water at the top of the reactor flows into a submerged recycle collector pipe and an effluent collector pipe under gravity. A portion of the fluid exits the system (i.e., the same volume as the feed to the system), while the balance is recycled back to the FBR vessel influent through the FBR Equipment Skid.

ACKNOWLEDGEMENTS

The authors would like to thank Casey Whittier, Gene Mazewski, and Chris Bryan of USFilter/Envirex Products, and Todd Webster from Envirogen for their assistance in the design, installation, and start-up of the FBR system. The authors would also like to thank Cyril Onewokae of the U.S. Army Operations Support Command, Rock Island, IL for his support of this project.

REFERENCES

Greene, M. R. and M. P. Pitre. 2000. "Treatment of Groundwater Containing Perchlorate Using Biological Fluidized Bed Reactors with GAC or Sand Media." In E. T. Urbansky (Ed.), *Perchlorate in the Environment*, pp. 241-256. Kluwer Academic/Plenum Publishers, New York, NY.

Hatzinger, P. B., M. R. Greene, S. Frisch, A. P. Togna, J. Manning, and W. J. Guarini. 2000. "Biological Treatment of Perchlorate-contaminated Groundwater Using Fluidized Bed Reactors." In G.B. Wickramanayake et al. (Eds.), *Case Studies in the Remediation of Chlorinated and Recalcitrant Compounds*, pp. 115-122. Battelle Press, Columbus, OH.

Sutton, P. M. and M. R. Greene. October 1999. "Perchlorate Treatment Utilizing GAC and Sand Based Biological Fluidized Bed Reactors." Proceedings of the 1999 Water Environment Federation Conference, New Orleans, LA.

U.S. Environmental Protection Agency. September 1993. "Nitrogen Control Manual." EPA/625/R-93/010. Washington, D.C., pp. 232-248.

In-Situ Bioremediation of Perchlorate-Contaminated Soils

James R. Kastner, K.C. Das, *Valentine A. Nzengung*, John Dowd, Jim Fields
(University of Georgia, Athens GA)

ABSTRACT: Previous research indicates that microbial communities in the rhizosphere are capable of transforming perchlorate to chloride, given an external carbon source such as acetate. Carbon source requirement in perchlorate transformation was independently confirmed in batch slurry experiments conducted with perchlorate-contaminated soils amended with acetate. Rapid degradation of perchlorate after an initial incubation period of about 7 days was observed in acetate-amended soils compared to controls. In subsequent batch studies, perchlorate contaminated soils from the Longhorn Army Ammunition Plant (LHAAP), were treated with organic amendments and liquid carbon sources to determine their ability to stimulate perchlorate biodegradation. The organic amendments tested included poultry and cow manure and cotton waste, and the liquid carbon sources included ethanol, molasses, and methanol. The different amendments stimulated the biodegradation of perchlorate in contaminated soils, with cotton waste resulting in slower rates compared to the other carbon sources. Based on our successful laboratory experiments, pilot tests were initiated at LHAAP. The pilot in-situ bioremediation started in October 2000 consisted of nine 1 x 1m treatment cells, three of which were amended with one of three different carbon sources; poultry manure, cow manure or ethanol. The maximum concentration of perchlorate in the selected treatment plots at the start of the pilot study was 400 mg/kg. Recent site analysis indicates that poultry manure and ethanol addition reduced perchlorate concentrations ranging from 80 to 250 ppm to below detection within 60 and 100 days, respectively, at depths of 24 and 36 inches.

INTRODUCTION

Most research on perchlorate bioremediation has centered on above ground pump and treat systems which are costly and do not treat source zones. Little research has focused on developing techniques to stimulate in-situ perchlorate remediation. However, previous research indicates that microbial communities in the rhizosphere are capable of transforming perchlorate to chloride (Nzengung et al., 1999; 2000). Microbial species capable of using perchlorate and chlorate as a terminal electron acceptor have been isolated from activated sludge, anaerobic digestors, biosolids, and soil (Logan et al., 2000; Rikken et al., 1996). These data suggest that subsurface microbial communities are capable of perchlorate transformation if given external carbon sources, under proper environmental conditions.

Past industrial operations, testing, or training activities at numerous Department of Defense (DoD) installations have resulted in the release of perchlorate into the soil, surface water, and groundwater. Perchlorate is a highly water soluble, potentially toxic oxy-anion (Urbansky, 1998). At the Longhorn Army Ammunition Plant (LHAAP), soil and groundwater perchlorate concentrations have significantly exceeded the action levels of 270 μg/kg and 66 μg/L respectively (Texas). Cost-effective technologies to remediate perchlorate is required to assist the U.S. military in meeting its stewardship goals.

A small body of literature suggests that perchlorate-reducing microorganisms are present in groundwater, soils, and sediments. Our group's recent research establishing perchlorate remediation via phytoremediation indicates that microbial systems in the rhizosphere contribute significantly to perchlorates transformation process. However, limited information is available on microbial communities in the subsurface (outside the rhizosphere) capable of transforming perchlorate.

Objectives. The primary purpose of this research was to determine if sub-surface microbial communities would transform perchlorate, thus demonstrating the feasibility of performing in-situ, perchlorate remediation.

METHODS

Site Description. The LHAAP site is located in a moist, sub-humid to humid, mild climate with an average annual rainfall of 46 inches and precipitation is fairly evenly distributed throughout the year. The depth to groundwater across the facility ranges from 1 ft to 70 ft below ground surface, with the depth to groundwater being typically 12 to 16 feet. LHAAP is presently inactive and scheduled to be excessed. A 1998 Remedial Investigation/Feasibility Study (RI/FS) for the LHAAP indicates that perchlorate has seriously impacted surface water, groundwater and soils at the site.

Batch Reactor Studies. Contaminated site soil was collected for batch treatability studies. A range of carbon sources, including cow manure, poultry manure, methanol, ethanol, molasses, and cotton gin trash were tested. Typically an extract of the agricultural residues was prepared and mixed with the soil and spiked with perchlorate when needed (e.g., 500g of dry cow manure was mixed with 1 L of DI water, 10 ml of this filtered extract mixed with 25 g of soil and spiked to achieve 34.3 ppm perchlorate). Methanol, ethanol and molasses, were diluted and mixed with soil as a slurry and perchlorate spiked into the reactors. Reactors were incubated statically at room temperature.

Modeling. HYDRUS-2D was used to model water and solute transport, and perchlorate biodegradation in the vadose zone (Simunek et al., 1999). The flow equation utilized is a two-dimensional variably saturated form of the Richard's Equation. Model parameters were estimated in the following manner. A Guelph permeameter was used to determine the in-situ saturated hydraulic conductivities of the soils beneath the pilot scale location. Soil samples were collected and transported to the University of Georgia. These soil samples were utilized to further refine the saturated hydraulic conductivities using falling head permeameter methods (Lambe, 1951). Other soil samples were used to determine the water retention curves for each of the soil horizons using Tempe cells. The relationship of pressure (ψ) versus volumetric soil moisture content (θ) and hydraulic conductivity (K_u) necessary to solve the flow and transport equations were determined from the water retention curves for each soil sample using Tempe cells (Richards, 1965). The soil samples were also used to determine size fraction, cation exchange capacity, bulk density, porosity, and percent organic carbon. Biodegradation rate constants were estimated from the batch treatability studies and K_d values for perchlorate and the carbon sources determined via batch partition studies.

Pilot Scale Studies. The pilot scale demonstration study was conducted at a former pilot scale wastewater treatment plant on the site and consisted of six 15 x 9 ft treatment plots and an 18 x 18 ft control plot. Previous soil analysis indicated that perchlorate concentration ranged from 36,200 to 144,000 μg/kg (0-2 ft). Perchlorate groundwater concentration was reported as 22,000 μg/L. Soil cores were obtained from the site (42) to determine spatial distribution of perchlorate in the soil and other soil parameters (e.g., TOC). Based on these data, cells to receive carbon source addition were identified. Each cell was tilled to ~12 inches and trenches installed to isolate each cell (24 ft). An attempt to hydraulically isolate each cell was made by installing plastic liners vertically inside the trenches. Liners were hung from a metal frame grid that was installed in place. Solid carbon sources were added to each of the cells and mixed with the tilled soil, and ethanol was added with the water source. Water was added in two stages to saturate the soil down to 12 and then 24 inches. Water saturation was monitored using tensiometers installed at 12, 24, and 36 inches below land surface (bls). Soil cores were periodically obtained at different depths for perchlorate analysis. In addition, oxidation-reduction potentials (ORP) were measured in multiple locations and depths in each cell. Each cell was covered during the incubation period.

Analytical Methods. From each soil core, six 10 g portions taken from different depths were extracted with deionized water and analyzed for perchlorate. The soil was extracted several times by homogenizing for 10 min with 100 mL of solution in a tissue homogenizer. For soils rich in organic matter (10% by weight), most of the sorbed perchlorate was desorbed using 1 mM NaOH solution. Each slurry was sonicated for 30 minutes and allowed to cool to room temperature, and separated from the aqueous-soil phase by centrifugation at 20,000 RPM for 30 minutes. The supernatant was passed through a 0.45 and 0.25 prewashed activated alumina (0.2 μm, Gelman Acrodisk ion membranes, Fisher Scientific, Fairlawn, NJ). These extracts were diluted as needed for IC analysis. A Dionex DX-100 IC and a Dionex DX500 with self-regenerating suppressor (ASRS) control were used for perchlorate and other anions analyses. The IC was equipped with a gradient pump run in isocratic mode, conductivity detector, autosampler, auto-injector, and an Ionpac AS 16 (4-mm x 250 mm) column and guard column. The column was operated at 30°C and the flow rate of solvent (sodium hydroxide 50 mM) was 1.0 ml/min. The injection loop volume was varied between 25 μL for high perchlorate concentrations in mg/L range and 1000 μL for low concentrations (μg/L). Water as a mobile phase was used for regeneration of ASRS and the conductivity was set below 10 μS. DI water (18 MΩ-cm) was used as a system blank sample to establish the baseline and to confirm the presence of or lack of contamination in the system. The detection limit for perchlorate was determined to be 2 μg/L. An Ionpac AS9-HC analytical column with an AG9-HC guard column and 9 mM Na_2CO_3 as eluent was used for the analysis of chloride, chlorite, chlorate, nitrate, and phosphate ions.

RESULTS AND DISCUSSION

The addition of an external carbon sources biostimulated the reduction of perchlorate in the site soil within 2 to 5 days. In the batch studies, perchlorate levels were typically reduced from 25-35 ppm to 0-5 ppm depending on the carbon source. Time course data for poultry manure and methanol indicated that these carbon sources gave the highest perchlorate transformation rates (Fig. 1-2). Subsequent analysis using ethanol and cow manure as a carbon source indicated similar results to methanol and chicken manure (data not shown). Ethanol was chosen as a carbon source due to its ability to stimulate perchlorate transformation and its ready availability on-site.

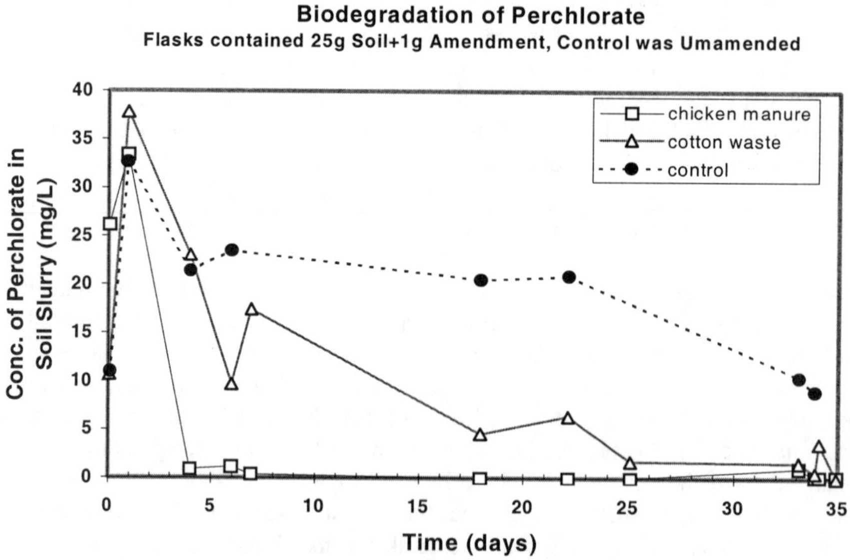

FIGURE 1: Effect of chicken manure and cotton waste on the transformation of perchlorate in batch reactors using site soil.

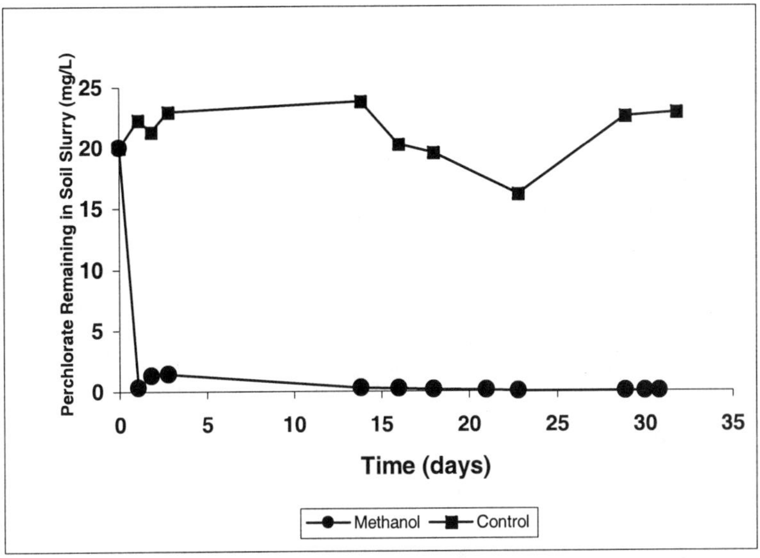

FIGURE 2: Effect of methanol on the transformation of perchlorate in batch reactors using site soil.

Based on these batch treatability results, poultry manure, ethanol, and cow manure were used in the pilot scale studies. Recent site analysis indicates that poultry manure and ethanol stimulated perchlorate biodegradation in the subsurface, relative to a control treatment (Fig. 3). Some perchlorate transport is indicated in the top section of the plots (0-12') due to the measured loss of perchlorate in the control (Fig. 3). However, biodegradation of perchlorate is indicated due to the complete exhaustion of perchlorate in the soil treated with ethanol and chicken manure at depths of 24 and 36 inches, relative to a constant perchlorate concentration in the control cell at these depths. Future research will focus on measuring biodegradation kinetics, microbial isolation, and bioremediation scale-up using vadoze zone modeling. The results of this pilot study demonstrate that perchlorate contaminated soils can be treated in-situ by applying the cost-effective techniques we have developed to deliver nutrients amendments to desired depths.

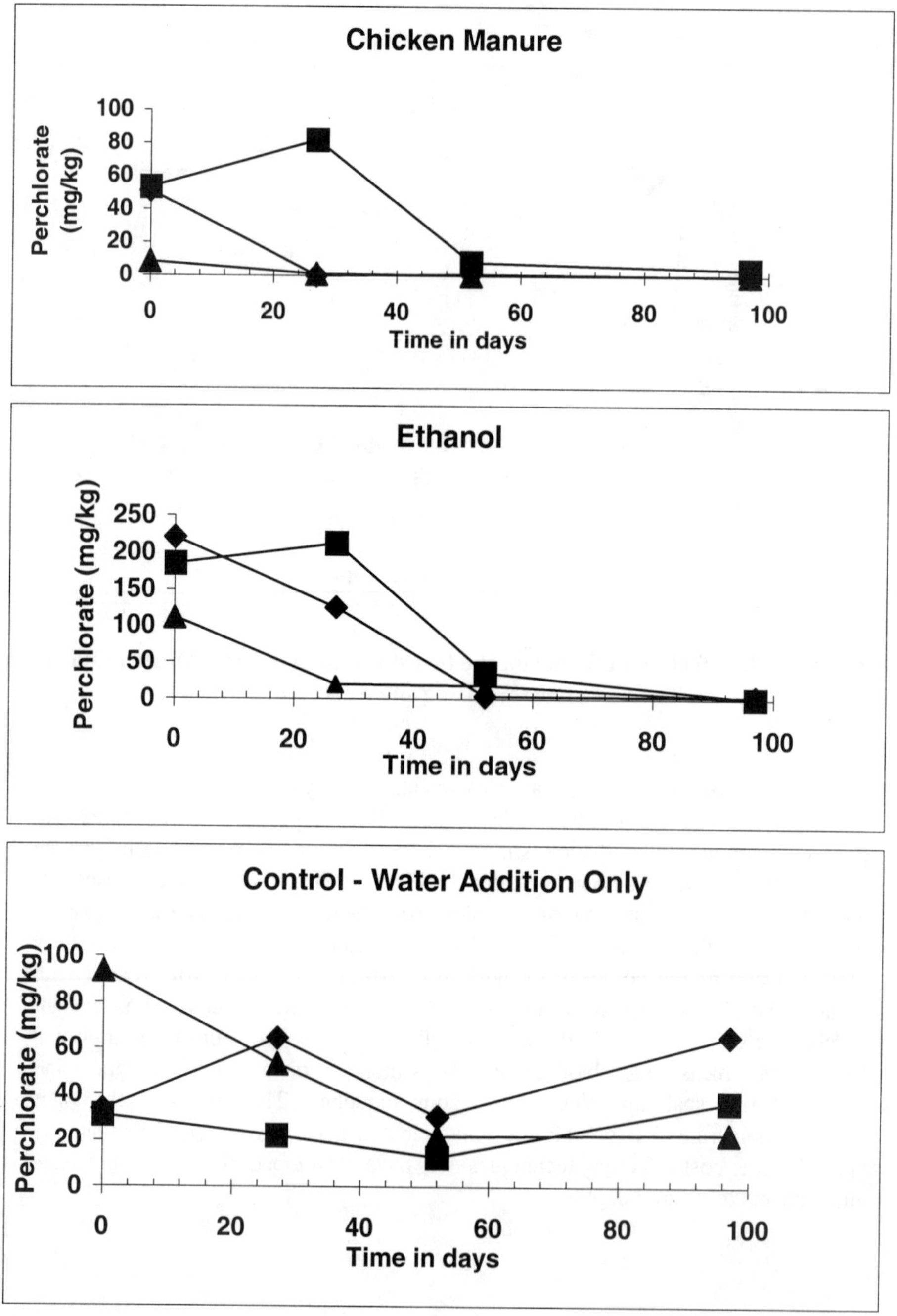

FIGURE 3: Effect of carbon source addition on in-situ biodegradation of perchlorate at different depths (0-12 in [▲]; 12-24 in [◆]; 24-36 in [■]). Pilot scale demonstration study at LHAAP in Karnack, Texas.

REFERENCES

Lambe, T.W. 1951. *Soil Testing for Engineers.* John Wiley & Sons, New York, 165 pp.

Logan, B.E., Zhang, H., Wu, and J., Unz, R. 2000. "The Potential for in situ perchlorate degradation." In Case studies in the remediation of chlorinated and recalcitrant compounds. The 2nd International Conference on Remediation of Chlorinated and Recalcitrant Compounds. Editors Wicramanayake, G.B., Gavaskar, A.R., Gibbs, J.T., and Means, J.L. Battelle Press. *C2*(7), 87-92.

Nzengung, V.A., and C. Wang. 2000. "Influences on phytoremediation of perchlorate contaminated Water." *American Chemical Society (ACS) Special Symposium Series: Perchlorate in the Environment. Editor: Urbansky. Kluwer Acad./Plenum Publ., NY.* Chapter 21, pp 219 - 229.

Nzengung, V.A., C. Wang, and G. Harvey. 1999. "Plant-mediated transformation of perchlorate into chloride." *Environ. Sci. Technol., 33*: 1470-1478.

Rikken, G.B., A.G.M. Kroon, and C.G. van Ginkel. 1996. "Transformation of (per)chlorate into chloride by a newly isolated bacterium: reduction and dismutation." *Appl. Microbiol. Biotechnol. 45*: 420-426.

Richards, L.A. 1965. "Physical condition of water in soil." *Methods of Soil Analysis, Part 1*, ed. C.A. Black. American Society of Agronomy, Madison, Wis., pp. 128-152.

Simunek, J., M. Sejna, and M. Th. Van Genuchten. 1999. *The HYDRUS-2D Software Package for Simulating the Two-Dimensional Movement of Water, Heat, and Multiple Solutes in Variably-Saturated Media, Version 2.0*, U.S. Salinity Laboratory, USDA, ARS, Riverside, California.

Urbansky, E.T. 1998. "Perchlorate Chemistry: Implications for Analysis and Remeidation." *Bioremediation Journal. 2*(2): 81-95.

SUCCESSFUL FIELD DEMONSTRATION OF IN SITU BIOREMEDIATION OF PERCHLORATE IN GROUNDWATER

Michaye L. McMaster and ***Evan E. Cox*** (GeoSyntec Consultants, Guelph, Ontario), Scott L. Neville and Laurence T. Bonsack (Aerojet, Sacramento, California)

ABSTRACT: A field demonstration was conducted to demonstrate the ability to accelerate in situ bioremediation of perchlorate in a deep aquifer (100 feet below ground surface) at the Aerojet Superfund Site in Sacramento, California. Perchlorate biodegradation was readily initiated, without apparent lag, through acetate addition. Perchlorate concentrations in the pilot test area groundwater declined from 12,000 μg/L to less than the Provisional Action Level (PAL) of 18 μg/L and even the method detection limit (MDL) of 4 μg/L within 15 feet of the electron donor delivery well. The in situ perchlorate biodegradation half-life ranged from 0.2 to 1.8 days, which is consistent with rates calculated from laboratory microcosm studies for this site. As a benefit, perchlorate biodegradation did not require highly reducing conditions, and appears to have minimal impact on secondary water quality parameters. These data confirm that in situ bioremediation will be a viable and cost-effective treatment technology for perchlorate-impacted groundwater.

INTRODUCTION

Groundwater contamination related to the production, handling and use of rocket propellants such as ammonium perchlorate has been identified as a widespread problem. It is estimated that perchlorate has been manufactured and/or used in 44 states, resulting in groundwater contamination in at least 14 of these states (Damian and Pontius, 1999). Perchlorate in groundwater and drinking water supplies causes concern due to possible human health effects, specifically related to thyroid function (Urbansky, 1997). While a national regulatory standard has yet to be set, the California Department of Health Services (CDHS) established a provisional action level (PAL) of 18 μg/L for perchlorate in drinking water.

Cost-effective remediation of perchlorate-contaminated groundwater is challenging due the physical and chemical properties of perchlorate. In groundwater, perchlorate is generally unreactive, it does not strongly sorb to soil materials and tends to behave like chloride, which is commonly considered a conservative tracer. Potential remedial technologies identified include: ion exchange, reverse osmosis, electrochemical reduction and bioremediation. Of the remedial technologies being developed, bioremediation is among the most promising because it has the potential to destroy perchlorate rather than transferring it to another waste stream (e.g., impacted resin or brine) requiring costly treatment or disposal. Accordingly, significant effort has been directed in recent years to the development of ex situ bioremediation treatment systems. While these ex situ bioreactors have been shown to effective in removing perchlorate from impacted groundwater they require costly long-term operations and maintenance (O&M). Passive and semi-passive in situ bioremediation may afford a less costly and less O&M-intensive approach to managing and remediating perchlorate-impacted groundwater. Specifically in situ bioremediation may aid in treating or controlling perchlorate source areas that serve as long-term contributors to groundwater.

Recently, GeoSyntec and others have been evaluating the applicability of in situ bioremediation for perchlorate-impacted groundwater. Our work has included bench-scale microcosm studies using soil and groundwater from more than a dozen different impacted sites around the nation. Results have been extremely promising, suggesting that bacteria capable of perchlorate-reduction are ubiquitous in subsurface environments, and that perchlorate can be rapidly biodegraded over a wide range of concentrations and starting conditions. This paper summarizes key results from bench-scale studies for multiple perchlorate-impacted sites and presents the results from our recent in situ field demonstration of perchlorate bioremediation at the Aerojet Superfund Site.

OVERVIEW OF PERCHLORATE BIODEGRADATION

Perchlorate biodegradation results from microbially-mediated redox reactions, whereby perchlorate serves as the electron acceptor, and is reduced via chlorate to chlorite. Chlorite then undergoes a biologically-mediated dismutation or disproportionation reaction, releasing chloride and oxygen. The oxygen is subsequently reduced, provided sufficient electron donors are available. Figure 1 shows the hypothesized pathway for perchlorate reduction.

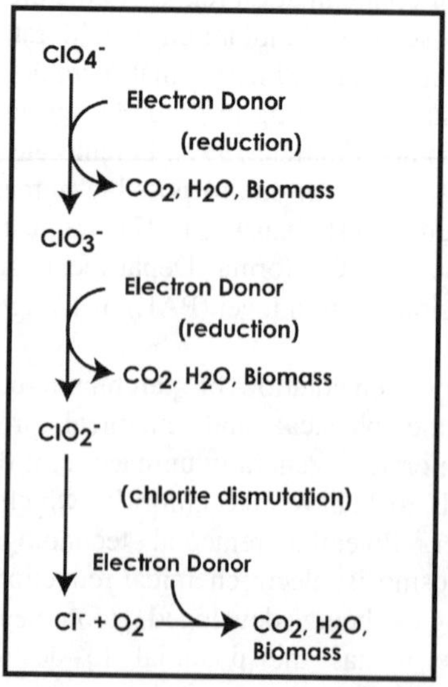

FIGURE 1. Pathway for the reduction of perchlorate.

A variety of electron donors have been used to stimulate perchlorate reduction using pure or mixed microbial cultures, including alcohols (e.g., ethanol, methanol), volatile fatty acids (e.g., acetate), edible oils (e.g., canola oil) and some sugar mixtures. To date, a number of microorganisms have been identified as possessing the ability to reduce perchlorate, and this information has been used to develop ex situ bioreactor processes to remove high perchlorate concentrations from rocket motor wash water (Attaway and Smith, 1993; Wallace et al., 1996) and low concentrations from groundwater (Greene and Pitre, 2000).

RESULTS FROM MICROCOSM STUDIES

In 1999, GeoSyntec was awarded a research grant under the U.S. DoD Strategic Environmental Research & Development Program (SERDP) to evaluate the ubiquity of perchlorate-degrading bacteria in differing geographical, geological and geochemical environments, and to assess the widespread applicability of in situ bioremediation as a remediation technology for perchlorate-impacted DoD sites. Laboratory microcosm studies were conducted for 9 test sites from 6 independent facilities around the nation. Perchlorate biodegradation was observed at all test sites, indicating that the distribution of perchlorate-biodegrading bacteria in subsurface environments is widespread. A range of inexpensive organic carbon-based electron donors, including acetate, sugars and canola oil stimulated perchlorate biodegradation over perchlorate concentrations ranging from 250 μg/L to in excess of 660,000 μg/L. Biodegradation typically reduced perchlorate concentrations below the PAL of 18 μg/L, making in situ bioremediation an appropriate technology for site remediation. Figure 2 presents results of from typical microcosm studies for a site in California.

SUCCESSFUL FIELD DEMONSTRATION

The field demonstration was initiated in May 2000 to demonstrate the ability to accelerate in situ bioremediation of perchlorate in a deep aquifer (100 feet below ground surface). The pilot test was implemented within the core of a perchlorate plume containing perchlorate concentrations in the range of 10,000 to 15,000 μg/L. The pilot test infrastructure consisted of a closed loop recirculation system, whereby groundwater was extracted from the aquifer, amended with electron donor (acetate; 4 one-hour pulses per day for a time weighted average concentration of 50 mg/L) and re-injected to the aquifer to promote perchlorate reduction in situ. Baseline groundwater redox conditions in the pilot test area (PTA) were aerobic and oxidizing, with dissolved oxygen (DO) concentrations typically ranging between 2 and 5 mg/L and ORP ranging from 143 to 263 mV. Consistent with the prevailing aerobic redox conditions, nitrate and sulfate were present in the groundwater at concentrations ranging up to 5 and 13 mg/L, respectively.

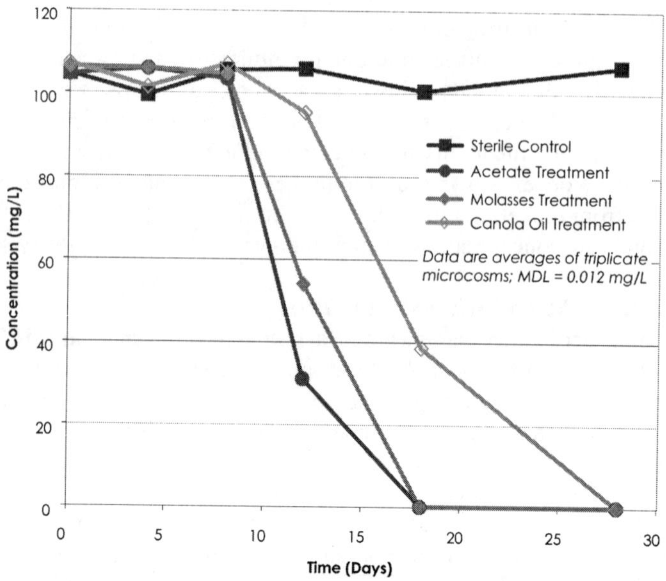

FIGURE 2. Perchlorate biodegradation in groundwater microcosms, California site

Performance monitoring in the PTA was accomplished using two monitoring wells screened across the target aquifer (85 to 100 feet bgs) and located at distances of 15 feet (well 3601) and 35 feet (well 3600) from the electron donor delivery well. PTA hydraulics (e.g., pore volume, residence time, travel times to biomonitoring wells) were estimated through conservative tracer (bromide) testing. Based on the tracer test results, the average travel times for non-retarded particles (similar to perchlorate) to reach these wells was estimated at 2.5 and 7 days, respectively.

Figure 3 presents perchlorate biodegradation results in the performance monitoring wells. Perchlorate biodegradation was readily initiated, without apparent lag, through acetate addition. Perchlorate concentrations in the PTA groundwater rapidly declined from 12,000 μg/L to less than the PAL of 18 μg/L and even the method detection limit (MDL) of 4 μg/L within 15 feet of the electron donor delivery well. The in situ perchlorate biodegradation half-life ranged from 0.2 to 1.8 days, which is consistent with rates calculated from laboratory microcosm studies for this site. As a benefit, perchlorate biodegradation did not require highly reducing conditions throughout the PTA. Specifically, DO concentrations in the performance monitoring wells (3600 and 3601) generally remained above 1 mg/L, while the oxidation-reduction potential was mildly reducing to slightly oxidizing ORP (19.7 to -25.6 mV at Day 60). These data have important implications for site remediation in that it may be possible to biodegrade perchlorate while not significantly impacting secondary water quality parameters (i.e., creating anaerobic groundwater with high dissolved metals, sulfide and methane). As an added benefit, nitrate concentrations in the PTA were reduced, while sulfate reduction was not initiated.

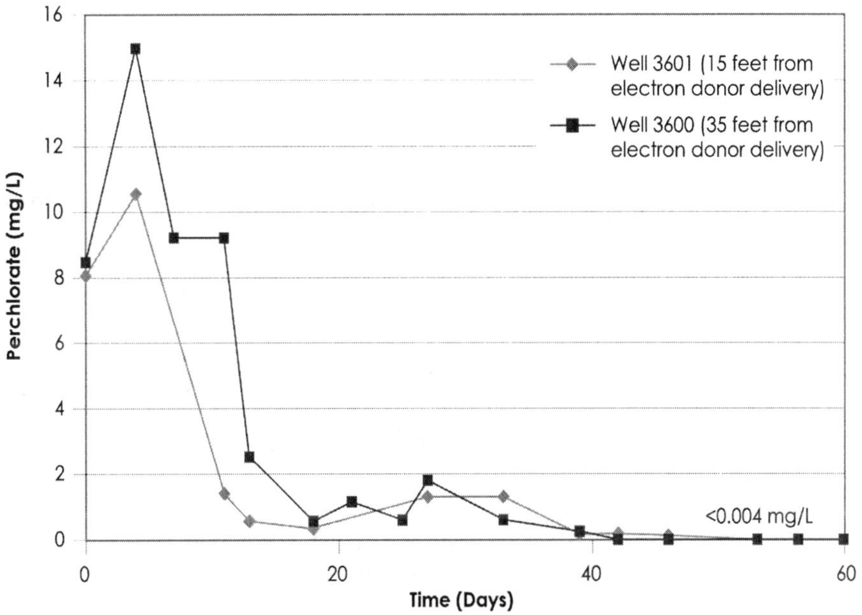

FIGURE 3. Results of in situ bioremediation of perchlorate, Aerojet Superfund site

CONCLUSIONS

The results of these studies and our on-going research suggest that bacteria capable of perchlorate reduction are ubiquitous in subsurface environments, and that perchlorate reduction in soil and groundwater can be stimulated through provision of a wide variety of electron donors. Given this, in situ bioremediation is likely to have a bright future as a low maintenance, cost-effective remedial approach for perchlorate-impacted groundwater in several ways. Firstly, the ability to stimulate rapid biodegradation of perchlorate at starting concentrations ranging between 100 to 1,000 mg/L indicates that in situ bioremediation may be effective in directly treating the perchlorate source areas that are the driving force behind the expansive perchlorate plumes at many sites. Secondly, the use of passive or semi-passive permeable bio-barriers should provide a cost-effective means of controlling plume migration and protecting receptors. Finally, the potential to jointly biodegrade a variety of common co-contaminants, including chlorinated solvents and by-products of hydrazine-based liquid rocket fuels (e.g., n-nitrosodimethylamine; NDMA) in the same in situ treatment system is of obvious benefit in reducing long-term O&M costs associated with conventional pump and treat based remedies.

ACKNOWLEDGMENTS

The authors would like to thank Dr. Elizabeth Edwards and Alison Waller at the University of Toronto and Gerald Swanick and Michael Girard at Aerojet for assistance in these and related perchlorate studies. The authors would also like to acknowledge SERDP funding of the laboratory microcosm studies.

REFERENCES

Attaway, H., and M. Smith. 1993. "Reduction of Perchlorate by an Anaerobic Enrichment Culture". *J. Ind. Microbiol.* 12:408-412.

Damian, P., and F.W. Pontius. 1999. "From Rockets to Remediation: The Perchlorate Problem." *Environmental Protection.* June 1999.

Greene, M.R., and M.P. Pitre. 2000. "Treatment of Groundwater Containing Perchlorate using Biological Fluidized Bed Reactors with GAC or Sand Media". In E.T. Urbansky (Ed.), *Perchlorate in the Environment.* Kluwer Academic Publishers: New York.

Urbansky, E.T. 1997. *Review and Discussion of Perchlorate Chemistry as Related to Analysis and Remediation.* U.S. EPA, December 1997.

Wallace, W., T. Ward. A. Breen, and H. Attaway. 1996. "Identification of an Anaerobic Bacterium which Reduces Perchlorate and Chlorate as *Wolinella succinogenes.*" *J. Ind. Microbiol.* 16:68-72.

PERCHLORATE DEGRADATION IN BENCH- AND PILOT-SCALE EX-SITU BIOREACTORS

Bruce E. Logan and Kijung Kim
(The Pennsylvania State University, University Park, Pennsylvania)
Steven Price (Camp, Dresser and McKee, Walnut Creek, California)

ABSTRACT: Perchlorate can be biologically degraded in fixed bed bioreactors, but there is little design information available on optimizing perchlorate removal rates in these reactors. During the past few years, we have been developing bioreactors to treat perchlorate-contaminated water, and analyzing perchlorate removal efficiencies in different bioreactor systems. In this paper, we review our previously published bioreactor work on mixed and pure-culture reactors for degrading perchlorate. In general, perchlorate removal rates were found to be proportional to the log mean perchlorate concentrations in the reactor. However, a fixed-bed reactor inoculated with a perchlorate-degrading bacterium, designated strain KJ (isolated in our laboratory) achieved perchlorate degradation rates nearly an order-of-magnitude higher than reported by others at similar perchlorate feed concentrations. Field tests are planned for treating perchlorate-contaminated water from the Redlands-Crafton plume during the summer of 2001.

INTRODUCTION

It is now well established that perchlorate can be degraded by microorganisms that use perchlorate as a terminal electron acceptor during the oxidation of both inorganic and organic substrates (Wallace et al., 1998; Logan, 1998; Herman and Frankenberger, 1998; Coates et al., 1999; Miller and Logan, 2000). Bioreactor systems were originally developed to remove high concentrations of perchlorate (grams per liter) present in wastewaters used to wash rocket casings to low concentrations (milligrams per liter). With the advent of ion chromatographic methods capable of detecting perchlorate to the low parts-per-billion range (Urbansky, 1998), it soon became clear that there was extensive low-level perchlorate contamination of drinking water sources, particularly in the arid Southwestern United States. Thus, bioreactor systems were needed to treat perchlorate down to the detection limit by ion chromatography (approximately 2-4 μg/L) (Logan and Kim, 1998; Urbansky, 1998; Herman and Frankenberger, 1999).

Research at Pennsylvania State University has concentrated on developing fixed-bed bioreactors to treat perchlorate to non-detectable levels (defined here as 4 μg/L). Our initial efforts were directed at proving fixed-bed bioreactor technology. Perchlorate-degrading enrichments were developed from biomass obtained from a wastewater treatment plan. These enrichments were used to inoculate sand columns. It was shown that perchlorate could be reduced from ~20 mg/L to <4 ug/L in these fixed bed reactors.

Our recent laboratory studies have been directed at optimizing the process and determining reactor detention times necessary to completely remove perchlorate from waters containing either high (~20 mg/L) or low (~80 μg/L) perchlorate concentrations. Here, we briefly summarize the findings obtained in our laboratory for the purposes of providing a consolidated review of our most recent work. Perchlorate removal rates obtained in our laboratory reactors are then compared to those obtained by others for the purposes of establishing design criteria of systems treating a variety of water and wastewaters containing different perchlorate concentrations. Based on own studies, we also discuss plans for field testing a unit to remove perchlorate from a contaminated groundwater in Southern California.

METHODS

Mixed and Pure Cultures. A perchlorate-degrading enrichment was developed using wastewater from a primary digester with perchlorate (1 g/L) and acetate (1 g/L) as described in Kim and Logan (2000). The perchlorate degrading isolate, KJ, was obtained by Mulvaney (1999) from the bioreactor described by Logan and Kim (1998). KJ has a maximum growth rate using acetate of 0.14 h^{-1} with perchlorate and 0.27 h^{-1} with oxygen as electron acceptors (Zhang, 2000).

Bioreactor Design. Bioreactors consisted of acrylic columns 28 cm long and 2.5 cm in diameter packed with silica sand (average diameter of 0.425 mm). The same reactor was used for both mixed and pure culture studies. These columns were inoculated and then left undisturbed for one day for the bacteria to adhere to the sand packing, and then the reactors were continuously fed perchlorate-containing solutions with acetate as a carbon source. Dissolved oxygen was not removed from the feed water, but in separate experiments (Kim, 1999) oxygen was found to be removed rapidly in the column from saturation levels (~8.5 mg/L) to below detection (≤0.1 mg/L). Additional details of these experiments are provided elsewhere (Kim and Logan, 2000; Kim and Logan, 2001).

Analysis of Degradation Rate Data. Perchlorate degradation rate data for organic feed reactors was obtained from studies reported by Wallace et al. (1998), Herman and Frankenberger (1999), Kim and Logan (2000) and Kim and Logan (2001). Perchlorate removal rates were calculated as

$$R = \frac{(Cin - Cout)}{\theta} \qquad (1)$$

where C_{in} and C_{out} are the influent and effluent perchlorate concentrations, and θ is the actual detention time. Removal rates were plotted as a function of the log-mean concentration of perchlorate in the reactor, C_{lm}, calculated as:

$$C_{lm} = \frac{C_{in} - C_{out}}{\ln (C_{in} / C_{out})} \qquad (2)$$

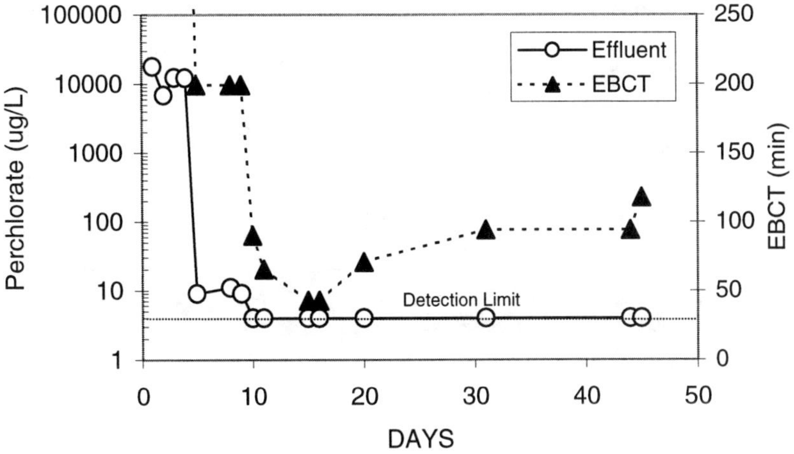

FIGURE 1. Effluent perchlorate concentrations and empty bed contact times (EBCT) for the packed bed reactor inoculated with a mixed culture. (Adapted from Figure 2 in Kim and Logan, 2000.)

A regression analysis of removal rate as a function of C_{lm} produces a straight line with a slope of unity (n=1) if the overall reaction is first order. Regression statistics were calculated using a spreadsheet at a 95% confidence interval.

RESULTS AND DISCUSSION

Mixed and Pure Culture Bioreactors. Using a perchlorate-degrading enrichment, a minimum empty-bed contact time (EBCT) was found to be 43 min, equal to an actual detention time of 18 min (Figure 1; data from Kim and Logan, 2000). Perchlorate concentrations in the feed averaged 24 mg/L, and perchlorate was removed to non-detectable levels over a 35 d period after a 10-d start up period. The maximum perchlorate removal rate in this reactor was 1.8 mg/L-min.

When a fixed bed reactor was inoculated with a pure culture (KJ), substantially lower detention times were needed to remove perchlorate from an average feed concentration of 19.6 mg/L (Figure 2; Kim and Logan, 2001). After a similar 10-d start-up period, perchlorate was removed to non-detectable concentrations for 56 days in 84% of samples (n=147). All effluent samples contained <0.146 mg/L of perchlorate. The minimum detention time to perchlorate breakthrough was found to be 1.08 min (EBCT of 2.1 min). The maximum perchlorate removal rate in the KJ reactor was 18.1 mg/L-min.

Fixed Film Bioreactor Performance Comparison. Perchlorate removal rates reported so far in different fixed bed bioreactors have varied from 0.0007 to 20 mg/L-min, or over six orders-of-magnitude (OOM) (Figure 3; Logan, 2001). While this range seems quite large, the average concentration of perchlorate in

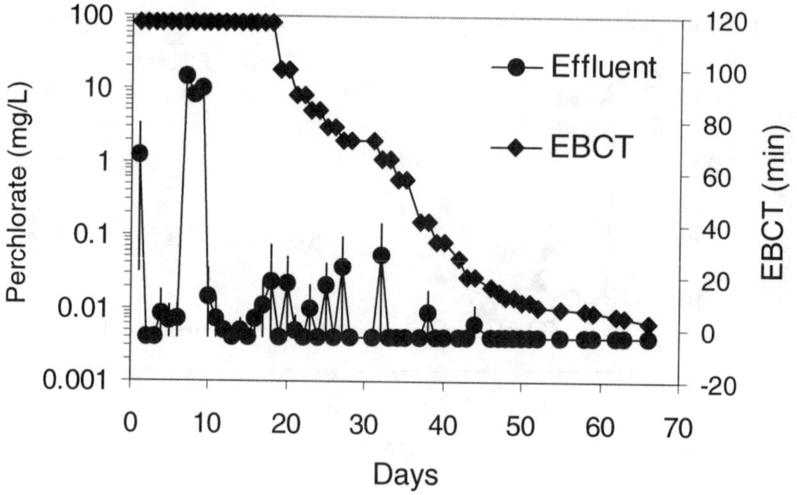

FIGURE 2. Effluent perchlorate concentrations and empty bed contact times (EBCT) for the packed bed reactor inoculated with a pure culture (KJ). (Adapted from Figure 1 in Kim and Logan, 2001.)

these reactors similarly ranged over many OOM. Only reactors using organic substrates (acetate, complex high-protein medium) have been included in this rate comparison.

Perchlorate removal rates were highly correlated (R^2=0.93) with log-mean perchlorate concentration. The slope obtained from this regression analysis was 1.03 ±0.13, which indicated overall first-order reaction kinetics with respect to perchlorate concentration (Figure 3; Logan, 2001). Data used for the regression analysis in Figure 3 included the organic feed reactor studies by others (mentioned above) and the mixed culture reactor data described above (from Kim and Logan, 2000). The pure culture reactor was excluded from this regression analysis as it was determined to have a rate that was significantly higher than other reactors at a comparable log-mean perchlorate concentration in the reactor (Logan, 2001).

Field Tests. Based on these results, packed bed reactors were concluded to be a feasible technology for the treatment of perchlorate contaminated water A pilot-scale, acetate-fed, packed-bed bioreactor, which has been designated as the Pennsylvania State University Perchlorate Treatment (PSU-O_4) system has been designed and is being field tested during the summer of 2001 using perchlorate-contaminated water from a site in Redlands, California. To fully evaluate scale up and operating considerations, two acetate fed reactors, one packed with sand and the other with plastic media, will be tested at this site.

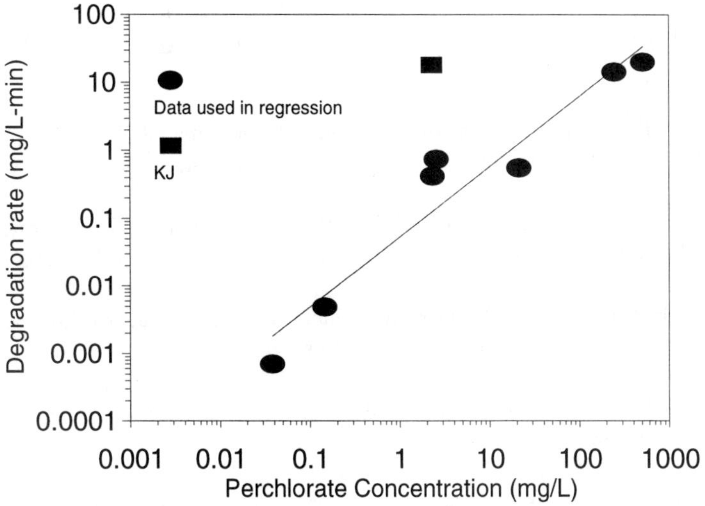

FIGURE 3. Perchlorate degradation rates as a function of log-mean perchlorate concentration. The regression line is based upon filled circle data. Adapted from Figure 1 in Logan (2001).

The fixed bed bioreactor technology developed here for perchlorate is sufficiently similar to existing anaerobic fixed bed filter (AFBF) technologies to make preliminary cost estimates in the range of $1 to $2 per gallon of capacity for a perchlorate based treatment system. Therefore, a 10 MGD system in Redlands, CA, for example, would cost about $10 to $20 million based on our current expectations of system performance. However, if extended side stream treatment is necessary, this might extend the costs to $3 to $5 per gallon of treated water. These cost estimates can be better refined following the conclusion of the field tests at the Redlands site.

ACKNOWLEDGMENTS

This research has been supported in part by grants from the National Science Foundation (BES-9714575) and the American Water Works Association Research Foundation (AWWARF, No. 2557).

REFERENCES

Coates, J. D., U. Michaelidou, R. A. Bruce, S. M. O'Conner, J. N. Crespi, and L. A. Achenbach. 1999. "Ubiquity and Diversity of Dissimilatory (Per)chlorate-Reducing Bacteria." *Applied and Environmental Microbiology.* *65*(12):5234-5241.

Herman D. C. and W. T. Frankenberger Jr. 1998. "Microbial-Mediated Reduction of Perchlorate in Groundwater." *Journal of Environmental Quality.* *27*:750-754.

Herman D. C. and W. T. Frankenberger Jr. 1999. "Bacterial Reduction of Perchlorate and Nitrate in Water." *Journal of Environmental Quality. 28*:1018-1024.

Kim K. 1999. "Microbial Treatment of Perchlorate-Contaminated Water." M.S. Thesis, The Pennsylvania State University, University Park, PA.

Kim, K. and B. E. Logan. 2000. "Fixed-Bed Bioreactor Treating Perchlorate-Contaminated Waters." *Environmental Engineering and Science. 17*(5):257-265.

Kim, K. and B. E. Logan. 2001. "Microbial Reduction of Perchlorate in Pure and Mixed Culture Packed-Bed Bioreactors." *Water Research.* In press.

Logan, B.E. 1998. "A Review of Chlorate and Perchlorate Respiring Microorganisms." *Bioremediation Journal. 2*(2):69-79.

Logan, B. E. 2001. "Analysis of Overall Perchlorate Removal Rates in Packed-Bed Reactors." *Journal of Environmental Engineering. 127*(5): In press.

Logan, B. E., and K. Kim. 1998. "Microbiological Treatment of Perchlorate Contaminated Groundwater." *Proceedings of the Southwest focused groundwater conference: Discussing the issue of MTBE and perchlorate in the ground water*, Anaheim, CA, June 3-4, pp. 87-90. National Ground Water Association, Columbus, OH.

Miller, J. P. and B. E. Logan. 2000. "Sustained Perchlorate Degradation in an Autotrophic, Gas phase, Packed Bed Bioreactor." *Environmental Science and Technology. 34*(14):3018-3022.

Mulvaney, P. 1999. "Perchlorate and Chlorate Reduction by Axenic Cultures." M.S. Thesis, The Pennsylvania State University, University Park, PA.

Urbansky, E. T. 1998. "Perchlorate Chemistry: Implications for Analysis and Remediation." *Bioremediation Journal. 2*(2):81-95.

Wallace, W. H., S. Beshear, D. Williams, W. Hospadar, and M. Owens. 1998. "Perchlorate Reduction by a Mixed Culture in an Up-Flow Anaerobic Fixed Bed Reactor." *Journal of Industrial and Microbiological Biotechnology. 20*:126-131.

Zhang, H. 2000. "Pure Culture Kinetics and Cell Yields of (Per)chlorate Reducing Bacteria." M.S. Thesis, The Pennsylvania State University, University Park, PA.

IN SITU REMOVAL OF PERCHLORATE FROM GROUNDWATER

W. J. Hunter (USDA-ARS, Fort Collins, Colorado, USA)

ABSTRACT: Laboratory column studies suggest that *in situ* barriers might be used to remove perchlorate from groundwater. Water containing 20 mg/L (0.2 mM) perchlorate was pumped through 1.5 by 30 cm columns containing sand. At the start of the study all columns were inoculated with bacteria from a soil extract and, after 14 days operation 0.47 mg of soybean oil was injected onto one group of columns, the treatment group, while a second group of columns, the control group, received no oil. Water containing perchlorate was pumped through both groups of columns and samples of the effluent water were collected at intervals and analyzed for chloride and perchlorate. In the control columns, perchlorate was present in the column effluents throughout the study. In the treatment columns, perchlorate in the effluent decreased by ~ 99 % (~ 0.2 mM) and the concentration of chloride increased by ~ 0.2 mM after the addition of the vegetable oil. In the control columns perchlorate levels remained high and chloride levels low throughout the study. Permeable barriers containing innocuous vegetable oils, other carbon substrates, or other electron donors might be used *in situ* to remove perchlorate from contaminated groundwater.

INTRODUCTION

Perchlorate is used in a number of industrial and agricultural applications, but is perhaps best known for its use as an oxidizer for solid rocket propellants and explosives. It is highly soluble in water, highly mobile, and remarkably persistent as a groundwater contaminant. It has been identified as a ground and surface water contaminant in over a dozen states (Logan et al., 1999; The Weinberg Group Inc., 1999). As a water contaminant in the Colorado River below Lake Mead perchlorate has the potential to affect almost 20 million people (Brechner et al., 2000). The presence of perchlorate in drinking water is a health concern because it resembles iodine in both charge and ionic radius (Renner, 1998) and, as a result, it interferes with the proper functioning of the thyroid gland by acting as a competitive inhibitor of iodine transport (Crooks and Wayne, 1960; Godley and Stanbury, 1954). This disruption in the thyroid gland in turn disrupts the thyroid-hypothalamus-pituitary feedback system (The Weinberg Group Inc., 1999). Perchlorate ingestion can cause goiter and hypothyroidism when iodine intake is at or below normal levels, in addition, perchlorate has the potential of influencing the neurological development of unborn babies by crossing the placenta (The Weinberg Group Inc., 1999; Brechner et al., 2000).

Presently, there is no federal drinking water standard for perchlorate. However, perchlorate is on the EPA's contaminants candidate list and unregulated contaminant monitoring list. A draft EPA proposal recommends that water consumed by humans contain less than 0.032 mg/L perchlorate

(Renner, 1999). In addition, California and Nevada have set "action levels" of 0.018 mg/L for perchlorate, Texas an "interim level" of 0.022 mg/L, and Arizona a "guidance level" of 0.031 mg/L (U.S. Environmental Protection Agency, 1999; The Weinberg Group Inc., 1999).

Conventional drinking water treatment processes do not remove perchlorate from water. Ion exchange, reverse osmosis, and nanofiltration can remove perchlorate from water but these processes do not destroy the anion. Biological treatment not only removes perchlorate from water but with biological treatment processes microorganisms destroy the perchlorate anion by converting it to chloride and oxygen. The chloride is released into the environment while the microorganisms use the oxygen for respiration. The biological treatment processes that have been evaluated so far have involved *ex situ* reactors (Logan et al., 1999; U. S. EPA, 1999).

Objective. For the present investigation laboratory columns, glass chromatography columns packed with sand, were used to evaluate the use of vegetable oil as a carbon substrate for the *in situ* treatment of groundwater contaminated with perchlorate. Previously, we demonstrated that vegetable oil could be used to form a stationary organic zone, barrier, or wall that was high in organic matter and permeable to water (Hunter et al., 1997; Hunter and Follett, 1997; Hunter, 1999). In these earlier studies, denitrifying bacteria were used to remove nitrate from flowing groundwater. The present study evaluated the same process with perchlorate rather than nitrate as the electron acceptor. In addition, a simple survey was conducted to determine the distribution of perchlorate reducing microorganisms in the environment. For this survey soil samples were collected from a number of different sites in northeastern Colorado and evaluated for the ability of these indigenous microorganisms to reduce perchlorate.

MATERIALS AND METHODS

Soil Survey. Surface soil samples were collected from nine different sites in the northeastern Colorado area. Sites included soil from beneath a cattle holding pen, pond sediment, creek sediment, six saline seeps, and a sludge sample from the Fort Collins, CO sewage treatment plant. Water extracts of the samples were placed in sealed 125 mL serum bottles containing soybean oil, phosphate buffer, nutrients and 20 mM perchlorate. Bottles were incubated in the dark under a helium atmosphere at 28°C for two weeks. After the incubation the amount of perchlorate remaining in the bottles was estimated using a perchlorate electrode.

Soil Columns. Columns were 1.5 by 30 cm glass chromatography columns filled with sand. At the start of the study, to assure the presence of a viable microbial population, columns were inoculated with 0.75 mL of a soil extract. After the addition of the soil extract a moderately-hard reconstituted water supplemented with iron, phosphate, NH_4^+-N, and 0.20 mM (20 mg/L) perchlorate was pumped through the columns at a rate of ~25 ml/day. No attempt was made to remove oxygen from the supply water. After 14 days of operation 0.47 mg of soybean oil was injected as an emulsion onto two of the columns, the treatment group, while two other columns, the control group, received no oil. Water containing perchlorate was pumped through both groups of columns for another 16 weeks. Samples of the effluent water were collected at regular intervals (3 times per week) and analyzed for perchlorate with an high pressure liquid chromatograph equipped with a methacrylate based quaternary amine column, a suppressor, and a conductivity detector.

RESULTS AND DISCUSSION

Soil survey. All soil samples were found to contain microorganisms capable of reducing perchlorate to chloride under anaerobic conditions. These results suggest that perchlorate-reducing organisms are ubiquitous or nearly ubiquitous in surface soils and that indigenous microorganisms can serve as the inoculum for the *in situ* remediation of perchlorate at most sites.

Column studies. Over a 20-week period water containing 0.2 mM (20 mg/L) perchlorate was pumped through the treatment and control columns. During the first 2 weeks of the study, before oil was applied to the treatment columns, no perchlorate was removed from the water pumped through any of the columns. However, after the addition of the oil emulsion to the treatment columns the amount of perchlorate in the effluents of these columns dropped rapidly and averaged 0.0007 mM (0.070 mg/L) between the 5th and 18th week of the study. This represents a decrease in the amount of perchlorate in the treated water of greater than 99 % in the period following the addition of the vegetable oil. At the same time the amount of chloride in the column fractions from the treatment columns increased rapidly (Figure 1) indicating the conversion of perchlorate to chloride by the treatment columns. In the control columns the amount of perchlorate in the effluent fractions remained high, averaging 0.14 mM (14 mg/L) over the 18-week study and the amount of chloride remained low (Figure 1) during the study.

CONCLUSIONS

These studies demonstrate the principles whereby a simple permeable barrier composed of sand, gravel and small amounts of an insoluble carbon substrate can be used to remove perchlorate from flowing ground water. In these studies the soil columns contained innocuous vegetable oils though other carbon substrates or electron donors might also be used *in situ* to remove perchlorate from contaminated groundwater.

Applications. Shallow permeable *in situ* barriers might be constructed by digging a trench and backfilling the trench with sand coated with vegetable oil or with a mixture of sand and an insoluble carbon substrate such as saw dust,

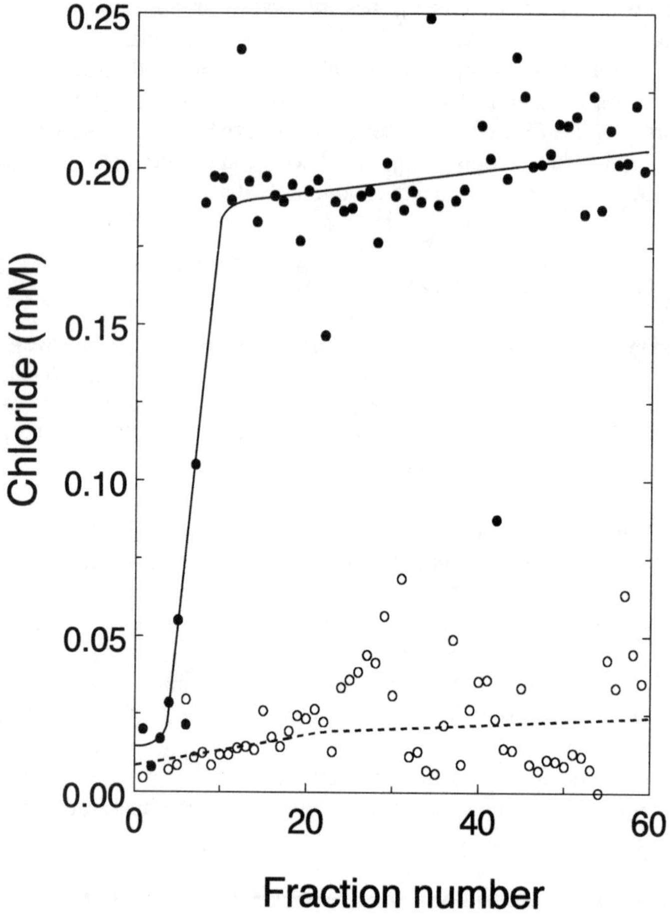

FIGURE 1. Chloride in treatment (•) and control (o) column effluent fractions over the course of the study. Each data point is an average of two readings.

crop residues, waste papers, etc. Such barriers can be quickly and cheaply constructed to remediate shallow aquifers or to protect an aquifer from runoff. However, the construction of deep trenches for deep barriers is much more difficult and expensive (Gavaskar et al., 1998). A simpler and less expensive approach for forming deeper barriers might involve the injection of vegetable oil emulsions across a section of a sand and gravel aquifer. Injection could reduce the cost of constructing deep barriers but additional research is needed to develop the methods and evaluate injection technologies.

REFERENCES

Brechner, R. J., G. D. Parkhurst, W. O. Humble, M. B. Brown, and W. H. Herman. 2000. "Ammonia Perchlorate Contamination of Colorado River Drinking Water Is Associated with Abnormal Thyroid Function in Newborns in Arizona." *J. Occup. Environ. Med.* *42*: 777 – 782.

Crooks, J. and E. Wayne. 1960. "A Comparison of Potassium Perchlorate, Methylthiouracil, and Carbamizole in the Treatment of Thyrotoxicosis." *Lancet*. *1*: 401 -- 404.

Gavaskar, A. R., N. Gupta, B. M. Sass, R. J. Janosy, and D. O'Sullivan. 1998. *Permeable Barriers for Groundwater Remediation: Design, Construction, and Monitoring.* Battelle Press, Columbus, OH.

Godley, A. and J. Stanbury. 1954. "Preliminary Experience in the Treatment of Hypothyroidism with Potassium Perchlorate." *J. Clin. Endocrinol.* *14*:470 – 478.

Hunter, W. J. 1999. "Use of Oil in a Pilot Scale Denitrifying Biobarrier: "In A. Leeson and B. C. Alleman (Eds). *Bioremediation of Metals and Inorganic Compounds.* Battelle Press. Columbus, OH. *5*(4):47-51.

Hunter, W. J. and Follett, R. F. 1997. "Removing Nitrate From Groundwater Using Innocuous Oils: Water Quality Studies." Pages 415 - 420 in B. C. Alleman and A. Leeson (Eds) *In Situ and On-Site Bioremediation: Volume 3.* Battelle Press, Columbus, OH. 531 pp.

Hunter, W. J., Follett, R. F. and Cary, J. W. 1997. "Use of Vegetable Oil to Stimulate Denitrification and Remove Nitrate from Flowing Water." *Trans. Am. Soc. Ag. Eng.* *40*(2):345-353.

Logan, B. to E., K. Kim, J. Miller, P. Mulvaney, and R. Unz. 1999. "Biological Treatment of Perchlorate Contaminated Waters." In A. Leeson and B. C. Alleman (Eds). *Bioremediation of Metals and Inorganic Compounds.* Battelle Press. Columbus, OH. *5*(4):147-151.

Renner, R. 1998. "Perchlorate-Tainted Wells Spur Government Action." *Environ. Sci. Technol.* *32*:210A.

Renner, R. 1999. EPA Draft Almost Doubles Safe Dose of Perchlorate in Water. *Environ. Sci. Technol.* *33*(5): 110 a -- 111a; EPA, 1999.

The Weinberg Group Inc. 1999. "Perchlorate Information Summaries Prepared for: the Fertilizer Institute." http://www.tfi.org/perchlorate-final.pdf (Confirmed 8/21/00).

U. S. Environmental Protection Agency, 1999. "EPA Region 9 Perchlorate Update." USEPA Region 9. San Francisco, CA. http://www.epa.gov/safewater/ccl/perchlor/r9699fac.pdf (Confirmed Aug 29, 2000).

IN SITU BIOTREATMENT OF PERCHLORATE AND CHROMIUM IN GROUNDWATER

Michael W. Perlmutter, Ronald Britto, James D. Cowan, and Alan K. Jacobs
(EnSafe Inc., Memphis, Tennessee)

ABSTRACT: A bench-scale treatability study was implemented to evaluate the feasibility of in situ biological treatment for perchlorate (ClO_4^-)- and hexavalent chromium [Cr(VI)]-contaminated groundwater at a solid rocket fuel manufacturing facility. Traditional batch and intermittent continual flow bioreactors were used to evaluate two media (sand and gravel) and four electron donors (acetate, molasses, composted manure, and concentrated fruit juice). Both acetate and molasses were effective electron donors during the bench-scale study. Because molasses costs less than acetate, molasses will be recommended for use in a planned pilot- or full-scale system. It was found that acetate may be required to initiate the treatment system, but molasses would be used as the long-term electron donor. No specialized consortia of microorganisms are required to innoculate the subsurface when initiating the in situ treatment system. Indigenous microorganisms can be acclimated to support an in situ ClO_4^- biotreatment system. Furthermore, there was no evidence of media plugging or fouling since the inception of the bench-scale study. ClO_4^- concentrations were regularly reduced at rates ranging from 200 to 600 mg/L/d. As a result, the bioreactive zone for ClO_4^--contaminated groundwater (1,500 mg/L) in an aquifer with a relatively high conductivity (1 ft/day) would only extend 2.5 to 7.5 feet downgradient of the amendment application. ClO_4^- reduction below method detection limits would take 2.5 to 7.5 days. Despite its potential toxic effects, Cr(VI) concentrations used in the study did not inhibit ClO_4^- reduction as it was routinely reduced to below method detection limits within minutes.

INTRODUCTION

A bench-scale study was implemented to evaluate the feasibility of in situ biological treatment for ClO_4^-- and Cr(VI)-contaminated groundwater at a solid rocket fuel manufacturing facility.

EXPERIMENTAL MATERIALS AND METHODS

Seven five-gallon bioreactors were packed with gravel or sand, filled with groundwater (two separate sources) and an acclimated perchlorate-reducing microbial (PRM) culture, and amended with a carbon source (acetate or molasses) and ammonium phosphate [$(NH_4)_2$ HPO_4] as shown in Table 1; each bioreactor was initially operated in batch mode. Aeration basin return line sludge (biomass) was collected from the local publicly owned treatment works (POTW) and acclimated to ClO_4^- for a week before being added to the bioreactors. A short polyvinyl chloride (PVC) pipe was installed through the cover of each sealed bioreactor to allow sampling access while maintaining anaerobic conditions.

TABLE 1. Bench-Scale Study Set-Up Summary

Reactor	Matrix	GW Volume	Carbon Source(s)	Acclimated POTW Sludge
1	Gravel	8 L	Acetate and molasses	no
2	Gravel	8 L	acetate and molasses	yes
3	Gravel	8 L	fruit juice concentrate	yes
4	Gravel	8 L	20% by volume composted manure	yes
5	Gravel	8 L	molasses	no
6	Gravel	8 L	acetate	no
7	Sand	5 L	acetate and molasses	no

Notes: Ammonium phosphate was used in all seven bioreactors. Bioreactors 5, 6, and 7 received acclimated PRM from Bioreactor 1.

SAMPLING AND ANALYTICAL PROTOCOLS

Samples were collected routinely and analyzed for ClO_4^-, Cr(VI), chlorides, dissolved oxygen (DO), and oxidation-reduction potential (ORP).

- ClO_4^- — *Onsite:* Orion ion-specific ion probe and 290A meter
  Lab: ion chromatography (Method 300.0)
- Cr(VI) — *Onsite:* Hach test kit; colorimetric method
  Lab: spectrophotometric analysis (Method 7196A)
- chloride — Hach test kit; colorimetric method
- DO/ORP — Horiba U-22 meter

Using a peristaltic pump, groundwater was extracted from the bioreactor through the PVC pipe and then re-infiltrated through a small opening in the cover. After several minutes to several hours of recirculation at a low flow rate, which distributed the contaminated groundwater throughout the bioreactor and simulated groundwater flow, the water was collected and sampled.

RESULTS AND DISCUSSION

Initially, Bioreactors 1 to 4 were evaluated using 10 gallons groundwater from an onsite location (Source 1) that had starting concentrations of 1,500 milligrams per liter (mg/L) ClO_4^- and 2.3 mg/L Cr(VI). A second groundwater source (Source 2) was used to continue Bioreactor 1 and start Bioreactors 5, 6, and 7; ClO_4^- and Cr(VI) concentrations were 1,500 and 8.0 mg/L, respectively. Bioreactor-specific results are discussed below.

Bioreactor 1. This bioreactor was established to assess whether indigenous microorganisms could stimulate in situ biotreatment of contaminated groundwater. Although it took nine weeks to initially reduce ClO_4^-, the results shown on Figure 1 indicate that in situ biotreatment is feasible without bioaugmentation.

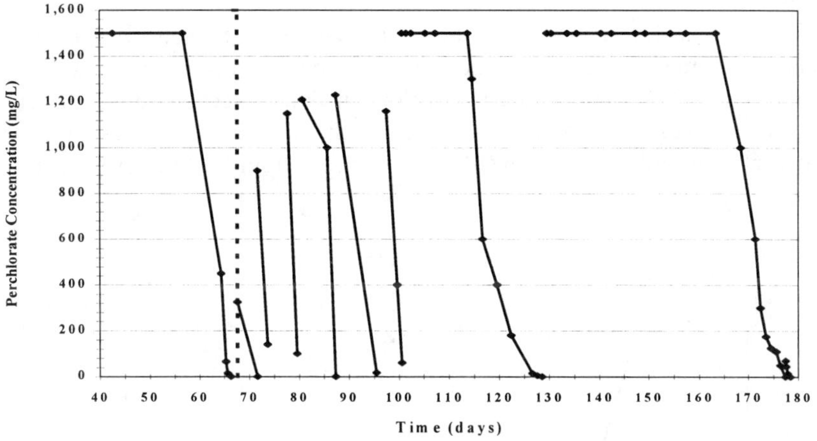

FIGURE 1. Perchlorate concentrations versus time in Bioreactor 1. Dashed line indicates when the carbon source was switched from acetate to molasses.

In addition to ClO_4^- degradation, Cr(VI) was reduced from 2.3 mg/L to below method detection limits within 24 hours.

After the groundwater from Source 1 was treated, sodium perchlorate was added to the bioreactor to simulate groundwater concentrations and, instead of acetate, molasses was added to the bioreactor as the carbon source to evaluate its viability as an electron donor for this site. Using molasses, ClO_4^- was reduced to below method detection limits in six successive perchlorate-dosing cycles, during which, the ClO_4^- removal rate was as a high as 600 milligrams per liter per day (mg/L/d).

On Day 100, groundwater from Source 2, molasses, and ammonium phosphate were added to the bioreactor after it was drained. Cr(VI) was reduced by 97% in approximately 20 hours; ClO_4^- reduction started after 2 weeks and was below method detection limits in 4 weeks. Because ClO_4^- reduction started promptly after an additional molasses dosing, the removal rate would likely have been faster if the bioreactor was not carbon limited. The bioreactor was drained and refilled with groundwater from Source 2, molasses, and ammonium phosphate on Day 129. After 5 weeks without measurable ClO_4^- reduction, the reactor was redosed with molasses following which, immediate reduction was observed.

During the bench-scale study, molasses to ClO_4^- ratios ranged from 0.8:1 to 2.6:1 on a mass basis.

Bioreactor 2. Unlike Bioreactor 1, this bioreactor was bioaugmented with an acclimated PRM culture. As a result, ClO_4^- was reduced to below method detection limits in six weeks rather than nine. As shown on Figure 2, the bioreactor was redosed with groundwater from Source 1 (diluted by 50%) two more times along with molasses (instead of acetate) and ammonium phosphate. ClO_4^- was reduced to below method detection limits in 3.7 and 2.2 days, respectively.

Similar to Bioreactor 1, after the groundwater from Source 1 was treated, sodium perchlorate was added to the bioreactor to simulate groundwater ClO_4^- concentrations. ClO_4^- was reduced to below method detection limits in four successive perchlorate-dosing cycles during which the ClO_4^- removal rate was as a high as 600 mg/L/d. Molasses to ClO_4^- ratios ranged from 1:1 to 5.2:1 on a mass basis.

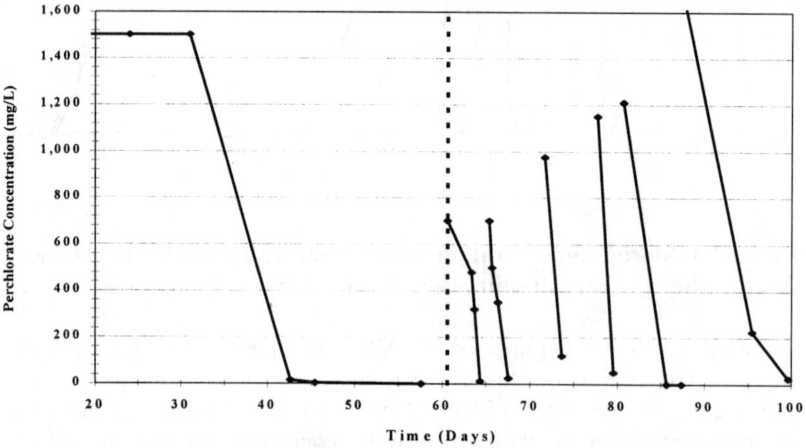

FIGURE 2. Perchlorate concentrations versus time in Bioreactor 2. Dashed line indicates when the carbon source was switched from acetate to molasses.

Bioreactors 3 and 4. Bioreactor 3 used fruit juice concentrate to evaluate whether a carbon source (electron donor) from an industrial source would be feasible. However, as discovered in previous bench-scale studies (EnSafe, 1999a and b), fermentation can inhibit ClO_4^- reduction. As a result, ClO_4^- concentrations remained unchanged for the entire experiment.

Bioreactor 4 was based on the promising findings of previous in situ bench-scale studies (EnSafe, 2000). However, the particular compost blend used may not have the appropriate type and/or quantities of organic carbon to act as the electron donor to treat the ClO_4^--contaminated groundwater. Similar to Bioreactor 3, ClO_4^- concentrations remained unchanged for the entire experiment. Bioreactors 3 and 4 were discontinued after 14 weeks.

Bioreactor 5. This bioreactor was initiated with groundwater from Source 2. Bioreactors 1 and 2 both successfully used acetate and molasses as carbon sources. However, neither one was acclimated with molasses. Groundwater, molasses, ammonium phosphate, and acclimated PRM from Bioreactor 1 were added to Bioreactor 5 when it was initiated.

Cr(VI) was reduced by more than 90% in approximately 45 hours; however, further reduction to method detection limits took two weeks. Molasses and acclimated PRM were added again after two weeks to reduce ORP and DO to levels indicative of anaerobic conditions; molasses was added again after eight weeks. After approximately 13 weeks, ClO_4^- concentrations remained unchanged and therefore, Bioreactor 5 monitoring was discontinued.

Bioreactor 6. Bioreactor 6 was initiated as a control (to distinguish between the feasibility of indigenous versus exogenous microbes) for Bioreactor 5. As such, Source 2 groundwater, acetate, ammonium phosphate, and acclimated PRM from Bioreactor 1 were added to the bioreactor when it was started.

Cr(VI) was reduced by approximately 95% in 45 hours; however further reduction to method detection limits took three weeks. Acetate and acclimated PRM were added again after two weeks to reduce ORP and DO to levels indicative of anaerobic conditions.

Coincident with a negative ORP measurement, ClO_4^- reduction began after seven weeks; the ClO_4^- concentration was below method detection limits one week later. The ClO_4^- removal rate was as a high as 250 mg/L/d during that week.

Bioreactor 7. This bioreactor was established to assess in situ biotreatment in an aquifer matrix more closely approximating site conditions at the solid rocket fuel manufacturing facility (sand versus gravel). Source 2 groundwater, acetate, ammonium phosphate, and acclimated PRM from Bioreactor 1 were added to sand-filled Bioreactor 7 when this bioreactor was initiated.

ClO_4^- reduction began after approximately two weeks; however, it took another two weeks before its concentration was below method detection limits (see Figure 3). During the two-week ClO_4^--reduction period, additional acetate was added to the bioreactor to accelerate ClO_4^- degradation; the ClO_4^- removal rate ranged from 50 to 300 mg/L/d during this period. Cr(VI) was reduced by approximately 98% in 2 days; however, further reduction to method detection limits took seven days.

The treatment strategy was modified after the groundwater from Source 2 was completely treated. Rather than adding an entire batch of contaminated groundwater, the bioreactor simulated a continuous flow system, which is more consistent with in situ conditions. In addition, molasses replaced acetate as the carbon source.

As shown on Figure 3, starting on Day 42, ClO_4^- concentrations introduced to the bioreactor were gradually increased each day. Groundwater was recirculated continuously with a peristaltic pump during this phase of the experiment to distribute the ClO_4^- and amendments more efficiently.

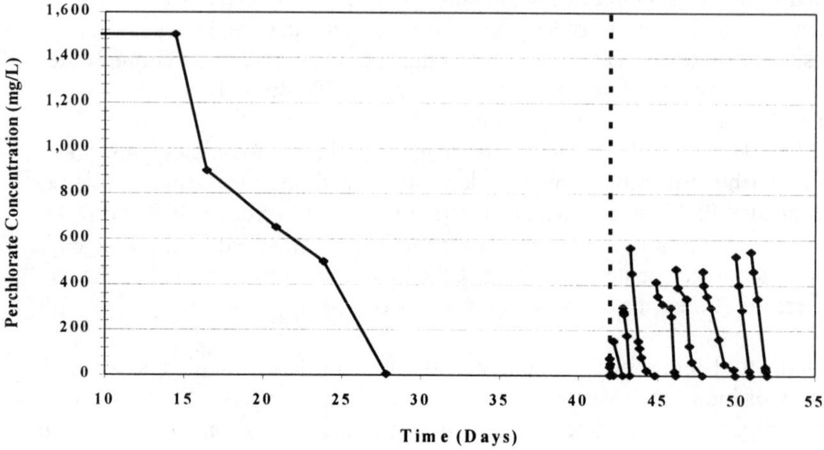

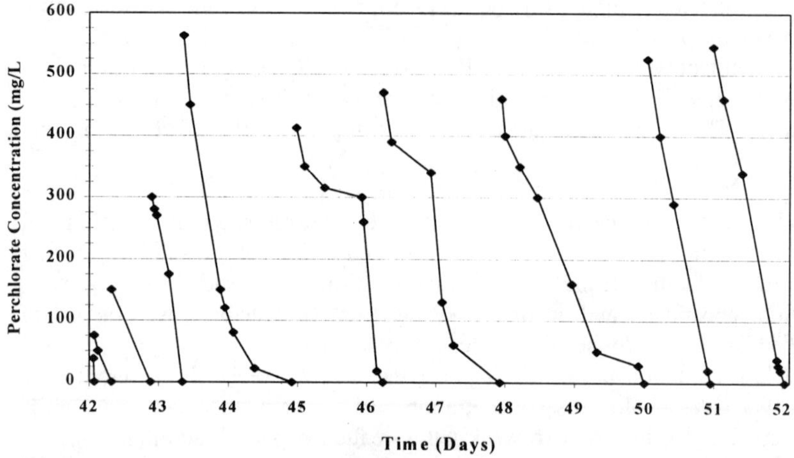

FIGURE 3. Perchlorate concentrations versus time in Bioreactor 7. Top: Dashed line indicates when the carbon source was switched from acetate to molasses. Bottom: starting on Day 42, ClO_4^- concentrations introduced to the bioreactor were gradually increased each day. Groundwater was recirculated continuously with a peristaltic pump during this phase of the experiment to distribute the ClO_4^- and amendments more efficiently.

At times, the ClO_4^- removal rate during this phase of the experiment was greater than 1,000 mg/L/d; however, removal rates generally ranged from 200 to 600 mg/L/d, which is reasonably consistent with the other bioreactors. Cr(VI) was routinely reduced to below method detection limits within minutes rather than hours. Molasses to ClO_4^- mass ratios ranged from 1.9:1 to 8.3:1 on a mass basis.

CONCLUSIONS

Carbon Source. Both acetate and molasses were effective electron donors during the bench-scale study. Because of considerable cost savings, molasses is recommended in a pilot- or full-scale system. It was found that acetate may be needed to initiate the treatment system, but molasses can be used as the long-term electron donor.

Biomass. As demonstrated in Bioreactors 1, 5, 6, and 7, no specialized microorganisms are required to inoculate the subsurface when initiating the in situ treatment system. Indigenous microorganisms are capable of supporting an in situ ClO_4^- and chromium biotreatment system.

The in situ treatment system is not expected to leave a "biochemical footprint." The carbon source (molasses) and the nutrients (ammonium phosphate or other) should be fully consumed in the aquifer. Furthermore, the aquifer's microbial population should return to pre-testing or pre-bioremediation conditions once amendment addition is discontinued. Once the concentration of available food is at a minimum, the microorganisms are forced to metabolize their own cell mass (otherwise known as endogenous respiration). Nutrients remaining in the dead cells diffuse out to furnish the remaining cells with food after remediation is complete (Tchobanoglous and Burton, 1991).

Finally, there was no evidence of aquifer matrix plugging or fouling in the bench-scale study. As expected with anaerobic systems, biomass generation is minimal, generally measured as a biofilm on the media with thicknesses from 10 to 70 microns for similar studies.

Perchlorate Reduction Rates. ClO_4^- concentrations were regularly reduced at rates ranging from 200 to 600 mg/L/d. As a result, the bioreactive zone for ClO_4^--contaminated groundwater (1,500 mg/L) in an aquifer with a relatively high conductivity (1 ft/day) would only extend 2.5 to 7.5 feet downgradient of the amendment application. ClO_4^- reduction below method detection limits would likely take 2.5 to 7.5 days.

Chromium Chemistry and Toxicity. In general, Cr(VI) is relatively mobile in the environment and acutely toxic, mutegenic, teratogenic, and carcinogenic. Because of their high solubility and their requirements for oxidizing conditions, natural Cr(VI) minerals are very scarce in nature. In contrast, Cr(III) has relatively low toxicity and is immobile under moderately alkaline to slightly acidic conditions. Despite its potential toxic effects, Cr(VI) concentrations used in the study did not inhibit ClO_4^- reduction.

Cr(VI) is expected to reduce completely within the time frame of ClO_4^- reduction in a pilot- or full-scale in situ anaerobic treatment system. As demonstrated in the bench-scale study, Cr(VI) was routinely reduced to below method detection limits within minutes, which was considerably faster than ClO_4^- reduction rates.

No attempt was made in this study to distinguish between biotic and abiotic mechanisms for Cr(VI) reduction. Further studies would be required to distinguish between these two mechanisms.

REFERENCES

EnSafe Inc. 1999a. *Draft Phase I Bench-Scale Study Status Report.* Naval Weapons Industrial Reserve Plant McGregor, McGregor, Texas. July 30, 1999.

EnSafe Inc. 1999b. *Draft Phase I Bench-Scale Study Status Report.* Naval Weapons Industrial Reserve Plant McGregor, McGregor, Texas. September 27, 1999.

EnSafe Inc. 2000. *Draft Interim Stabilization Measuress Report.* Naval Weapons Industrial Reserve Plant McGregor, McGregor, Texas. April 21, 2000.

Tchobanoglous, G., and F.L. Burton. 1991. *Wastewater Engineering: Treatment, Disposal, and Reuse.* Third Edition. McGraw-Hill.

ON-SITE EVALUATION OF SELENIUM TREATMENT TECHNOLOGIES

Jonathan C. Cherry, P.E. (Kennecott Utah Copper Corporation, Magna, Utah)
Jennifer Saran (Kennecott Utah Copper Corporation, Magna, Utah)

Abstract: A detailed selenium treatment technology review, based on the site specific chemistry of selenium-contaminated artesian flows, identified eight treatment technologies for bench and pilot scale testing during 1999 and 2000. The purpose of the selenium treatment technology evaluation was to determine the most efficient and cost effective technology that could be implemented at the Kennecott Utah Copper Corporation (KUCC) project site. Results of the evaluation and testing showed two promising options: 1) recycling the artesian flows into the Kennecott process water circuit for industrial use, and 2) reducing the selenium biologically.

INTRODUCTION

The aquatic toxicity of selenium was described in the U.S. Environmental Protection Agency's (U.S. EPA) Ambient Water Quality Criteria document, and the freshwater chronic criterion was derived to be 0.005 mg/L (U.S. EPA, 1988). At the Kennecott site, selenium-contaminated artesian flows (approximately 500 gpm (1893 L/min)) are surfacing in wetlands adjacent to the southern shore of Great Salt Lake (GSL) at concentrations of approximately 1.8 mg/L, or over 300 times the established freshwater quality standard. The source of the contamination is the underlying ground water plume containing elevated selenium.

As part of a voluntary cleanup response with federal and state oversight, Kennecott was tasked with completing a Remedial Investigation and Feasibility Study (RI/FS) to define the nature and extent of surface and ground water contamination, and to develop and evaluate remedial alternatives. The two-year FS screened over 53 potential selenium treatment technologies, tested 20 technologies at bench scale, and determined that eight had the potential for successful treatment of selenium at the Kennecott site. The eight selenium treatment technologies that were evaluated included process water recycle, biological selenium removal, ferrihydrite adsorption, catalyzed cementation, iron filing packed columns, iron powder reductive precipitation, permeable reactive barriers and ferrous iron treatment.

Objective. The objective of the FS was to determine and evaluate the efficiency and cost effectiveness of treating selenium-contaminated artesian flows. The goal was to achieve a selenium effluent concentration of 0.05 mg/L or less for a cost that would not be prohibitive at full-scale implementation.

Site Description. The Kennecott site is located on the northern end of the Oquirrh Mountain range adjacent to the southern shore of the Great Salt Lake. A precious metals refinery operated in this area from the early 1950s through 1995. In addition to refining copper, gold and silver, other rare and precious metals were recovered including platinum, palladium, tellurium and selenium. During approximately 40 years of operation, varying concentrations and quantities of selenium-laden process solutions were released from the refinery and subsequently migrated into the underlying ground water system. The refinery was demolished in 1995, and soils removal and capping of the site removed the source of contamination. However, a selenium plume remains at and down gradient of the site. The ground water travels northward from the refinery for approximately one mile (1609 m) where it is discharged under artesian conditions at Kessler springs. Sampling and monitoring of the ground water have shown that over 95 percent of the selenium in the plume is in the form of selenate (Se^{+6}) (Shepherd Miller, Inc., 1998). Other typical analytical parameters for the Kessler springs water are shown in Table 1.

TABLE 1. Kessler springs water quality.

Analyte	Concentration (mg/L)	Analyte	Concentration (mg/L)
pH	7.0 – 7.4 (s.u.)	Arsenic	0.025
TDS	1500	Cadmium	<0.001
TSS	3.0	Chromium	<0.01
Conductivity	2800 (umho/cm)	Copper	<0.02
Sulfate	200- 300	Iron	<0.3
Calcium	100-150	Lead	<0.005
Chloride	100	Mercury	<0.005
Magnesium	40	Selenium	1.75
Potassium	10	Selenate	1.70
Sodium	400	Selenite	0.05
		Zinc	<0.01

INITIAL TECHNOLOGY SCREENING

Typically, the removal of metals from water is accomplished by raising the pH to cause the precipitation of metallic hydroxides, sulfates or carbonates. Metals such as copper, zinc and chromium are removed because their soluble cations form insoluble hydroxides, sulfates and carbonate compounds. This does not work for selenium or arsenic because they are anions. Furthermore, because selenate salts tend to be more soluble than the corresponding selenite (Se^{+4}) salts, many technologies effective at removing selenite are ineffective at removing selenate.

After an extensive literature review, the 53 selenium treatment technologies were separated into the following general treatment categories: metals coprecipitation, biological treatment, conventional water treatment, ion exchange, chemical reduction, membrane separation, phytoextraction, metal oxide adsorption, electrochemical separation, evaporation, stripping, activated carbon adsorption, polymer coagulation and solvent extraction (Montgomery Watson, 1999). Specific technologies were listed within each category, and general advantages and limitations were described for the Kennecott site. Based on this general screening, groups of selenium removal technologies were recommended for further evaluation. The groups of technologies that were carried forward were evaluated for selenium removal efficiency, potential capital cost, potential operating and maintenance (O&M) cost, potential waste disposal costs and anticipated cost per 1000 gallons (3785 L) of water treated. From this technology screening, the key advantages, limitations, state of technology development, and selenate removal efficiencies were listed for the remaining individual selenium treatment technologies, along with a recommendation of which technologies should be considered for bench or pilot scale testing at the Kennecott site (Montgomery Watson, 1999). Twenty technologies were tested in bench scale experiments. From these 20, eight were selected for further testing: process water recycle, biological selenium removal, ferrihydrite adsorption, catalyzed cementation, iron filing packed columns, iron powder reductive precipitation, permeable reactive barriers and ferrous iron treatment.

ON-SITE BENCH AND PILOT SCALE TESTING

The eight technologies that were selected for on-site bench and pilot scale testing were evaluated as part of the FS. Brief descriptions of each technology, and corresponding bench or pilot scale test results are described below.

Process Water Recycle. All selenium-contaminated flows have been captured and pumped into the Kennecott process water circuit since May 2000. Water quality samples and concentrator recovery experiments have shown this option to be viable at full scale. Samples from various locations within the process water circuit have consistently shown that selenium is not concentrating within the circuit to levels that exceed permit discharge limits (KUCC, 2001). The fairly steady concentration of selenium within the circuit can be attributed to reactions that occur when these flows contact the tailings slurry/mine leachate mixture, which contains high concentrations of iron. Bench scale experiments conducted in 1999 and 2000 showed that up to 49 percent of dissolved selenium is removed when the selenium-contaminated water contacts the mixture (KUCC, 2000).

Experiments were conducted to determine if the added selenium-contaminated flows in the circuit would negatively effect recoveries at the Copperton concentrator. Testing at the concentrator using various combinations of Kessler springs water and process water showed that the selenium had no effect on the recoveries of copper and molybdenum (KUCC, 2001).

Biological Selenium Removal. In a cooperative effort with Applied Biosciences Corporation, site-specific selenate reducing microbes were collected from the Kessler springs system. These microbes were combined with other cataloged selenate reducing microbes to create a site-specific microbial cocktail that reduced selenate to elemental selenium in an anaerobic solids bed reactor. After a series of successful bench scale tests that removed over 99 percent of selenium from the water, an on-site pilot scale system was designed and subsequently operated at Kessler springs (MSE, 2000).

The pilot scale testing used six, 500-gallon (1893 L) tanks configured first as a series of five and later as two series of three anaerobic solids beds. Laboratory field reactors, started in advance of the project, were used to help optimize the field reactors. Reactor retention times of 12, 11, 8 and 5.5 hours were tested to evaluate selenium removal efficiencies at flow rates up to 1 gpm (3.78 L/min).

This system operated for over six months. The biological reduction process effectively removed selenium from the water at an initial concentration of approximately 1.8 mg/L to less than 0.002 mg/L. At no time during the testing did reactor effluents exceed 0.019 mg/L selenium. Nutrient costs for the reactor operation at the various retention times ranged from $0.51 per 1000 gallons (3785 L) to $0.58 per 1000 gallons (3785 L) (MSE, 2000).

Bench scale testing of aquifer bedrock and alluvial materials for in-situ selenium microbial reduction began in October 2000 and continues in 2001 (Applied Biosciences Corporation, 2000). Initial results for this option appear promising, and long term cost estimates for full-scale in-situ microbial reduction would be lower than a full-scale ex-situ biological treatment system (KUCC 2001).

Ferrihydrite Adsorption. As part of the U.S. EPA's Mine Waste Technology Program, Kennecott provided site access and laboratory services for MSE Technology Applications, Inc. to evaluate this and other selenium treatment technologies. The ferrihydrite adsorption technology is the U.S. EPA's Best Demonstrated Available Technology (BDAT) for selenium treatment. Selenium is removed in the process when ferrihydrite is precipitated with concurrent adsorption of selenium onto the ferrihydrite surface. In order for coprecipitation of selenium using ferrihydrite to occur, the ferric ion must be present in the water. Selenate is optimally removed from the water at a pH below 4.0.

This technology was tested on site at a flow rate of 5 gpm (19 L/min). From the bulk storage tank, Kessler springs water was pumped to a static mixer where ferric chloride was added to the water immediately prior to the mixer. From the static mixer, the solution was pumped to an 80-gallon (303 L) tank where a lime slurry was injected to increase the pH of the process water. From the pH adjustment tank, the process water was pumped to a 1000 gallon (3785 L) thickener where flocculent was added to assist with solids separation. The treated water was removed from the top of the thickener and adjusted once more to bring the water to a neutral pH.

Depending on the different test runs and iron additions, the mean selenium effluent concentrations ranged from 0.090 to 0.631 mg/L. The iron consumption of this technology ranged from $14.31 – $15.17 per 1000 gallons (3785 L) (MSE, 2000).

Catalyzed Cementation. This technology also was tested as part of the Mine Waste Technology Program by MSE at the Kennecott site. Catalyzed cementation is based on the ability to remove heavy metals from water by adding proprietary catalysts that enhance the cementation of the contaminant on the surface of the iron particle.

A catalyst reagent was added to Kessler springs water from a bulk storage tank at 1 gpm (3.78 L/min) prior to a static mixer. Sulfuric acid was added prior to a second static mixer to achieve the desired pH. Once the desired pH was achieved, the process water was pumped into an elemental iron reactor that consisted of a fluidized bed of elemental iron particles. From the iron reactor the process water was sent to an 80-gallon (303 L), secondary reactor where the pH was adjusted upward with a lime slurry and an oxidizer to complete the reaction. Flocculent also was added to enhance solids separation.

Based on the different variations tested, the mean effluent selenium concentration ranged from 0.035 mg/L to 0.834 mg/L. The reagent consumption costs for this technology were estimated to be $8.11 per 1000 gallons (3785 L), of which $5.81 was attributed to oxidizing reagents (MSE, 2000).

Iron Filing Packed Columns (IFPC). Developed for Kennecott by Harding Lawson and Associates (HLA), IFPC involves the flow of selenium-contaminated water through a column filled with iron filings (8x50 mesh). This process was tested at bench scale and small pilot scale. The bench scale testing involved twelve, 1- and 2-inch (2.54 cm and 5.08 cm) diameter columns, 30 inches (76.2 cm) tall, that were filled with three different iron media types. Kessler springs water was pumped into the bottom of the columns at various flow rates to determine the minimal contact time required. Results indicated selenium removal was consistently greater than 98 percent for two of the media types at contact times of 200 and 300 minutes (HLA, 1999). However, after approximately 50 days of operation, most of the columns were showing increased pressure readings near the influent ports. Further study indicated that the columns were most likely plugging due to calcite precipitation at the influent end of the columns.

The small, pilot scale IFPC involved six, 1.83-foot (0.558 m) diameter HDPE columns with an effective bed depth of five feet (1.52 m). The operational parameters that were tested included contact time, and chemical and physical methods for preventing calcite precipitation in the columns, which included pH adjustment of influent water, and a mixture of iron filings with sand. The longer contact time (pertaining to a flow rate of 0.67 gpm or 2.54 L/min) was the most effective in all of the columns, with 74 to 92 percent of the selenium removed (Harding ESE, 2000a). The iron filings/sand mixture proved to be an effective physical method for preventing flow blockage due to calcite precipitation, but the mixture lowered selenium removal efficiencies. Adjusting the pH of the influent water was an effective chemical method to prevent calcite precipitation within the columns, and did not affect the removal efficiencies. Reagent costs for the IFPC were estimated to be $1.32 per 1000 gallons (3785 L) (Harding ESE, 2000b).

Iron Powder Reductive Precipitation (IPRP). Based on promising results from beaker experiments performed by HLA involving the ability of iron powder to remove selenium from solution, a flow through process (IPRP) was developed and tested at both bench scale and small pilot scale. The bench scale system involved the addition of Kessler springs water to a continually mixed reaction beaker containing iron powder. Sulfuric acid was added to lower and maintain the pH at 4.0. From the reaction beaker, the water flowed by gravity to the pH adjust beaker, where sodium hydroxide was added to raise the pH to various levels. Iron dose and final pH parameters were optimized, and high intensity mixing showed the most promising selenium removal efficiencies of 66 to 83 percent during a five day run time (HLA, 1999).

Small pilot scale testing of the IPRP system involved a rectangular, stainless steel reaction tank designed for a retention time of 150 minutes at a flow rate of 1 gpm (3.78 L/min). Kessler springs water was pumped into the continually mixed tank containing the iron powder, where the pH was maintained at a particular level. The water then flowed by gravity to a clarifier through an in-line mixer, where caustic was added to raise the pH of the water. Iron flocs settled to the bottom of the clarifier, and the treated and clarified water was recycled in the Kennecott process water circuit. Various pH levels in the reaction tank and clarifier were tested. Selenium removal efficiencies ranged from 29 to 85 percent (Harding ESE, 2000a). Reagent costs to treat 1000 gallons (3785 L) for IPRP were estimated to be $6.54 (Harding ESE, 2000b).

Permeable Reactive Barriers (PRB). A PRB is an in-situ zone of reactive media, through which contaminated water passes and is treated. At the request of Kennecott, the University of Waterloo performed a series of column tests on water that was similar to Kessler springs water to evaluate the potential to remove selenate in ground water through a PRB system.

Two acrylic columns measuring 15.8 inches (40 cm) in length by 1.97 inches (5 cm) in diameter were constructed with an internal volume of 0.810 L and multiple sample ports. One column was filled with a reactive media consisting of zero-valent iron, and the other was filled with mixture of zero-valent iron and pyrite. The influent water was introduced to the columns at a flow rate similar to ground water flow velocities at the Kennecott site.

The column tests ran for 63 days during which time all effluent selenium concentrations were less than 0.003 mg/L (University of Waterloo, 2000). However, corresponding decreases in calcium and silica concentrations were noted. Geochemical modeling suggests the precipitation of carbonate, silicate and sulfide minerals within the reactive media is likely to occur. This was observed in the columns as a decrease in porosity. Reagent costs for this option were estimated to be $13.00 per 1000 gallons (3785 L) (University of Waterloo, 2000).

Ferrous Iron Treatment. At the request of Kennecott, the University of Utah tested a ferrous iron treatment system for the removal of selenate from Kessler springs water. The experiments, conducted at the University, involved the placement of Kessler springs water into a small reactor, which was purged with argon for approximately 15 minutes. Once the reactor had been purged, a dry ferrous powder sulfate was added to the reactor and allowed to dissolve for 5 minutes. After 5 minutes, sodium hydroxide was added to the reactor to raise the pH of the solution in the reactor to between 9.0 and 9.5. Samples of the water were collected at predetermined time intervals beginning at one minute after the addition of sodium hydroxide.

This technology provided the fastest removal of selenate from Kessler springs water. Effluent selenium concentrations ranged from less than 0.002 mg/L to 0.01 mg/L after 10 minutes or less of treatment (University of Utah, 2000). The estimated costs of reagents for this process are approximately $2.50 to $3.00 per 1000 gallons (3785 L).

Technology Comparison. Of the eight selenium treatment technologies that Kennecott elected to study on-site, selenium effluent concentrations ranged from less than 0.002 mg/L to 0.834 mg/L, with reagent consumption costs ranging from $0.51 to $15.17 per 1000 gallons (3785 L). Table 2 summarizes the percent selenium removed and estimated costs for each technology.

TABLE 2. Summary of selenium treatment technologies.

Selenium Treatment Technology	Percent Selenium Removed	Reagents: $/1000 gallons (3785 L)	20 Year O&M: $/1000 gallons (3785 L)
Process Water Recycle	49	0	0.10 – 0.15
Biological Reduction	98.9–99.9	0.51 – 0.58	2.34
Ferrihydrite (EPA BDAT)	65 – 95	14.31 – 15.17	50.37
Catalyzed Cementation	54 – 98	8.11	10.36
Iron Filing Packed Columns	74 – 92	1.32	2.95
Iron Reductive Precipitation	29 – 85	6.54	10.37
Permeable Reactive Barrier	99.8–99.9	13.00	6.62
Ferrous Iron	99.4–99.9	2.50 – 3.00	4.80

CONCLUSION

Over the past two years, Kennecott has evaluated eight potential selenium (mainly in the form of selenate) treatment technologies through on-site bench and pilot scale testing. The testing showed that EPA's BDAT (ferrihydrite adsorption) for selenium treatment did not achieve selenium reductions to primary drinking water standards, and was cost prohibitive. The preferred technology for the Kennecott site is the process water recycle option, provided there will be a process in which to recycle the water for the next 30 years. If the process recycle option becomes unavailable, Kennecott will likely utilize the biological selenium removal process. Investigations are currently underway to evaluate the potential for the selenate reducing microbes to treat the ground water plume in-situ.

Of the eight treatment technologies studied, at least three appear to have the potential for cost effective, beneficial application to other waste water, ground water or surface water which contain elevated levels of selenium in the form of selenate: biological reduction, iron filing packed columns and ferrous iron treatment.

REFERENCES

Applied Biosciences Corporation. 2000. "Testing of KUCC's Aquifer Bedrock and Alluvial Materials for In-situ Selenium Reduction Potential." Contractor report to KUCC, December.

Harding ESE (formerly HLA). 2000a. "Preliminary Selenium Treatment Pilot Scale Testing: Iron Filing Packed Columns (IFPC) and Iron Powder Reductive Precipitation (IPRP) Processes." Contractor report to KUCC, October.

————2000b. "Cost Estimate for Kessler Springs Selenium Treatment Options." Contractor report to KUCC, December.

Harding Lawson and Associates (HLA). 1999. "Draft Report: Selenium Treatment Bench Testing, KUCC, Magna UT." Contractor report to KUCC, August.

Kennecott Utah Copper Corporation (KUCC). 2000. *North Facilities Remedial Investigation Report.* August.

————2001. *North Facilities Feasibility Study Report.* March; in-progress.

Montgomery Watson. 1999. "Technology Screening Study for Selenate." Contractor report to KUCC, April.

MSE Technology Applications, Inc. 2000. "Interim Report – Selenium Treatment/Removal Alternatives." Contractor report to KUCC, November.

Shepherd Miller, Inc. 1998. "Geochemical Characterization of Sources of Selenium, Historic Precious Metal Refinery Area, Kennecott Utah Copper." Contractor report to KUCC, February.

University of Utah (Trujillo, F., J. Turner, and J.Cantrell). 2000. "Investigation of the Removal of Selenium From a Waste Water Using Ferrous Iron: Contractor report to KUCC, February.

University of Waterloo (Smyth, D., and D. Blowes). 2000. "Removal of Dissolved Selenium From Groundwater Using Reactive Materials: Laboratory Column Treatability Testing." Contractor report to KUCC, September.

United States Environmental Protection Agency. 1988. *Ambient Water Quality Criteria for Selenium – 1988.* Office of Water Regulations and Standards, Washington, D.C. NTIS PB88-142237.

PILOT-SCALE NITRATE, SELENIUM AND CYANIDE REMOVAL

D.J. Adams (Applied Biosciences Corporation, Salt Lake City, UT, USA and Weber State University, Center for Bioremediation, Ogden, UT, USA)
T. Pickett (Applied Biosciences Corporation, Salt Lake City, UT, USA)

ABSTRACT: The Zortman – Landusky, MT mining properties contain 225 million gallons of water in storage, with an additional contribution of 150 million gallons annually. The bulk of the water is ponded under the heap material in lined heap leach pads. The site is currently dewatered through pumping and land application. The 8° C, pH ~7.4 site water contains ~250 mg/L Nitrate-N, ~0.7 mg/L Se and ~0.8 mg/L cyanide. Treatment targets are currently defined at 10 mg/L Nitrate-N, 0.05 mg/L Se, and 0.02 mg/L CN at ~300 gal/min for year-round treatment. An initial treatability assessment optimized non-pathogenic, proprietary denitrifying, selenium reducing and cyanide degrading microbes supplemented with indigenous denitrifying bacteria to develop a robust microbial biofilm. Three separate, but complementary microbial mixes were prepared in the laboratory and scaled up at the site in an existing 3,500 tank and carbon columns. Removal of all three contaminants to or below target discharge criteria was accomplished in staged treatment system at site water temperatures; no system or water heating was required. Nutrient costs from pilot-scale projections are <$0.55/1,000 gallons. A full-scale biotreatment system, consisting of three staged bioreactors, is being constructed at the Landusky 87 Pad. Based on pilot-scale testing, the biotreatment system is expected to run 15^{+}-years before harvesting of reduced selenium and bioreactor biofilm regeneration is required.

INTRODUCTION

At the Zortman and Landusky site, two contaminants are of primary concern, nitrate and cyanide. Additionally, selenium contamination is an important secondary concern. Currently, from the Landusky side of the mountain all lower pads go into the barren pond then into the 87 Pad and then over the hill to the Zortman side for treatment. Landusky Pads 87 and 91 contain the bulk of the contaminated waters requiring treatment for nitrate, cyanide and selenium. Most of the water going to Goslin Flats Land Application Disposal (LAD) comes from Landusky Pads 87 and 91. Cyanide containing waters from Landusky Pads 87 and 91 is being pumped to the Zortman 82 Pond and is being treated with H_2O_2 before going to the LAD.

Treatment goals for full scale are ~150 million gallon of leach pad water per year, corresponding to ~300 gpm for year around treatment. Contaminant effluent goals are to meet target discharge criteria for nitrate – 10 mg/L, cyanide – 0.02 mg/L and selenium 0.05 mg/L. A fully successful pilot-scale biotreatment of Landusky 87 Pad waters used three different flow rates to optimize nutrient composition and amount and to demonstrate nitrate, cyanide and selenium removal to target discharge criteria. Data was gathered to develop full-scale plant criteria, operational parameters and nutrient costs. This report presents the pilot-scale test methods and results.

METHODS

Bioreactor Configuration. Pilot-scale testing was configured to use existing site equipment to minimize testing costs. Applied Biosciences used a rectangular tank of ~3500 gallons and five 7' x 8' carbon tanks in carbon plant #3 to implement a site specific, staged treatment process. The reactors were filled with ~60,000 lbs of 8 x 30-mesh coconut shell activated carbon.

Microbial Inocula. Applied Biosciences' microbial inocula for nitrate, selenium, and cyanide was supplemented with indigenous site microbes and delivered to the site for scale up in existing tanks.

Sample Collection. Samples were collected from each step in the treatment system. Samples were collected in sterile containers using methods to stabilize samples for the particular microorganism or analyte. Samples for nitrate, nitrite, selenium, total cyanide and weak acid dissociable (WAD) cyanide analysis were prepared by Spectrum Engineering personnel for shipping. Samples were collected in sterile containers, filtered (0.45 μm), preserved and submitted to an EPA certified laboratory for analysis. Submitted samples included the following controls: a field duplicate, field cross contamination blank and calibration certification controls prepared with a known concentration of analyte. To help evaluate bioreactor performance, pH, dissolved oxygen (DO) and REDOX were measured at each sample point on a daily basis using calibrated instruments.

Sample Point	Description
1	Feed
2	After 1st rectangular tank
3	After 1st carbon tank
4	After 2nd carbon tank
5	After 3rd carbon tank
6	After 4th carbon tank
7	After 5th carbon tank - Effluent

Microbial Analysis/Characterization. Applied Biosciences completed all microbial analysis and characterization. All samples were stored at 4° C to inhibit microbial growth until analysis was performed. Baseline microbial characterization included microbial isolations and plate counts. Isolates were characterized by colony morphology and gram stain, and isolates were slanted on appropriate media for future testing. Where appropriate, MIDI fatty acid profiles were used to fingerprint the microbial population to evaluate maintenance of appropriate microbial populations within the bioreactor.

Process Evaluation. The following process evaluations were made at three flow rates 2.5, 7.5 and 15 gpm using, a balanced C:N:P ratio, trace elements and vitamin mix prepared by Applied Biosciences. Other less optimized nutrient mixtures were also evaluated.

- Optimum process operation and efficiency
- Different nutrient mixtures
- Nutrient amounts
- System limitations

RESULTS AND DISCUSSION

Nitrate, selenium and cyanide were removed from Landusky 87 Pad waters to or below discharge criteria during pilot-scale tests. Nitrite levels were normally below 1.0 mg/L in both feed and effluent samples during testing using the Applied Biosciences nutrient mix. Nitrite levels above 1.0 mg/L were noted with other nutrient mixes. No tracer studies were conducted to determine fates of contaminant components, but from past laboratory testing, using balanced nutrient sources, it is expected that the bulk of the nitrate was converted to nitrogen gas. The selenium was reduced to elemental selenium that remained in the bioreactor system and the cyanide was converted to carbon dioxide and ammonia. Based on this and other laboratory and pilot-scale testing, the biotreatment system is expected to run 15^{+}-years before harvesting of reduced selenium and bioreactor biofilm regeneration is required.

Pilot-Scale Configuration. Site equipment was used throughout the test and included a ~3500 gal rectangular tank situated on the south-facing side of the carbon plant and painted black to take advantage of solar heating. 87 Pad waters were fed directly into this tank and then through existing carbon tanks. This produced a very workable, but less than optimal treatment system. Nitrate and selenium removal bioprocesses require anaerobic conditions and cyanide biodegradation requires oxygen. Through on-site and laboratory testing/evaluation, it is estimated that the overall biotreatment system effectiveness was reduced by ~35 percent.

System effectiveness losses were primarily due to two factors, 1) malfunction of existing equipment and 2) unseasonably cold temperatures that interfered with biofilm production/establishment. Malfunction of existing on-site equipment, including the portion of the carbon plant system used for nutrient distribution. The system was partially plugged and did not deliver the correct amount of nutrient to the bioreactors during pilot-scale tests. Additionally, the cascading water system inherent to the carbon tanks used for the bioreactor system introduced oxygen and cooled the water at each stage. An unseasonably cold temperature hampered optimal biofilm production and establishment and is discussed below.

Inocula/Biofilm Preparation. Applied Biosciences' microbial inocula for the ***BDN***™, ***BCND***™ and ***BSeR***™ processes were delivered to the site for scale up in existing 1,000-gallon heated tanks for scale-up into the bioreactor system. The inocula was augmented with nitrate reducing microbial isolates obtained from Landusky Pad 87 waters following demonstrated establishment of a robust biofilm in site waters. Optimal biofilm were not reached for pilot-scale tests due to unseasonably cold temperatures. Measured biofilm densities established for these tests were ~1.5 logs less than optimal. Performance reductions were measured by calculation of observed versus known nitrate, selenium and cyanide removal rates per kilogram of biofilm material in field and laboratory systems.

The biofilm establishment process had to be temporarily shut down because the pipes and most of the bioreactor system had frozen. Although the biofilms did not reach the desired density, laboratory tests conducted during biofilm production and on-site tests indicated that the biofilm had reached a level that would produce satisfactory pilot-scale results.

Pilot-Scale Testing. Table 1 provides a general summary of the treatment conditions used during pilot-scale testing. Detailed discussions are provided in each of the treatment flow rate and nutrient sections. No heating was used during pilot-scale process operation.

TABLE 1. Treatment conditions

Process	Contaminant	Influent Level	Drinking Water MCL	Retention Time (hr) at Flow Rates (gpm)		
				2.5	7.5	15
Applied Biosciences Process System (*BDN, BSeR, BCND*)	NO_3-N	~250 mg/L	10 mg/L	~49	~16	~8
	Se	~0.8 mg/l	0.05 mg/L			
	CN	~0.7 mg/L	0.02 mg/L			

Water temperatures. Water temperature in the pilot tests ranged from 6.4° C to 9.5° C in the feed waters and from 5° C to 16° C in the different stages of the bioreactor system. The water temperature in the system increases until it enters the carbon columns then decreases again, exiting the reactor system at the same or lower temperature as when it entered the system. An optimized full-scale biotreatment system will better maintain heat.

2.5 gpm flow rate. At 2.5 gpm, the pilot treatment system removed nitrate, selenium and cyanide from Landusky Pad 87 waters to below drinking water criteria, Figures 1-3. Throughout the 12 days of operation, nitrate was removed to below 10 mg/L by the fourth reactor of the treatment system, Figure 1. Selenium was removed to below 50 ppb by the third bioreactor, sample point #4 and to below detection in the final reactor, Figure 2. Cyanide was removed to below 0.10 mg/L by the fourth reactor and to below 0.001 mg/L by the last reactor, Figure 3. The 2.5 gpm flow rate was continued for an extended time to determine the base rate for nutrient addition using a complete nutrient mix – a balanced C:N:P ratio, trace elements and vitamins mix prepared by Applied Biosciences.

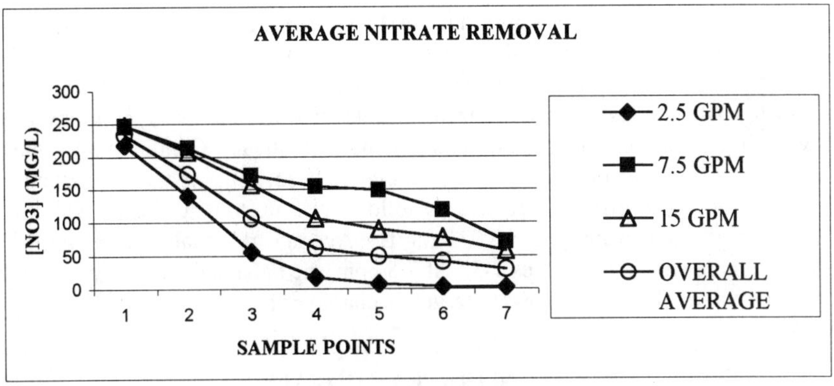

FIGURE 1. Average nitrate removal.

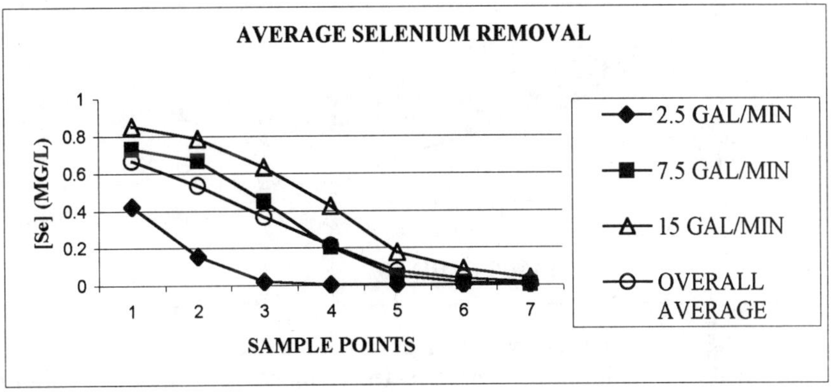

FIGURE 2. Average selenium concentration.

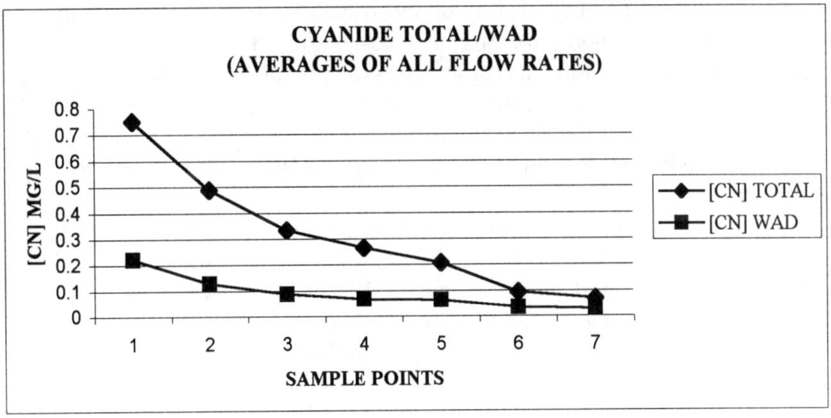

FIGURE 3. Average Total and WAD cyanide concentrations.

7.5 gpm flow rate. At 7.5 gpm, Figure 4, the treatment system nutrient amount and composition was varied to determine the minimal nutrient composition required to treat site waters. It was found that removal of all contaminants was affected when a less than optimized nutrient mixture was used; nitrate removal was affected more than cyanide and selenium removal. Contaminant removals varied depending on the nutrient mixture used, but reduced contaminant concentrations in effluent samples as follows: nitrate to <50 mg/L, selenium to <0.02 mg/L and cyanide to <0.071 mg/L. Average removals for the entire 7.5 gpm run are shown in Figures 1-3. At 7.5 gpm, the balanced nutrient mix resulted in removal of all contaminants to target discharge criteria.

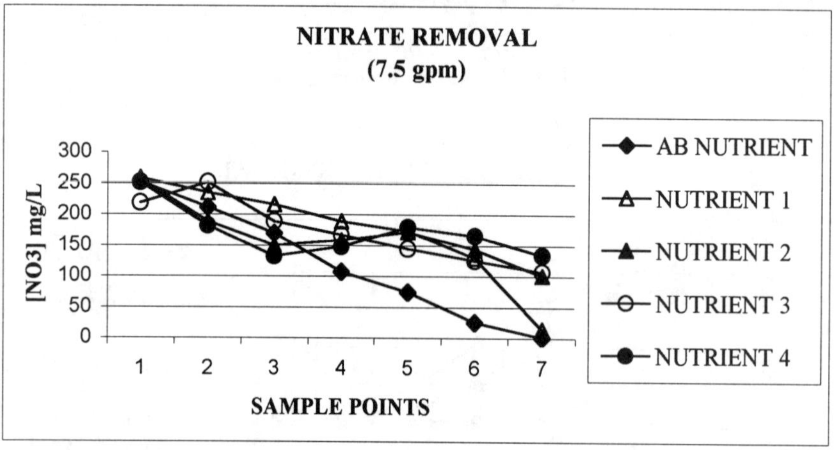

FIGURE 4. Nitrate removal using different nutrient components. Nitrate removal was significantly affected by less complete nutrient mixes with the Applied Biosciences complete nutrient resulting in the best nitrate removal.

15 gpm flow rate. This flow rate, Figures 1-3, & 5, was used to further examine nutrient amounts required, to help define system limitations, and in evaluation of trade-off between bioreactor size and cost. For this flow rate, the balanced nutrient mix was used in varying amounts to determine optimal nutrient amounts required for contaminant removal, the nutrient increase required with increased flow rates and the nutrient requirements at the lower ambient temperatures. The data shown in the Figures 1-3 are average effluent values for contaminant removal.

At the 15 gpm flow rate, effluent nitrate concentrations ranged from ~18 mg/L to ~79 mg/L, selenium ranged from 0.004 mg/L to 0.09 mg/L and cyanide range from 0.027 mg/L to 0.046 mg/L. When the nutrient feeding rates were close to optimal, the 15-gpm flow rate contaminant concentrations in the bioreactor effluents were near target discharge criteria.

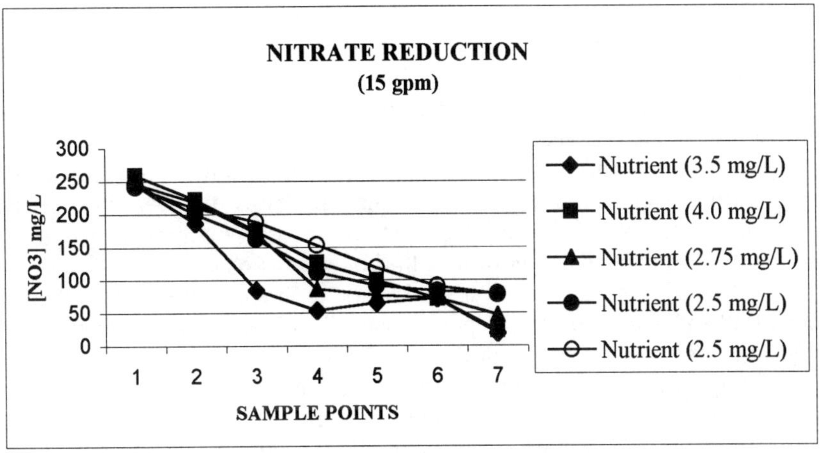

FIGURE 5. Nitrate reduction at 15 gpm using various nutrient concentrations.

Nutrient Costs. The Applied Biosciences' balanced nutrient mixture provided the best contaminant removal. Nitrate in the Landusky waters was the most susceptible of the three primary contaminants to nutrient composition and amount. Laboratory tests indicate that removals of selenium and cyanide would also be affected, but these contaminants do not have the short response time shown for nitrate. Nitrate removal was best using a balanced nutrient mix, Figures 4-6. A less than balanced formulation, such as just sugar or methanol, caused nitrate removal to fall substantially.

Costs for the balanced nutrient mix ranged from ~$0.35/1,000 gal at 2.5 gpm to ~$0.74/1,000 at 15 gpm. Calculated nutrient costs do not generally have this broad a range. Lower biofilm densities and malfunction of old site equipment contributed to lower bioreactor efficiencies, this would be amplified at the higher flow rates. This conclusion was also supported by laboratory biofilm analysis.

In theory, nutrient costs to treat 1,000 gal should be the same, that is, the same volume of contaminant is present no matter how long the retention time. However, in practice this is not usually the case, we have observed that the higher the flow rate or the shorter the contact time the larger the amount of nutrient required to maintain consistent contaminant removal. This increase is often site or contaminant mixture dependent and not often substantial unless there are great differences in contaminant contact time. When dealing with high-density biofilms in a semi-steady metabolism state and low growth rate state, nutrient requirements have been observed to be related to contaminant contact time. As observed in laboratory and field-testing, the longer the contact time the lower the amount of nutrient required.

Only a portion of the energy supplied by the nutrients provides energy needed to remove contaminants, the balance is used to main the biofilm. With a fixed number of cells, shorter contaminant contact times require that the biofilm's metabolic state be maintained at a higher level, which requires more energy, thus, larger nutrient amounts. Nutrients are also required for synthesis of new cellular materials and at higher flow rates there is a requirement for higher metabolic rates to synthesize new cellular materials.

While bioconversions can often be achieved, in short term applications, with less complex nutrients, these simpler nutrient formulations or compounds do not provide the proper C:N:P ratio and other micronutrients required for long-term biofilm stability and optimal contaminant removal. With less than balanced nutrients, a portion of the biofilm must die in order to provide the missing nutrients to the living cells performing the bioremediation.

The 7.5 gpm flow rate was used to test different nutrient mixes that were both less and more costly per pound, but all proved to be significantly less effective for contaminant removal. The 7.5 gpm test period was not used in calculating nutrient costs. Three main factors played a role in the nutrient costs observed, retention time, malfunctioning nutrient distribution systems and low biofilm density. Full-scale treatment costs are expected to be <$0.55/1,000 gallons.

CONCLUSIONS/RECOMMENDATIONS

- Pilot-scale tests of the Applied Biosciences biotreatment system using Landusky 87 Pad waters was fully successful
- Nitrate, cyanide and selenium removal to target discharge criteria was demonstrated at ambient site water temperatures
- A biotreatment time long enough to remove nitrate will also remove cyanide and selenium to target discharge criteria
- A three-stage biotreatment system with an 18+ hr retention time is recommended to meet discharge criteria at full scale

2001 AUTHOR INDEX

This index contains names, affiliations, and volume/page citations for all authors who contributed to the ten-volume proceedings of the Sixth International In Situ and On-Site Bioremediation Symposium (San Diego, California, June 4-7, 2001). Ordering information is provided on the back cover of this book.

The citations reference the ten volumes as follows:

6(1): Magar, V.S., J.T. Gibbs, K.T. O'Reilly, M.R. Hyman, and A. Leeson (Eds.), *Bioremediation of MTBE, Alcohols, and Ethers*. Battelle Press, Columbus, OH, 2001. 249 pp.

6(2): Leeson, A., M.E. Kelley, H.S. Rifai, and V.S. Magar (Eds.), *Natural Attenuation of Environmental Contaminants*. Battelle Press, Columbus, OH, 2001. 307 pp.

6(3): Magar, V.S., G. Johnson, S.K. Ong, and A. Leeson (Eds.), *Bioremediation of Energetics, Phenolics, and Polycyclic Aromatic Hydrocarbons*. Battelle Press, Columbus, OH, 2001. 313 pp.

6(4): Magar, V.S., T.M. Vogel, C.M. Aelion, and A. Leeson (Eds.), *Innovative Methods in Support of Bioremediation*. Battelle Press, Columbus, OH, 2001. 197 pp.

6(5): Leeson, A., E.A. Foote, M.K. Banks, and V.S. Magar (Eds.), *Phytoremediation, Wetlands, and Sediments*. Battelle Press, Columbus, OH, 2001. 383 pp.

6(6): Magar, V.S., F.M. von Fahnestock, and A. Leeson (Eds.), *Ex Situ Biological Treatment Technologies*. Battelle Press, Columbus, OH, 2001. 423 pp.

6(7): Magar, V.S., D.E. Fennell, J.J. Morse, B.C. Alleman, and A. Leeson (Eds.), *Anaerobic Degradation of Chlorinated Solvents*. Battelle Press, Columbus, OH, 2001. 387 pp.

6(8): Leeson, A., B.C. Alleman, P.J. Alvarez, and V.S. Magar (Eds.), *Bioaugmentation, Biobarriers, and Biogeochemistry*. Battelle Press, Columbus, OH, 2001. 255 pp.

6(9): Leeson, A., B.M. Peyton, J.L. Means, and V.S. Magar (Eds.), *Bioremediation of Inorganic Compounds*. Battelle Press, Columbus, OH, 2001. 377 pp.

6(10): Leeson, A., P.C. Johnson, R.E. Hinchee, L. Semprini, and V.S. Magar (Eds.), *In Situ Aeration and Aerobic Remediation*. Battelle Press, Columbus, OH, 2001. 391 pp.

Aagaard, Per (University of Oslo/NORWAY) *6*(2):181
Aarnink, Pedro J.P. (Tauw BV/THE NETHERLANDS) *6*(10):253
Abbott, James E. (Battelle/USA) *6*(5):231, 237
Accashian, John V. (Camp Dresser & McKee, Inc./USA) *6*(7):133
Adams, Daniel J. (Camp Dresser & McKee, Inc./USA) *6*(8):53
Adams, Jack (Applied Biosciences Corporation/USA) *6*(9):331
Adriaens, Peter (University of Michigan/USA) *6*(8):19, 193
Adrian, Neal R. (U.S. Army Corps of Engineers/USA) *6*(6):133
Agrawal, Abinash (Wright State University/USA) *6*(5):95
Aiken, Brian S. (Parsons Engineering Science/USA) *6*(2): *6*5, 189

Aitchison, Eric (Ecolotree, Inc./USA) *6*(5):121
Al-Awadhi, Nader (Kuwait Institute for Scientific Research/KUWAIT) *6*(6):249
Alblas, B. (Logisticon Water Treatment/THE NETHERLANDS) *6*(8):11
Albores, A. (CINVESTAV-IPN/MEXICO) *6*(6):219
Al-Daher, Reyad (Kuwait Institute for Scientific Research/KUWAIT) *6*(6):249
Al-Fayyomi, Ihsan A. (Metcalf & Eddy, Inc./USA) *6*(7):173
Al-Hakak, A. (McGill University/CANADA) *6*(9):139
Allen, Harry L. (U.S. EPA/USA) *6*(3):259
Allen, Jeffrey (University of Cincinnati/USA) *6*(9):9
Allen, Mark H. (Dames & Moore/USA) *6*(10):95
Allende, J.L. (Universidad Complutense/SPAIN) *6*(4):29
Alonso, R. (Universidad Politecnica/SPAIN) *6*(6):377
Alphenaar, Arne (TAUW bv/THE NETHERLANDS) *6*(7):297
Alvarez, Pedro J. J. (University of Iowa/USA) *6*(1):195; *6*(3):1; *6*(8):147, 175
Alvestad, Kimberly R. (Earth Tech/USA) *6*(3):17
Ambert, Jack (Battelle Europe/SWITZERLAND) *6*(6):241
Amezcua-Vega, Claudia (CINVESTAV-IPN/MEXICO) *6*(3):243
Amy, Penny (University of Nevada Las Vegas/USA) *6*(9):257
Andersen, Peter F. (GeoTrans, Inc./USA) *6*(10):163
Anderson, Bruce (Plan Real AG/AUSTRALIA) *6*(2):223
Anderson, Jack W. (RMT, Inc./USA) *6*(10):201
Anderson, Todd (Texas Tech University/USA) *6*(9):273
Andreotti, Giorgio (ENI Sop.A.) *6*(5):41
Andretta, Massimo (Centro Ricerche Ambientali Montecatini/ITALY) *6*(4):131
Andrews, Eric (Environmental Management, Inc./USA) *6*(10):23
Andrews, John (SHN Consulting Engineers & Geologists, Inc./USA) *6*(3):83
Archibald, Brent B. (Exxon Mobil Environmental Remediation/USA) *6*(8):87
Archibold, Errol (Spelman College/USA) *6*(9):53
Aresta, Michele (Universita di Catania/ITALY) *6*(3):149
Arias, Marianela (PDVSA Intevep/VENEZUELA) *6*(6):257
Atagana, Harrison I. (Mangosuthu Technikon/REP OF SOUTH AFRICA) *6*(6):101
Atta, Amena (U.S. Air Force/USA) *6*(2):73
Ausma, Sandra (University of Guelph/CANADA) *6*(6):185
Autenrieth, Robin L. (Texas A&M University/USA) *6*(5): 17, 25
Aziz, Carol E. (Groundwater Services, Inc./USA) *6*(7):19; *6*(8):73
Azizian, Mohammad (Oregon State University/USA) *6*(10): 145, 155

Babel, Wolfgang (UFZ Center for Environmental Research/GERMANY) *6*(4):81
Bae, Bumhan (Kyungwon University/REPUBLIC OF KOREA) *6*(6):51
Baek, Seung S. (Kyonggi University/REPUBLIC OF KOREA) *6*(1):161
Bagchi, Rajesh (University of Cincinnati/USA) *6*(5):243, 253, 261
Baiden, Laurin (Clemson University/USA) *6*(7):109
Bakker, C. (IWACO/THE NETHERLANDS) *6*(7):141
Balasoiu, Cristina (École Polytechnique de Montreal/CANADA) *6*(9):129
Balba, M. Talaat (Conestoga-Rovers & Associates/USA) *6*(1):99; *6*(6):249; *6*(10):131

Banerjee, Pinaki (Harza Engineering Company, Inc./USA) *6*(7):157
Bankston, Jamie L. (Camp Dresser and McKee Inc./USA) *6*(5):33
Barbé, Pascal (Centre National de Recherche sur les Sites et Sols Pollués/FRANCE) *6*(2):129
Barcelona, Michael J. (University of Michigan/USA) *6*(8):19, 193
Barczewski, Baldur (Universitat Stuttgart/GERMANY) *6*(2):137
Barker, James F. (University of Waterloo/CANADA) *6*(8):95
Barnes, Paul W. (Earth Tech, Inc./USA) *6*(3): 17, 25
Basel, Michael D. (Montgomery Watson Harza/USA) *6*(10):41
Baskunov, Boris B. (Russian Academy of Sciences/RUSSIA) *6*(3):75
Bastiaens, Leen (VITO/BELGIUM) *6*(4):35; *6*(9):87
Batista, Jacimaria (University of Nevada Las Vegas/USA) *6*(9): 257, 265
Bautista-Margulis, Raul G. (Centro de Investigacion en Materiales Avanzados/MEXICO) *6*(6):361
Becker, Paul W. (Exxon Mobil Refining & Supply/USA) *6*(8):87
Beckett, Ronald (Monash University/AUSTRALIA) *6*(4):1
Beckwith, Walt (Solutions Industrial & Environmental Services/USA) *6*(7):249
Beguin, Pierre (Institut Pasteur/FRANCE) *6*(1):153
Behera, N. (Sambalpur University/INDIA) *6*(9):173
Bell, Nigel (Imperial College London/UK) *6*(10):123
Bell, Mike (Coats North America/USA) *6*(7):213
Beller, Harry R. (Lawrence Livermore National Laboratory/USA) *6*(1):195
Belloso, Claudio (Facultad Catolica de Quimica e Ingenieria/ARGENTINA) *6*(6): 235, 303
Benner, S. G. (Stanford University/USA) *6*(9):71
Bensch, Jeffrey C. (GeoTrans, Inc/USA) *6*(7):221
Béron, Patrick (Université du Québec à Montréal/CANADA) *6*(3):165
Berry, Duane F. (Virginia Polytechnic Institute & State University/USA) *6*(2):105
Betts, W. Bernard (Cell Analysis Ltd./UK) *6*(6):27
Billings, Bradford G. (Brad) (Billings & Associates, Inc./USA) *6*(1):115
Bingler, Linda (Battelle Sequim/USA) *6*(5):231, 237
Birkle, M. (Fraunhofer Institute/GERMANY) *6*(2):137
Bitter, Paul (URS Corporation./USA) *6*(2):261
Bittoni, A. (EniTecnologie/ITALY) *6*(6):173
Bjerg, Poul L (Technical University of Denmark/DENMARK) *6*(2):11
Blanchet, Denis (Institut Français du Pétrole/FRANCE) *6*(3):227
Bleckmann, Charles A. (Air Force Institute of Technology/USA) *6*(2):173
Blokzijl, R. (DHV Environment and Infrastructure/THE NETHERLANDS) *6*(8):11
Blowes, David (University of Waterloo/CANADA) *6*(9):71
Bluestone, Simon (Montgomery Watson/ITALY) *6*(10):41
Boben, Carolyn (Williams/USA) *6*(1):175
Böckle, Karin (Technologiezentrum Wasser/GERMANY) *6*(8):105
Boender, H. (Logisticon Water Treatment/THE NETHERLANDS) *6*(8):11
Böhler, Anja (BioPlanta GmbH/GERMANY) *6*(3):67
Bonner, James S. (Texas A&M University/USA) *6*(5):17, 25
Bononi, Vera Lucia Ramos (Instituto de Botânica/BRAZIL) *6*(3):99
Bonsack, Laurence T. (Aerojet/USA) *6*(9):297
Borazjani, Abdolhamid (Mississippi State University/USA) *6*(5):329; *6*(6):279

Borden, Robert C. (Solutions Industrial & Environmental Services/USA) *6*(7):249
Bornholm, Jon (U.S. EPA/USA) *6*(6):81
Bosco, Francesca (Politecnico di Torino/ITALY) *6*(3):211
Bosma, Tom N.P. (TNO Environment/THE NETHERLANDS) *6*(7):61
Bourquin, Al W. (Camp Dresser & McKee Inc./USA) *6*(5):33; *6*(6):81; *6*(7):133,
Bouwer, Edward J. (Johns Hopkins University/USA) *6*(2):19
Bowman, Robert S. (New Mexico Institute of Mining & Technology/USA) *6*(8):131
Boyd, Sian (CEFAS Laboratory/UK) *6*(10):337
Boyd-Kaygi, Patricia (Harding ESE/USA) *6*(10):231
Boyle, Susan L. (Haley & Aldrich, Inc./USA) *6*(7):27, 281
Brady, Warren D. (IT Corporation/USA) *6*(9):215
Breedveld, Gijs (University of Oslo/NORWAY) *6*(2):181
Bregante, M. (Istituto di Cibernetica e Biofisica/ITALY) *6*(5):157
Brenner, Richard C. (U.S. EPA/USA) *6*(5):231, 237
Breteler, Hans (Oostwaardhoeve Co./THE NETHERLANDS) *6*(6):59
Bricka, Mark R. (U.S. Army Corps of Engineers/USA) *6*(9):241
Brickell, James L. (Earth Tech, Inc./USA) *6*(10):65
Brigmon, Robin L. (Westinghouse Savannah River Co/USA) *6*(7):109
Britto, Ronnie (EnSafe, Inc./USA) *6*(9):315
Brossmer, Christoph (Degussa Corporation/USA) *6*(10):73
Brown, Bill (Dunham Environmental Services/USA) *6*(6):35
Brown, Kandi L. (IT Corporation/USA) *6*(1):51
Brown, Richard A. (ERM, Inc./USA) *6*(7):45, 213
Brown, Stephen (Queen's University/CANADA) *6*(2):121
Brown, Susan (National Water Research Institute/CANADA) *6*(7):321, 333, 341
Brubaker, Gaylen (ThermoRetec North Carolina Corp./USA) *6*(7):1
Bruce, Cristin (Arizona State University/USA) *6*(8):61
Bruce, Neil C. (University of Cambridge/UK) *6*(5):69
Buchanan, Gregory (Tait Environmental Management, Inc./USA) *6*(10):267
Bucke, Christopher (University of Westminster/UK) *6*(3):75
Bulloch, Gordon (BAE Systems Properties Ltd./UK) *6*(6):119
Burckle, John (U.S. EPA/USA) *6*(9):9
Burden, David S. (U.S. EPA/USA) *6*(2):163
Burdick, Jeffrey S. (ARCADIS Geraghty & Mills/USA) *6*(7):53
Burgos, William (The Pennsylvania State University/USA) *6*(8):201
Burken, Joel G. (University of Missouri-Rolla/USA) *6*(5):113, 199
Burkett, Sharon E. (ENVIRON International Corp./USA) *6*(7):189
Burnell, Daniel K. (GeoTrans, Inc./USA) *6*(2):163
Burns, David A. (ERM, Inc./USA) *6*(7):213
Burton, Christy D. (Battelle/USA) *6*(1):137; *6*(10):193
Buscheck, Timothy E. (Chevron Research & Technology Co/USA) *6*(1): 35, 203
Buss, James A. (RMT, Inc./USA) *6*(2):97
Butler, Adrian P. (Imperial College London/UK) *6*(10):123
Butler, Jenny (Battelle/USA) *6*(7):13
Büyüksönmez, Fatih (San Diego State University/USA) *6*(10):301

Caccavo, Frank (Whitworth College/USA) *6*(8):1
Callender, James S. (Rockwell Automation/USA) *6*(7):133
Calva-Calva, G. (CINVESTAV-IPN/MEXICO) *6*(6):219
Camper, Anne K. (Montana State University/USA) *6*(7):117

Camrud, Doug (Terracon/USA) *6*(10):15
Canty, Marietta C. (MSE Technology Applications/USA) *6*(9):35
Carman, Kevin R. (Louisiana State University/USA) *6*(5):305
Carrera, Paolo (Ambiente S.p.A./ITALY) *6*(6):227
Carson, David A. (U.S. EPA/USA) *6*(2):247
Carvalho, Cristina (Clemson University/USA) *6*(7):109
Case, Nichole L. (Haley & Aldrich, Inc./USA) *6*(7):27, 281
Castelli, Francesco (Universita di Catania/ITALY) *6*(3):149
Cha, Daniel K. (University of Delaware/USA) *6*(6):149
Chaney, Rufus L. (U.S. Department of Agriculture/USA) *6*(5):77
Chang, Ching-Chia (National Chung Hsing University/TAIWAN) *6*(10):217
Chang, Soon-Woong (Kyonggi University/REPUBLIC OF KOREA) *6*(1):161
Chang, Wook (University of Maryland/USA) *6*(3):205
Chapuis, R. P. (École Polytechnique de Montréal/CANADA) *6*(4):139
Charrois, Jeffrey W.A. (Komex International, Ltd./CANADA) *6*(4):7
Chatham, James (BP Exploration/USA) *6*(2):261
Chekol, Tesema (University of Maryland/USA) *6*(5):77
Chen, Abraham S.C. (Battelle/USA) *6*(10):245
Chen, Chi-Ruey (Florida International University/USA) *6*(10):187
Chen, Zhu (The University of New Mexico/USA) *6*(9):155
Cherry, Jonathan C. (Kennecott Utah Copper Corp/USA) *6*(9):323
Child, Peter (Investigative Science Inc./CANADA) *6*(2):27
Chino, Hiroyuki (Obayashi Corporation/JAPAN) *6*(6):249
Chirnside, Anastasia E.M. (University of Delaware/USA) *6*(6):9
Chiu, Pei C. (University of Delaware/USA) *6*(6):149
Cho, Kyung-Suk (Ewha University/REPUBLIC OF KOREA) *6*(6):51
Choung, Youn-kyoo (Yonsei University/REPUBLIC OF KOREA) *6*(6):51
Clement, Bernard (École Polytechnique de Montréal/CANADA) *6*(9):27
Clemons, Gary (CDM Federal Programs Corp./USA) *6*(6):81
Cocos, Ioana A. (École Polytechnique de Montréal/CANADA) *6*(9):27
Cocucci, M. (Universita' degli Studi di Milano/ITALY) *6*(5):157
Coelho, Rodrigo O. (CSD-GEOLOCK/BRAZIL) *6*(1):27
Collet, Berto (TAUW bv/THE NETHERLANDS) *6*(10):253
Compton, Joanne C. (REACT Environmental Engineers/USA) *6*(3):25
Connell, Doug (Barr Engineering Company/USA) *6*(5):105
Connor, Michael A. (University of Melbourne/AUSTRALIA) *6*(10):329
Cook, Jim (Beazer East, Inc./USA) *6*(2):239
Cooke, Larry (NOVA Chemicals Corporation/USA) *6*(4):117
Coons, Darlene (Conestoga-Rovers & Associates/USA) *6*(1):99; *6*(10):131
Costley, Shauna C. (University of Natal/REP OF SOUTH AFRICA) *6*(9):79
Cota, Jennine L. (ARCADIS Geraghty & Miller, Inc./USA) *6*(7):149
Covell, James R. (EG&G Technical Services, Inc./USA) *6*(10):49
Cowan, James D. (Ensafe Inc./USA) *6*(9):315
Cox, Evan E. (GeoSyntec Consultants/CANADA) *6*(8):27, *6*(9):297
Cox, Jennifer (Clemson University/USA) *6*(7):109
Craig, Shannon (Beazer East, Inc./USA) *6*(2):239
Crawford, Donald L. (University of Idaho/USA) *6*(3):91; *6*(9):147

Crecelius, Eric (Battelle/USA) *6*(5): 231, 237
Crotwell, Terry (Solutions Industrial & Environmental Services/USA) *6*(7):249
Cui, Yanshan (Chinese Academy of Sciences/CHINA) *6*(9):113
Cunningham, Al B. (Montana State University/USA) *6*(7):117; *6*(8):1
Cunningham, Jeffrey A. (Stanford University/USA) *6*(7):95
Cutright, Teresa J. (The University of Akron/USA) *6*(3):235

da Silva, Marcio Luis Busi (University of Iowa/USA) *6*(1):195
Daly, Daniel J. (Energy & Environmental Research Center/USA) *6*(5):129
Daniel, Fabien (AEA Technology Environment/UK) *6*(10):337
Daniels, Gary (GeoTrans/USA) *6*(8):19
Das, K.C. (University of Georgia/USA) *6*(9):289
Davel, Jan L. (University of Cincinnati/USA) *6*(6):133
Davis, Gregory A. (Microbial Insights Inc./USA) *6*(2):97
Davis, Jeffrey L. (U.S. Army/USA) *6*(3): 43, 51
Davis, John W. (The Dow Chemical Company/USA) *6*(2):89
Davis-Hoover, Wendy J. (U.S. EPA/USA) *6*(2):247
De'Ath, Anna M. (Cranfield University/UK) *6*(6):329
Dean, Sean (Camp Dresser & McKee. Inc/USA) *6*(7):133
DeBacker, Dennis (Battelle/USA) *6*(10):145
DeHghi, Benny (Honeywell International Inc./USA) *6*(2):39;*6*(10):283
de Jong, Jentsje (TAUW BV/THE NETHERLANDS) *6*(10):253
Del Vecchio, Michael (Envirogen, Inc./USA) *6*(9):281
Delille, Daniel (CNRS/FRANCE) *6*(2):57
DeLong, George (AIMTech/USA) *6*(7):321, 333, 341
Demers, Gregg (ERM/USA) *6*(7):45
De Mot, Rene (Catholic University of Leuven/BELGIUM) *6*(4):35
Deobald, Lee A. (University of Idaho/USA) *6*(9):147
Deschênes, Louise (École Polytechnique de Montréal/CANADA) *6*(3):115; *6*(9):129
Dey, William S. (Illinois State Geological Survey/USA) *6*(9):179
Díaz-Cervantes, Dolores (CINVESTAV-IPN/MEXICO) *6*(6):369
Dick, Vincent B. (Haley & Aldrich, Inc./USA) *6*(7):27, 281
Diehl, Danielle (The University of New Mexico/USA) *6*(9):155
Diehl, Susan V. (Mississippi State University/USA) *6*(5):329
Diels, Ludo (VITO/BELGIUM) *6*(9):87
DiGregorio, Salvatore (University della Calabria/ITALY) *6*(4):131
Di Gregorio, Simona (Universita degli Studi di Verona/ITALY) *6*(3):267
Dijkhuis, Edwin (Bioclear/THE NETHERLANDS) *6*(5):289
Di Leo, Cristina (EniTecnologie/ITALY) *6*(6):173
Dimitriou-Christidis, Petros (Texas A& M University) *6*(5):17
Dixon, Robert (Montgomery Watson/ITALY) *6*(10):41
Dobbs, Gregory M. (United Technologies Research Center/USA) *6*(7):69
Doherty, Amy T. (GZA GeoEnvironmental, Inc./USA) *6*(7):165
Dolan, Mark E. (Oregon State University/USA) *6*(10):145, 155, 179
Dollhopf, Michael (Michigan State University/USA) *6*(8):19
Dondi, Giovanni (Water & Soil Remediation S.r.l./ITALY) *6*(6):179
Dong, Yiting (Chinese Academy of Sciences/CHINA) *6*(9):113
Dooley, Maureen A. (Regenesis/USA) *6*(7):197
Dottridge, Jane (Komex Europe Ltd./UK) *6*(4):17
Dowd, John (University of Georgia/USA) *6*(9):289
Doughty, Herb (U.S. Navy/USA) *6*(10):1

Doze, Jacco (RIZA/THE NETHERLANDS) *6*(5):289
Dragich, Brian (California Polytechnic State University/USA) *6*(2):1
Drake, John T. (Camp Dresser & McKee Inc./USA) *6*(7):273
Dries, Victor (Flemish Public Waste Agency/BELGIUM) *6*(7):87
Du, Yan-Hung (National Chung Hsing University/TAIWAN) *6*(6):353
Dudal, Yves (École Polytechnique de Montréal/CANADA) *6*(3):115
Duffey, J. Tom (Camp Dresser & McKee Inc./USA) *6*(5):33
Duffy, Baxter E. (Inland Pollution Services, Inc./USA) *6*(7):313
Duijn, Rik (Oostwaardhoeve Co./THE NETHERLANDS) *6*(6):59
Durant, Neal D. (GeoTrans, Inc./USA) *6*(2):19, 163
Durell, Gregory (Battelle Ocean Sciences/USA) *6*(5):231
Dworatzek, S. (University of Toronto/CANADA) *6*(8):27
Dwyer, Daryl F. (University of Minnesota/USA) *6*(3):219
Dzantor, E. K. (University of Maryland/USA) *6*(5):77

Ebner, R. (GMF/GERMANY) *6*(2):137
Ederer, Martina (University of Idaho/USA) *6*(9):147
Edgar, Michael (Camp Dresser & McKee Inc./USA) *6*(7):133
Edwards, Elizabeth A. (University of Toronto/CANADA) *6*(8):27
Edwards, Grant C. (University of Guelph/CANADA) *6*(6):185
Eggen, Trine (Jordforsk Centre for Soil and Environmental Research/NORWAY) *6*(6):157
Eggert, Tim (CDM Federal Programs Corp./USA) *6*(6):81
Elberson, Margaret A. (DuPont Co./USA) *6*(8):43
Elliott, Mark (Virginia Polytechnic Institute & State University/USA) *6*(5):1
Ellis, David E. (Dupont Company/USA) *6*(8):43
Ellwood, Derek C. (University of Southampton/UK) *6*(9):61
Else, Terri (University of Nevada Las Vegas/USA) *6*(9):257
Elväng, Annelie M. (Stockholm University/SWEDEN) *6*(3):133
England, Kevin P. (USA) *6*(5):105
Ertas, Tuba Turan (San Diego State University/USA) *6*(10):301
Escalon, Lynn (U.S. Army Corps of Engineers/USA) *6*(3):51
Esparza-Garcia, Fernando (CINVESTAV-IPN/MEXICO) *6*(6):219
Evans, Christine S. (University of Westminster/UK) *6*(3):75
Evans, Patrick J. (Camp Dresser & McKee, Inc./USA) *6*(2):113, 199; *6*(8):209

Fabiani, Fabio (EniTecnologie S.p.A./ITALY) *6*(6):173
Fadullon, Frances Steinacker (CH2M Hill/USA) *6*(3):107
Fang, Min (University of Massachusetts/USA) *6*(6):73
Faris, Bart (New Mexico Environmental Department/USA) *6*(9):223
Farone, William A. (Applied Power Concepts, Inc./USA) *6*(7):103
Fathepure, Babu Z. (Oklahoma State University/USA) *6*(8):19
Faust, Charles (GeoTrans, Inc./USA) *6*(2):163
Fayolle, Françoise (Institut Français du Pétrole/FRANCE) *6*(1):153
Feldhake, David (University of Cincinnati/USA) *6*(2):247
Felt, Deborah (Applied Research Associates, Inc./USA) *6*(7):125
Feng, Terry H. (Parsons Engineering Science, Inc./USA) *6*(2):39; *6*(10):283
Fenwick, Caroline (Aberdeen University/UK) *6*(2):223
Fernandez, Jose M. (University of Iowa/USA) *6*(1):195
Fernández-Sanchez, J. Manuel (CINVESTAV-IPN/MEXICO) *6*(6):369

Ferrer, E. (Universidad Complutense de Madrid/SPAIN) *6*(4):29
Ferrera-Cerrato, Ronald (Colegio de Postgraduados/MEXICO) *6*(6):219
Fiacco, R. Joseph (Environmental Resources Management) *6*(7):45
Fields, Jim (University of Georgia/USA) *6*(9):289
Fields, Keith A. (Battelle/USA) *6*(10):1
Fikac, Paul J. (Jacobs Engineering Group, Inc./USA) *6*(6):35
Fischer, Nick M. (Aquifer Technology/USA) *6*(8):157, *6*(10):15
Fisher, Angela (The Pennsylvania State University/USA) *6*(8):201
Fisher, Jonathan (Environment Agency/UK) *6*(4):17
Fitch, Mark W. (University of Missouri-Rolla/USA) *6*(5):199
Fleckenstein, Janice V. (USA) *6*(6):89
Fleischmann, Paul (ZEBRA Environmental Corp./USA) *6*(10):139
Fletcher, John S. (University of Oklahoma/USA) *6*(5):61
Foget, Michael K. (SHN Consulting Engineers & Geologists, Inc./USA) *6*(3):83
Foley, K.L. (U.S. Army Engineer Research & Development Center/USA) *6*(5):9
Follner, Christina G. (University of Leipzig/GERMANY) *6*(4):81
Fontenot, Martin M. (Syngenta Crop Protection, Inc./USA) *6*(6):35
Foote, Eric A. (Battelle/USA) *6*(1):137; *6*(7):13
Ford, James (Investigative Science Inc./CANADA) *6*(2):27
Forman, Sarah R. (URS Corporation/USA) *6*(7):321, 333, 341
Fortman, Tim J. (Battelle Marine Sciences Laboratory/USA) *6*(3):157
Francendese, Leo (U.S. EPA/USA) *6*(3):259
Francis, M. McD. (NOVA Research & Technology Center/CANADA) *6*(4):117; *6*(5):53,
François, Alan (Institut Français du Pétrole/FRANCE) *6*(1):153
Frankenberger, William T. (University of California/USA) *6*(9):249
Freedman, David L. (Clemson University/USA) *6*(7):109
French, Christopher E. (University of Cambridge/UK) *6*(5):69
Friese, Kurt (UFZ Center for Environmental Research/GERMANY) *6*(9):43
Frisbie, Andrew J. (Purdue University/USA) *6*(3):125
Frisch, Sam (Envirogen Inc./USA) *6*(9):281
Frömmichen, René (UFZ Centre for Environmental Research/GERMANY) *6*(9):43
Fuierer, Alana M. (New Mexico Institute of Mining & Technology/USA) *6*(8):131
Fujii, Kensuke (Obayashi Corporation/JAPAN) *6*(10):239
Fujii, Shigeo (Kyoto University/JAPAN) *6*(4):149
Furuki, Masakazu (Hyogo Prefectural Institute of Environmental Science/JAPAN) *6*(5):321

Gallagher, John R. (University of North Dakota/USA) *6*(5):129; *6*(6):141
Gambale, Franco (Istituto di Cibernetica e Biofisica/ITALY) *6*(5):157
Gambrell, Robert P. (Louisiana State University/USA) *6*(5):305
Gandhi, Sumeet (University of Iowa/USA) *6*(8):147
Garbi, C. (Universidad Complutense de Madrid/SPAIN) *6*(4):29; *6*(6):377
García-Arrazola, Roeb (CINVESTAV-IPN/MEXICO) *6*(6):369
García-Barajas, Rubén Joel (ESIQIE-IPN/MEXICO) *6*(6):369
Garrett, Kevin (Harding ESE/USA) *6*(7):205
Garry, Erica (Spelman College/USA) *6*(9):53
Gavaskar, Arun R. (Battelle/USA) *6*(7):13
Gavinelli, Marco (Ambiente S.p.A./ITALY) *6*(6):227
Gebhard, Michael (GeoTrans/USA) *6*(8):19

Gec, Bob (Degussa Canada Ltd./CANADA) *6*(10):73
Gehre, Matthias (UFZ - Centre for Environmental Research/GERMANY) *6*(4):99
Gemoets, Johan (VITO/BELGIUM) *6*(4):35; *6*(9):87
Gent, David B. (U.S. Army Corps of Engineers/USA) *6*(9):241
Gentry, E. E. (Science Applications International Corporation/USA) *6*(8):27
Georgiev, Plamen S. (University of Mining & Geology/BULGARIA) *6*(9):97
Gerday, Charles (Université de Liège/BELGIUM) *6*(2):57
Gerlach, Robin (Montana State University/USA) *6*(8):1
Gerritse, Jan (TNO Environmental Sciences/THE NETHERLANDS) *6*(2):231; *6*(7):61
Gerth, André (BioPlanta GmbH/GERMANY) *6*(3):67; *6*(5):173
Ghosh, Upal (Stanford University/USA) *6*(3):189; *6*(6):89
Ghoshal, Subhasis (McGill University/CANADA) *6*(9):139
Gibbs, James T. (Battelle/USA) *6*(1):137
Gibello, A. (Universidad Complutense/SPAIN) *6*(4):29
Giblin, Tara (University of California/USA) *6*(9):249
Gilbertson, Amanda W. (University of Missouri-Rolla/USA) *6*(5):199
Gillespie, Rick D. (Regenesis/USA) *6*(1):107
Gillespie, Terry J. (University of Guelph/CANADA) *6*(6):185
Glover, L. Anne (Aberdeen University /UK) *6*(2):223
Goedbloed, Peter (Oostwaardhoeve Co./THE NETHERLANDS) *6*(6):59
Golovleva, Ludmila A. (Russian Academy of Sciences/RUSSIA) *6*(3):75
Goltz, Mark N. (Air Force Institute of Technology/USA) *6*(2):173
Gong, Weiliang (The University of New Mexico/USA) *6*(9):155
Gossett, James M. (Cornell University/USA) *6*(4):125
Govind, Rakesh (University of Cincinnati/USA) *6*(5):269; *6*(8):35; *6*(9):1, 9, 17
Gozan, Misri (Water Technology Center/GERMANY) *6*(8):105
Grainger, David (IT Corporation/USA) *6*(1):51; *6*(2):73
Grandi, Beatrice (Water & Soil Remediation S.r.l./ITALY) *6*(6):179
Granley, Brad A. (Leggette, Brashears, & Graham/USA) *6*(10):259
Grant, Russell J. (University of York/UK) *6*(6):27
Graves, Duane (IT Corporation/USA) *6*(2):253; *6*(4):109; *6*(9):215
Green, Chad E. (University of California/USA) *6*(10):311
Green, Donald J. (USAG Aberdeen Proving Ground/USA) *6*(7):321, 333, 341
Green, Robert (Alcoa/USA) *6*(6):89
Green, Roger B. (Waste Management, Inc./USA) *6*(2):247; *6*(6):127
Gregory, Kelvin B. (University of Iowa/USA) *6*(3):1
Griswold, Jim (Construction Analysis & Management, Inc./USA) *6*(1):115
Groen, Jacobus (Vrije Universiteit/THE NETHERLANDS) *6*(4):91
Groenendijk, Gijsbert Jan (Hoek Loos bv/THE NETHERLANDS) *6*(7):297
Grotenhuis, Tim (Wageningen Agricultural University/THE NETHERLANDS) *6*(5):289
Groudev, Stoyan N. (University of Mining & Geology/BULGARIA) *6*(9):97
Guarini, William J. (Envirogen, Inc./USA) *6*(9):281
Guieysse, Benoît (Lund University/SWEDEN) *6*(3):181
Guiot, Serge R. (Biotechnology Research Institute/CANADA) *6*(3):165
Gunsch, Claudia (Clemson University/USA) *6*(7):109
Gurol, Mirat (San Diego State University/USA) *6*(10):301

Ha, Jeonghyub (University of Maryland/USA) *6*(10):57
Haak, Daniel (RMT, Inc./USA) *6*(10):201
Haas, Patrick E. (Mitretek Systems/USA) *6*(7):19, 241, 249; *6*(8):73
Haasnoot, C. (Logisticon Water Treatment/THE NETHERLANDS) *6*(8):11
Habe, Hiroshi (The University of Tokyo/JAPAN) *6*(4):51; *6*(6):111
Haeseler, Frank (Institut Français du Pétrole/FRANCE) *6*(3):227
Haff, James (Meritor Automotive, Inc./USA) *6*(7):173
Haines, John R. (U.S. EPA/USA) *6*(9):17
Håkansson, Torbjörn (Lund University/SWEDEN) *6*(9):123
Halfpenny-Mitchell, Laurie (University of Guelph/CANADA) *6*(6):185
Hall, Billy (Newfields, Inc./USA) *6*(5):189
Hampton, Mark M. (Groundwater Services/USA) *6*(8):73
Hannick, Nerissa K. (University of Cambridge/UK) *6*(5):69
Hannigan, Mary (Mississippi State University) *6*(5):329; *6*(6):279
Hannon, LaToya (Spelman College/USA) *6*(9):53
Hansen, Hans C. L. (Hedeselskabet /DENMARK) *6*(2):11
Hansen, Lance D. (U.S. Army Corps of Engineers/USA) *6*(3):9, 43, 51; *6*(4):59; *6*(6):43; *6*(7):125; *6*(10):115
Haraguchi, Makoo (Sumitomo Marine Research Institute/JAPAN) *6*(10):345
Hardisty, Paul E. (Komex Europe, Ltd./ENGLAND) *6*(4):17
Harmon, Stephen M. (U.S. EPA/USA) *6*(9):17
Harms, Hauke (Swiss Federal Institute of Technology/SWITZERLAND) *6*(3):251
Harmsen, Joop (Alterra, Wageningen University and Research Center/THE NETHERLANDS) *6*(5):137, 279; *6*(6):1, 59
Harper, Greg (TetraTech EM Inc./USA) *6*(3):259
Harrington-Baker, Mary Ann (MSE, Inc./USA) *6*(9):35
Harris, Benjamin Cord (Texas A&M University/USA) *6*(5):17, 25
Harris, James C. (U.S. EPA/USA) *6*(6):287, 295
Harris, Todd (Mason and Hanger Corporation/USA) *6*(3):35
Harrison, Patton B. (American Airlines/USA) *6*(1):121
Harrison, Susan T.L. (University of Cape Town/REP OF SOUTH AFRICA) *6*(6):339
Hart, Barry (Monash University/AUSTRALIA) *6*(4):1
Hartzell, Kristen E. (Battelle/USA) *6*(1):137; *6*(10):193
Harwood, Christine L. (Michael Baker Corporation/USA) *6*(2):155
Hassett, David J. (Energy & Environmental Research Center/USA) *6*(5):129
Hater, Gary R. (Waste Management Inc./USA) *6*(2):247
Hausmann, Tom S. (Battelle Marine Sciences Laboratory/USA) *6*(3):157
Hawari, Jalal (National Research Council of Canada/CANADA) *6*(9):139
Hayes, Adam J. (Triple Point Engineers, Inc./USA) *6*(1):183
Hayes, Dawn M. (U.S. Navy/USA) *6*(3):107
Hayes, Kim F. (University of Michigan/USA) *6*(8):193
Haynes, R.J. (University of Natal/REP OF SOUTH AFRICA) *6*(6):101
Heaston, Mark S. (Earth Tech/USA) *6*(3):17, 25
Hecox, Gary R. (University of Kansas/USA) *6*(4):109
Heebink, Loreal V. (Energy & Environmental Research Center/USA) *6*(5):129
Heine, Robert (EFX Systems, Inc./USA) *6*(8):19
Heintz, Caryl (Texas Tech University/USA) *6*(3):9

Hendrickson, Edwin R. (DuPont Co./USA) *6*(8):27, 43
Hendriks, Willem (Witteveen+Bos Consulting Engineers/THE NETHERLANDS) *6*(5):289
Henkler, Rolf D. (ICI Paints/UK) *6*(2):223
Henny, Cynthia (University of Maine/USA) *6*(8):139
Henry, Bruce M. (Parsons Engineering Science, Inc/USA) *6*(7):241
Henssen, Maurice J.C. (Bioclear Environmental Biotechnology/THE NETHERLANDS) *6*(8):11
Herson, Diane S. (University of Delaware/USA) *6*(6):9
Hesnawi, Rafik M. (University of Manitoba/CANADA) *6*(6):165
Hetland, Melanie D. (Energy & Environmental Research Center/USA) *6*(5):129
Hickey, Robert F. (EFX Systems, Inc./USA) *6*(8):19
Hicks, Patrick H. (ARCADIS/USA) *6*(1):107
Hiebert, Randy (MSE Technology Applications, Inc./USA) *6*(8):79
Higashi, Teruo (University of Tsukuba/JAPAN) *6*(9):187
Higgins, Mathew J. (Bucknell University/USA) *6*(2):105
Higinbotham, James H. (ExxonMobil Environmental Remediation/USA) *6*(8):87
Hines, April (Spelman College/USA) *6*(9):53
Hinshalwood, Gordon (Delta Environmental Consultants, Inc./USA) *6*(1):43
Hirano, Hiroyuki (The University of Tokyo/JAPAN) *6*(6):111
Hirashima, Shouji (Yakult Pharmaceutical Industry/JAPAN) *6*(10):345
Hirsch, Steve (Environmental Protection Agency/USA) *6*(5):207
Hiwatari, Takehiko (National Institute for Environmental Studies/JAPAN) *6*(5):321
Hoag, Rob (Conestoga-Rovers & Associates/USA) *6*(1):99
Hoelen, Thomas P. (Stanford University/USA) *6*(7):95
Hoeppel, Ronald E. (U.S. Navy/USA) *6*(10):245
Hoffmann, Johannes (Hochtief Umwelt GmbH/GERMANY) *6*(6):227
Hoffmann, Robert E. (Chevron Canada Resources/CANADA) *6*(6):193
Höfte, Monica (Ghent University/BELGIUM) *6*(5):223
Holder, Edith L. (University of Cincinnati/USA) *6*(2):247
Holm, Thomas R. (Illinois State Water Survey/USA) *6*(9):179
Holman, Hoi-Ying (Lawrence Berkeley National Laboratory/USA) *6*(4):67
Holoman, Tracey R. Pulliam (University of Maryland/USA) *6*(3):205
Hopper, Troy (URS Corporation/USA) *6*(2):239
Hornett, Ryan (NOVA Chemicals Corporation/USA) *6*(4):117
Hosangadi, Vitthal S. (Foster Wheeler Environmental Corp./USA) *6*(9):249
Hough, Benjamin (Tetra Tech EM, Inc./USA) *6*(10):293
Hozumi, Toyoharu (Oppenheimer Biotechnology/JAPAN) *6*(10):345
Huang, Chin-I (National Chung Hsing University/TAIWAN) *6*(10):217
Huang, Chin-Pao (University of Delaware/USA) *6*(6):9, 149
Huang, Hui-Bin (DuPont Co./USA) *6*(8):43
Huang, Junqi (Air Force Institute of Technology/USA) *6*(2):173
Huang, Wei (University of Sheffield/UK) *6*(2):207
Hubach, Cor (DHV Noord Nederland/THE NETHERLANDS) *6*(8):11
Huesemann, Michael H. (Battelle/USA) *6*(3):157
Hughes, Joseph B. (Rice University/USA) *6*(5):85; *6*(7):19
Hulsen, Kris (University of Ghent/BELGIUM) *6*(5):223
Hunt, Jonathan (Clemson University/USA) *6*(7):109

Hunter, William J. (U.S. Dept of Agriculture/USA) *6*(9):209, 309
Hwang, Sangchul (University of Akron/USA) *6*(3):235
Hyman, Michael R. (North Carolina State University/USA) *6*(1): 83, 145

Ibeanusi, Victor M. (Spelman College/USA) *6*(9):53
Ickes, Jennifer (Battelle/USA) *6*(5):231, 237
Ide, Kazuki (Obayashi Corporation Ltd./JAPAN) *6*(6):111; *6*(10):239
Igarashi, Tsuyoshi (Nippon Institute of Technology/JAPAN) *6*(5):321
Infante, Carmen (PDVSA Intevep/VENEZUELA) *6*(6):257
Ingram, Sherry (IT Corporation/USA) *6*(4):109
Ishikawa, Yoji (Obayashi Corporation/JAPAN) *6*(6):249; *6*(10):239

Jackson, W. Andrew (Texas Tech University/USA) *6*(5):207, 313; *6*(9):273
Jacobs, Alan K. (EnSafe, Inc./USA) *6*(9):315
Jacques, Margaret E. (Rowan University/USA) *6*(5):215
Jahan, Kauser (Rowan University/USA) *6*(5):215
James, Garth (MSE Inc./USA) *6*(8):79
Jansson, Janet K. (Södertörn University College/SWEDEN) *6*(3):133
Japenga, Jan (Alterra/THE NETHERLANDS) *6*(5):137
Jauregui, Juan (Universidad Nacional Autonoma de Mexico/MEXICO) *6*(6):17
Jensen, James N. (State University of New York at Buffalo/USA) *6*(6):89
eon, Mi-Ae (Texas Tech University/USA) *6*(9):273
Jerger, Douglas E. (IT Corporation/USA) 6(3):35
Jernberg, Cecilia (Södertörn University College/SWEDEN) *6*(3):133
Jindal, Ranjna (Suranaree University of Technology/THAILAND) *6*(4):149
Johnson, Dimitra (Southern University at New Orleans/USA) *6*(5):151
Johnson, Glenn (University of Utah/USA) *6*(5):231
Johnson, Paul C. (Arizona State University/USA) *6*(1):11; *6*(8):61
Johnson, Richard L. (Oregon Graduate Institute/USA) *6*(10):293
Jones, Antony (Komex H_2O Science, Inc./USA) *6*(2):223; *6*(3):173; *6*(10):123
Jones, Clay (University of New Mexico/USA) *6*(9):223
Jones, Triana N. (University of Maryland/USA) *6*(3):205
Jonker, Hendrikus (Vrije Universiteit/THE NETHERLANDS) *6*(4):91
Ju, Lu-Kwang (The University of Akron/USA) *6*(6):319

Kaludjerski, Milica (San Diego State University/USA) *6*(10):301
Kamashwaran, S. Ramanathen (University of Idaho/USA) *6*(3):91
Kambhampati, Murty S. (Southern University at New Orleans/USA) *6*(5):145, 151
Kamimura, Daisuke (Gunma University/JAPAN) *6*(8):113
Kang, James J. (URS Corporation/USA) *6*(1):121; *6*(10):223
Kappelmeyer, Uwe (UFZ Centre for Environmental Research/GERMANY) 6(5):337
Karamanev, Dimitre G. (University of Western Ontario/CANADA) *6*(10):171
Karlson, Ulrich (National Environmental Research Institute) *6*(3):141
Kastner, James R. (University of Georgia/USA) *6*(9):289
Kästner, Matthias (UFZ Centre for Environmental Research/GERMANY) *6*(4):99; *6*(5):337
Katz, Lynn E. (University of Texas/USA) *6*(8):139
Kavanaugh, Rathi G. (University of Cincinnati/USA) *6*(2):247

Kawahara, Fred (U.S. EPA/USA) *6*(9):9
Kawakami, Tsuyoshi (University of Tsukuba/JAPAN) *6*(9):187
Keefer, Donald A. (Illinois State Geological Survey/USA) *6*(9):179
Keith, Nathaniel (Texas A&M University/USA) *6*(5):25
Kelly, Laureen S. (Montana Department of Environmental Quality/USA) *6*(6):287
Kempisty, David M. (U.S. Air Force/USA) *6*(10):145, 155
Kerfoot, William B. (K-V Associates, Inc./USA) *6*(10):33
Keuning, S. (Bioclear Environmental Technology/THE NETHERLANDS) *6*(8):11
Khan, Tariq A. (Groundwater Services, Inc./USA) *6*(7):19
Khodadoust, Amid P. (University of Cincinnati/USA) *6*(5):243, 253, 261
Kieft, Thomas L. (New Mexico Institute of Mining and Technology/USA) *6*(8):131
Kiessig, Gunter (WISMUT GmbH/GERMANY) *6*(5):173; *6*(9):155
Kilbride, Rebecca (CEFAS Laboratory/UK) *6*(10):337
Kim, Jae Young (Seoul National University/REPUBLIC OF KOREA) *6*(9):195
Kim, Jay (University of Cincinnati/USA) *6*(6):133
Kim, Kijung (The Pennsylvania State University/USA) *6*(9):303
Kim, Tae Young (Ewha University/REPUBLIC OF KOREA) *6*(6):51
Kinsall, Barry L. (Oak Ridge National Laboratory/USA) *6*(4):73
Kirschenmann, Kyle (IT Corp/USA) *6*(4):109
Klaas, Norbert (University of Stuttgart/GERMANY) *6*(2):137
Klecka, Gary M. (The Dow Chemical Company/USA) *6*(2):89
Klein, Katrina (GeoTrans, Inc./USA) *6*(2):163
Klens, Julia L. (IT Corporation/USA) *6*(2):253; *6*(9):215
Knotek-Smith, Heather M. (University of Idaho/USA) *6*(9):147
Koch, Stacey A. (RMT, Inc./USA) *6*(7):181
Koenen, Brent A. (U.S. Army Engineer Research & Development Center/USA) *6*(5):9
Koenigsberg, Stephen S. (Regenesis Bioremediation Products/USA) *6*(7):197, 257; *6*(8):209; *6*(10):9, 87
Kohata, Kunio (National Institute for Environmental Studies/JAPAN) *6*(5):321
Kohler, Keisha (ThermoRetec Corporation/USA) *6*(7):1
Kolhatkar, Ravindra V. (BP Corporation/USA) *6*(1):35, 43
Komlos, John (Montana State University/USA) *6*(7):117
Komnitsas, Kostas (National Technical University of Athens/GREECE) *6*(9):97
Kono, Masakazu (Oppenheimer Biotechnology/JAPAN) *6*(10):345
Koons, Brad W. (Leggette, Brashears & Graham, Inc./USA) *6*(1):175
Koschal, Gerard (PNG Environmental/USA) *6*(1):203
Koschorreck, Matthias (UFZ Centre for Environmental Research/GERMANY) *6*(9):43
Koshikawa, Hiroshi (National Institute for Environmental Studies/JAPAN) *6*(5):321
Kramers, Jan D. (University of Bern/SWITZERLAND) *6*(4):91
Krooneman, Jannneke (Bioclear Environmental Biotechnology/THE NETHERLANDS) *6*(7):141
Kruk, Taras B. (URS Corporation/USA) *6*(10):223
Kuhwald, Jerry (NOVA Chemicals Corporation/CANADA) *6*(5):53
Kuschk, Peter (UFZ Centre for Environmental Research Leipzig/GERMANY) *6*(5):337

Laboudigue, Agnes (Centre National de Recherche sur les Sites et Sols Pollués/FRANCE) *6*(2):129
LaFlamme, Brian (Engineering Management Support, Inc./USA) *6*(10):231
Lafontaine, Chantal (École Polytechnique de Montréal/CANADA) *6*(10):171
Laha, Shonali (Florida International University/USA) *6*(10):187
Laing, M.D. (University of Natal/REP OF SOUTH AFRICA) *6*(9):79
Lamar, Richard (EarthFax Development Corp/USA) *6*(6):263
Lamarche, Philippe (Royal Military College of Canada/CANADA) *6*(8):95
Lamb, Steven R. (GZA GeoEnvironmental, Inc./USA) *6*(7):165
Landis, Richard C. (E.I. du Pont de Nemours & Company/USA) *6*(8):185
Lang, Beth (United Technologies Corp./USA) *6*(10):41
Langenhoff, Alette (TNO Institute of Environmental Science/THE NETHERLANDS) *6*(7):141
LaPat-Polasko, Laurie T. (Parsons Engineering Science, Inc./USA) *6*(2):65, 189
Lapus, Kevin (Regenesis/USA) *6*(7):257; *6*(10):9
LaRiviere, Daniel (Texas A&M University/USA) *6*(5):17, 25
Larsen, Lars C. (Hedeselskabet/DENMARK) *6*(2):11
Larson, John R. (TranSystems Corporation/USA) *6*(7):229
Larson, Richard A. (University of Illinois at Urbana-Champaign/USA) *6*(5):181
Lauzon, Francois (Dept of National Defence/CANADA) *6*(8):95
Leavitt, Maureen E. (Newfields Inc./USA) *6*(1):51; *6*(5):189
Lebron, Carmen A. (U.S. Navy/USA) *6*(7):95
Lee, B. J. (Science Applications International Corporation) *6*(8):27
Lee, Brady D. (Idaho National Engineering & Environmental Laboratory/USA) *6*(7):77
Lee, Chi Mei (National Chung Hsing University/TAIWAN) *6*(6):353
Lee, Eun-Ju (Louisiana State University/USA) *6*(5):313
Lee, Kenneth (Fisheries & Oceans Canada/CANADA) *6*(10):337
Lee, Michael D. (Terra Systems, Inc./USA) *6*(7):213, 249
Lee, Ming-Kuo (Auburn University/USA) *6*(9):105
Lee, Patrick (Queen's University/CANADA) *6*(2):121
Lee, Seung-Bong (University of Washington/USA) *6*(10):211
Lee, Si-Jin (Kyonggi University/REPUBLIC OF KOREA) *6*(1):161
Lee, Sung-Jae (ChoongAng University/REPUBLIC OF KOREA) *6*(6):51
Leeson, Andrea (Battelle/USA) *6*(10):1, 145, 155, 193
Lehman, Stewart E. (California Polytechnic State University/USA) *6*(2):1
Lei, Li (University of Cincinnati/USA) *6*(5):243, 261
Leigh, Daniel P. (IT Corporation/USA) *6*(3):35
Leigh, Mary Beth (University of Oklahoma/USA) *6*(5):61
Lendvay, John (University of San Francisco/USA) *6*(8):19
Lenzo, Frank C. (ARCADIS Geraghty & Miller/USA) *6*(7):53
Leon, Nidya (PDVSA Intevep/VENEZUELA) *6*(6):257
Leong, Sylvia (Crescent Heights High School/CANADA) *6*(5):53
Leontievsky, Alexey A. (Russian Academy of Sciences/RUSSIA) *6*(3):75
Lerner, David N. (University of Sheffield/UK) *6*(1):59; *6*(2):207
Lesage, Suzanne (National Water Research Institute/CANADA) *6*(7):321, 333, 341

Leslie, Jolyn C. (Camp Dresser & McKee, Inc./USA) *6*(2):113
Lewis, Ronald F. (U.S. EPA/USA) *6*(5):253, 261
Li, Dong X. (USA) *6*(7):205
Li, Guanghe (Tsinghua University/CHINA) *6*(7):61
Li, Tong (Tetra Tech EM Inc./USA) *6*(10):293
Librando, Vito (Universita di Catania/ITALY) *6*(3):149
Lieberman, M. Tony (Solutions Industrial & Environmental Services/USA) *6*(7):249
Lin, Cindy (Conestoga-Rovers & Associates/USA) *6*(1):99; *6*(10):131
Lipson, David S. (Blasland, Bouck & Lee, Inc./USA) *6*(10):319
Liu, Jian (University of Nevada Las Vegas/USA) *6*(9):265
Liu, Xiumei (Shandong Agricultural University/ CHINA) *6*(9):113
Livingstone, Stephen (Franz Environmental Inc./CANADA) *6*(6):211
Lizzari, Daniela (Universita degli Studi di Verona/ITALY) *6*(3):267
Llewellyn, Tim (URS/USA) *6*(7):321, 333, 341
Lobo, C. (El Encin IMIA/SPAIN) *6*(4):29
Loeffler, Frank E. (Georgia Institute of Technology/USA) *6*(8):19
Logan, Bruce E. (The Pennsylvania State University/USA) *6*(9):303
Long, Gilbert M. (Camp Dresser & McKee Inc./USA) *6*(6):287
Longoni, Giovanni (Montgomery Watson/ITALY) *6*(10):41
Lorbeer, Helmut (Technical University of Dresden/GERMANY) *6*(8):105
Lors, Christine (Centre National de Recherche sur les Sites et Sols Pollués /FRANCE) *6*(2):129
Lorton, Diane M. (King's College London/UK) *6*(2):223; *6*(3):173
Losi, Mark E. (Foster Wheeler Environ. Corp./USA) *6*(9):249
Loucks, Mark (U.S. Air Force/USA) *6*(2):261
Lu, Chih-Jen (National Chung Hsing University/TAIWAN) *6*(6):353; *6*(10):217
Lu, Xiaoxia (Tsinghua University/CHINA) *6*(7):61
Lubenow, Brian (University of Delaware/USA) *6*(6):149
Lucas, Mary (Parsons Engineering Science, Inc./USA) *6*(10):283
Lundgren, Tommy S. (Sydkraft SAKAB AB/SWEDEN) *6*(6):127
Lundstedt, Staffan (Umeå University/SWEDEN) *6*(3):181
Luo, Xiaohong (NRC Research Associate/USA) *6*(8):167
Luthy, Richard G. (Stanford University/USA) *6*(3):189
Lutze, Werner (University of New Mexico/USA) *6*(9):155
Luu, Y.-S. (Queen's University/CANADA) *6*(2):121
Lynch, Regina M. (Battelle/USA) *6*(10):155

Macek, Thomáš (Institute of Chemical Technology/Czech Republic) *6*(5):61
MacEwen, Scott J. (CH2M Hill/USA) *6*(3):107
Machado, Kátia M. G. (Fund. Centro Tecnológico de Minas Gerais/BRAZIL) *6*(3):99
Maciel, Helena Alves (Aberdeen University/UK) *6*(1):1
Mack, E. Erin (E.I. du Pont de Nemours & Co./USA) *6*(2):81; *6*(8):43
Macková, Martina (Institute of Chemical Technology/Czech Republic) *6*(5):61
Macnaughton, Sarah J. (AEA Technology/UK) *6*(5):305; *6*(10):337
Macomber, Jeff R. (University of Cincinnati/USA) *6*(6):133
Macrae, Jean (University of Maine/USA) *6*(8):139
Madden, Patrick C. (Engineering Consultant/USA) *6*(8):87
Madsen, Clint (Terracon/USA) *6*(8):157; *6*(10):15
Magar, Victor S. (Battelle/USA) *6*(1):137; *6*(5):231, 237; *6*(10):145, 155

Mage, Roland (Battelle Europe/SWITZERLAND) *6*(6):241; *6*(10):109
Magistrelli, P. (Istituto di Cibernetica e Biofisica/ITALY) *6*(5):157
Maierle, Michael S. (ARCADIS Geraghty & Miller, Inc./USA) *6*(7):149
Major, C. Lee (Jr.) (University of Michigan/USA) *6*(8):19
Major, David W. (GeoSyntec Consultants/CANADA) *6*(8):27
Maki, Hideaki (National Institute for Environmental Studies/JAPAN) *6*(5):321
Makkar, Randhir S. (University of Illinois-Chicago/USA) *6*(5):297
Malcolm, Dave (BAE Systems Properties Ltd./UK) *6*(6):119
Manabe, Takehiko (Hyogo Prefectural Fisheries Research Institute/JAPAN) *6*(10):345
Maner, P.M. (Equilon Enterprises, LLC/USA) *6*(1):11
Maner, Paul (Shell Development Company/USA) *6*(8):61
Manrique-Ramírez, Emilio Javier (SYMCA, S.A. de C.V./MEXICO) *6*(6):369
Marchal, Rémy (Institut Français du Pétrole/FRANCE) *6*(1):153
Maresco, Vincent (Groundwater & Environmental Srvcs/USA) *6*(10):101
Marnette, Emile C. (TAUW BV/THE NETHERLANDS) *6*(7):297
Marshall, Timothy R. (URS Corporation/USA) *6*(2):49
Martella, L. (Istituto di Cibernetica e Biofisica/ITALY) *6*(5):157
Martin, C. (Universidad Politecnica/SPAIN) *6*(4):29
Martin, Jennifer P. (Idaho National Engineering & Environmental Laboratory/USA) *6*(7):265
Martin, John F. (U.S. EPA/USA) *6*(2):247
Martin, Margarita (Universidad Complutense de Madrid/SPAIN) *6*(4):29; *6*(6):377
Martinez-Inigo, M.J. (El Encin IMIA/SPAIN) *6*(4):29
Martino, Lou (Argonne National Laboratory/USA) *6*(5):207
Mascarenas, Tom (Environmental Chemistry/USA) *6*(8):157
Mason, Jeremy (King's College London/UK) *6*(2):223; *6*(3):173; *6*(10):123
Massella, Oscar (Universita degli Studi di Verona/ITALY) *6*(3):267
Matheus, Dacio R. (Instituto de Botânica/BRAZIL) *6*(3):99
Matos, Tania (University of Puerto Rico at Rio Piedras/USA) *6*(9):179
Matsubara, Takashi (Obayashi Corporation/JAPAN) *6*(6):249
Mattiasson, Bo (Lund University/SWEDEN) *6*(3):181; *6*(6):65; *6*(9):123
McCall, Sarah (Battelle/USA) *6*(10):155, 245
McCarthy, Kevin (Battelle Duxbury Operations/USA) *6*(5):9
McCartney, Daryl M. (University of Manitoba/CANADA) *6*(6):165
McCormick, Michael L. (The University of Michigan/USA) *6*(8):193
McDonald, Thomas J. (Texas A&M University) *6*(5):17
McElligott, Mike (U.S. Air Force/USA) *6*(1):51
McGill, William B. (University of Northern British Columbia/CANADA) *6*(4):7
McIntosh, Heather (U.S. Army/USA) *6*(7):321, 333
McLinn, Eugene L. (RMT, Inc./USA) *6*(5):121
McLoughlin, Patrick W. (Microseeps Inc./USA) *6*(1):35
McMaster, Michaye (GeoSyntec Consultants/CANADA) *6*(8):27, 43, *6*(9):297
McMillen, Sara J. (Chevron Research & Technology Company/USA) *6*(6):193
Meckenstock, Rainer U. (University of Tübingen/GERMANY) *6*(4):99
Mehnert, Edward (Illinois State Geological Survey/USA) *6*(9):179

Meigio, Jodette L. (Idaho National Engineering & Environmental Laboratory/USA) *6*(7):77
Meijer, Harro A.J. (University of Groningen/THE NETHERLANDS) *6*(4):91
Meijerink, E. (Province of Drenthe/THE NETHERLANDS) *6*(8):11
Merino-Castro, Glicina (Inst Technol y de Estudios Superiores/MEXICO) *6*(6):377
Messier, J.P. (U.S. Coast Guard/USA) *6*(1):107
Meyer, Michael (Environmental Resources Management/BELGIUM) *6*(7):87
Meylan, S. (Queen's University/CANADA) *6*(2):121
Miles, Victor (Duracell Inc./USA) *6*(7):87
Millar, Kelly (National Water Research Institute/CANADA) *6*(7):321, 333, 341
Miller, Michael E. (Camp Dresser & McKee, Inc./USA) *6*(7):273
Miller, Thomas Ferrell (Lockheed Martin/USA) *6*(3):259
Mills, Heath J. (Georgia Institute of Technology/USA) *6*(9):165
Millward, Rod N. (Louisiana State University/USA) *6*(5):305
Mishra, Pramod Chandra (Sambalpur University/INDIA) *6*(9):173
Mitchell, David (AEA Technology Environment/UK) *6*(10):337
Mitraka, Maria (Serres/GREECE) *6*(6):89
Mocciaro, PierFilippo (Ambiente S.p.A./ITALY) *6*(6):227
Moeri, Ernesto N. (CSD-GEOKLOCK/BRAZIL) *6*(1):27
Moir, Michael (Chevron Research & Technology Co./USA) *6*(1):83
Molinari, Mauro (AgipPetroli S.p.A/ITALY) *6*(6):173
Mollea, C. (Politecnico di Torino/ITALY) *6*(3):211
Mollhagen, Tony (Texas Tech University/USA) *6*(3):9
Monot, Frédéric (Institut Français du Pétrole/FRANCE) *6*(1):153
Moon, Hee Sun (Seoul National University/REPUBLIC OF KOREA) *6*(9):195
Moosa, Shehnaaz (University of Cape Town/REP OF SOUTH AFRICA) *6*(6):339
Morasch, Barbara (University Konstanz/GERMANY) *6*(4):99
Moreno, Joanna (URS Corporation/USA) *6*(2):239
Morgan, Scott (URS - Dames & Moore/USA) *6*(7):321
Morrill, Pamela J. (Camp, Dresser, & McKee, Inc./USA) *6*(2):113
Morris, Damon (ThermoRetec Corporation/USA) *6*(7):1
Mortimer, Marylove (Mississippi State University/USA) *6*(5):329
Mortimer, Wendy (Bell Canada/CANADA) *6*(2):27; *6*(6):185, 203, 211,
Mossing, Christian (Hedeselskabet/DENMARK) *6*(2):11
Mossmann, Jean-Remi (Centre National de Recherche sur les Sites et Sols Pollués/FRANCE) *6*(2):129
Moteleb, Moustafa A. (University of Cincinnati/USA) *6*(6):133
Mowder, Carol S. (URS/USA) *6*(7):321, 333, 341
Moyer, Ellen E. (ENSR International./USA) *6*(1):75
Mravik, Susan C. (U.S. EPA/USA) *6*(1):167
Mueller, James G. (URS Corporation/USA) *6*(2):239
Müller, Axel (Water Technology Center/GERMANY) *6*(8):105
Müller, Beate (Umweltschutz Nord GmbH/GERMANY) *6*(4):131
Müller, Klaus (Battelle Europe/SWITZERLAND) *6*(5):41; *6*(6):241
Muniz, Herminio (Hart Crowser Inc./USA) *6*(10):9
Murphy, Sean M. (Komex International Ltd./CANADA) *6*(4):7
Murray, Cliff (United States Army Corps of Engineers/USA) *6*(9):281
Murray, Gordon Bruce (Stella-Jones Inc./CANADA) *6*(3):197

Murray, Willard A. (Harding ESE/USA) *6*(7):197
Mutch, Robert D. (Brown and Caldwell/USA) *6*(2):145
Mutti, Francois (Water & Soil Remediation S.r.l./ITALY) *6*(6):179
Myasoedova, Nina M. (Russian Academy of Sciences/RUSSIA) *6*(3):75

Nadolishny, Alex (Nedatek, Inc./USA) *6*(10):139
Nagle, David P. (University of Oklahoma/USA) *6*(5):61
Nam, Kyoungphile (Seoul National University/REPUBLIC OF KOREA) *6*(9):195
Narayanaswamy, Karthik (Parsons Engineering Science/USA) *6*(2):65
Nelson, Mark D. (Delta Environmental Consultants, Inc./USA) *6*(1):175
Nelson, Yarrow (California Polytechnic State University/USA) *6*(10):311
Nemati, M. (University of Cape Town/REP OF SOUTH AFRICA) *6*(6):339
Nestler, Catherine C. (Applied Research Associates, Inc./USA) *6*(4):59, *6*(6):43
Nevárez-Moorillón, G.V. (UACH/MEXICO) *6*(6):361
Neville, Scott L. (Aerojet General Corp./USA) *6*(9):297
Newell, Charles J. (Groundwater Services, Inc./USA) *6*(7):19
Nieman, Karl (Utah State University/USA) *6*(4):67
Niemeyer, Thomas (Hochtief Umwelt Gmbh/GERMANY) *6*(6):227
Nies, Loring (Purdue University/USA) *6*(3):125
Nipshagen, Adri A.M. (IWACO/THE NETHERLANDS) *6*(7):141
Nishino, Shirley (U.S. Air Force/USA) *6*(3):59
Nivens, David E. (University of Tennessee/USA) *6*(4):45
Noffsinger, David (Westinghouse Savannah River Company/USA) *6*(10):163
Noguchi, Takuya (Nippon Institute of Technology/JAPAN) *6*(5):321
Nojiri, Hideaki (The University of Tokyo/JAPAN) *6*(4):51; *6*(6):111
Noland, Scott (NESCO Inc./USA) *6*(10):73
Nolen, C. Hunter (Camp Dresser & McKee/USA) *6*(6):287
Norris, Robert D. (Eckenfelder/Brown and Caldwell/USA) *6*(2):145; *6*(7):35
North, Robert W. (Environ Corporation./USA) *6*(7):189
Novak, John T. (Virginia Polytechnic Institute & State University/USA) *6*(2):105; *6*(5):1
Novick, Norman (Exxon/Mobil Oil Corp/USA) *6*(1):35
Nuttall, H. Eric (The University of New Mexico/USA) *6*(9): 155, 223
Nuyens, Dirk (Environmental Resources Management/BELGIUM) *6*(7):87; *6*(9):87
Nzengung, Valentine A. (University of Georgia/USA) *6*(9):289

Ochs, L. Donald (Regenesis/USA) *6*(10):139
O'Connell, Joseph E. (Environmental Resolutions, Inc./USA) *6*(1):91
Odle, Bill (Newfields, Inc./USA) *6*(5):189
O'Donnell, Ingrid (BAE Systems Properties, Ltd./UK) *6*(6):119
Ogden, Richard (BAE Systems Properties Ltd./UK) *6*(6):119
Oh, Byung-Taek (The University of Iowa/USA) *6*(8):147, 175
Oh, Seok-Young (University of Delaware/USA) *6*(6):149
Omori, Toshio (The University of Tokyo/JAPAN) *6*(4):51; *6*(6):111
O'Neal, Brenda (ARA/USA) *6*(3):43
Oppenheimer, Carl H. (Oppenheimer Biotechnology/USA) *6*(10):345
O'Regan, Gerald (Chevron Products Company/USA) *6*(1):203
O'Reilly, Kirk T. (Chevron Research & Technology Co/USA) *6*(1):83, 145, 203
Oshio, Takahiro (University of Tsukuba/JAPAN) *6*(9):187

Ozdemiroglu, Ece (EFTEC Ltd./UK) *6*(4):17

Padovani, Marco (Centro Ricerche Ambientali/ITALY) *6*(4):131
Paganetto, A. (Istituto di Cibernetica e Biofisica/ITALY) *6*(5):157
Pahr, Michelle R. (ARCADIS Geraghty & Miller/USA) *6*(1):107
Pal, Nirupam (California Polytechnic State University/USA) *6*(2):1
Palmer, Tracy (Applied Power Concepts, Inc./USA) *6*(7):103
Palumbo, Anthony V. (Oak Ridge National Laboratory/USA) *6*(4):73; *6*(9):165
Panciera, Matthew A. (University of Connecticut/USA) *6*(7):69
Pancras, Tessa (Wageningen University/THE NETHERLANDS) *6*(5):289
Pardue, John H. (Louisiana State University/USA) *6*(5): 207, 313; *6*(9):273
Park, Kyoohong (ChoongAng University/REPUBLIC OF KOREA) *6*(6):51
Parkin, Gene F. (University of Iowa/USA) *6*(3):1
Paspaliaris, Ioannis (National Technical University of Athens/GREECE) *6*(9):97
Paton, Graeme I. (Aberdeen University/UK) *6*(1):1
Patrick, John (University of Reading/UK) *6*(10):337
Payne, Frederick C. (ARCADIS Geraghty & Miller/USA) *6*(7):53
Payne, Jo Ann (DuPont Co./USA) *6*(8):43
Peabody, Jack G. (Regenesis/USA) *6*(10):95
Peacock, Aaron D. (University of Tennessee/USA) *6*(4):73; *6*(5):305
Peargin, Tom R. (Chevron Research & Technology Co/USA) *6*(1):67
Peeples, James A. (Metcalf & Eddy, Inc./USA) *6*(7):173
Pehlivan, Mehmet (Tait Environmental Management, Inc./USA) *6*(10):267, 275
Pelletier, Emilien (ISMER/CANADA) *6*(2):57
Pennie, Kimberley A. (Stella-Jones, Inc./CANADA) *6*(3):197
Peramaki, Matthew P. (Leggette, Brashears, & Graham, Inc./USA) *6*(10):259
Perey, Jennie R. (University of Delaware/USA) *6*(6):149
Perez-Vargas, Josefina (CINVESTAV-IPN/MEXICO) *6*(6):219
Perina, Tomas (IT Corporation/USA) *6*(1):51; *6*(2):73
Perlis, Shira R. (Rowan University/USA) *6*(5):215
Perlmutter, Michael W. (EnSafe, Inc./USA) *6*(9):315
Perrier, Michel (École Polytechnique de Montréal/CANADA) *6*(4):139
Perry, L.B. (U.S. Army Engineer Research & Development Center/USA) *6*(5):9
Persico, John L. (Blasland, Bouck & Lee, Inc./USA) *6*(10):319
Peschong, Bradley J. (Leggette, Brashears & Graham, Inc./USA) *6*(1):175
Peters, Dave (URS/USA) *6*(7):333
Peterson, Lance N. (North Wind Environmental, Inc./USA) *6*(7):265
Petrovskis, Erik A. (Geotrans Inc./USA) *6*(8):19
Peven-McCarthy, Carole (Battelle Ocean Sciences/USA) *6*(5):231
Pfiffner, Susan M. (University of Tennessee/USA) *6*(4):73
Phelps, Tommy J. (Oak Ridge National Laboratory/USA) *6*(4):73
Pickett, Tim M. (Applied Biosciences Corporation/USA) *6*(9):331
Pickle, D.W. (Equilon Enterprises LLC/USA) *6*(8):61
Pierre, Stephane (École Polytechnique de Montréal/CANADA) *6*(10):171
Pijls, Charles G.J.M. (TAUW BV/THE NETHERLANDS) *6*(10):253
Pirkle, Robert J. (Microseeps, Inc./USA) *6*(1):35
Pisarik, Michael F. (New Fields/USA) *6*(1):121

Piveteau, Pascal (Institut Français du Pétrole/FRANCE) *6*(1):153
Place, Matthew (Battelle/USA) *6*(10):245
Plata, Nadia (Battelle Europe/SWITZERLAND) *6*(5):41
Poggi-Varaldo, Hector M. (CINVESTAV-IPN/MEXICO) *6*(3):243; *6*(6):219
Pohlmann, Dirk C. (IT Corporation/USA) *6*(2):253
Pokethitiyook, Prayad (Mahidol University/THAILAND) *6*(10):329
Polk, Jonna (U.S. Army Corps of Engineers/USA) *6*(9):281
Pope, Daniel F. (Dynamac Corp/USA) *6*(1):129
Porta, Augusto (Battelle Europe/SWITZERLAND) *6*(5):41; *6*(6):241; *6*(10):109
Portier, Ralph J. (Louisiana State University/USA) *6*(5):305
Powers, Leigh (Georgia Institute of Technology/USA) *6*(9):165
Prandi, Alberto (Water & Soil Remediation S.r.1/ITALY) *6*(6):179
Prasad, M.N.V. (University of Hyderabad/INDIA) *6*(5):165
Price, Steven (Camp Dresser & McKee, Inc./USA) *6*(9):303
Priester, Lamar E. (Priester & Associates/USA) *6*(10):65
Pritchard, P. H. (Hap) (U.S. Navy/USA) *6*(7):125
Profit, Michael D. (CDM Federal Programs Corporation/USA) *6*(6):81
Prosnansky, Michal (Gunma University/JAPAN) *6*(9):201
Pruden, Amy (University of Cincinnati/USA) *6*(1):19
Ptacek, Carol J. (University of Waterloo/CANADA) *6*(9):71

Radosevich, Mark (University of Delaware/USA) *6*(6):9
Radtke, Corey (INEEL/USA) *6*(3):9
Raetz, Richard M. (Global Remediation Technologies, Inc./USA) *6*(6):311
Rainwater, Ken (Texas Tech University/USA) *6*(3):9
Ramani, Mukundan (University of Cincinnati/USA) *6*(5):269
Raming, Julie B. (Georgia-Pacific Corp./USA) *6*(1):183
Ramírez, N. E. (ECOPETROL-ICP/COLOMBIA) *6*(6):319
Ramsay, Bruce A. (Polyferm Canada Inc./CANADA) *6*(2):121; *6*(10):171
Ramsay, Juliana A. (Queen's University/CANADA) *6*(2):121; *6*(10):171
Rao, Prasanna (University of Cincinnati/USA) *6*(9):1
Ratzke, Hans-Peter (Umweltschutz Nord GMBH/GERMANY) *6*(4):131
Reardon, Kenneth F. (Colorado State University/USA) *6*(8):53
Rectanus, Heather V. (Virginia Polytechnic Institute & State University/USA) *6*(2):105
Reed, Thomas A. (URS Corporation/USA) *6*(8):157; *6*(10):15, 95
Rees, Hubert (CEFAS Laboratory/UK) *6*(10):337
Rehm, Bernd W. (RMT, Inc./USA) *6*(2):97; *6*(10):201
Reinecke, Stefan (Franz Environmental Inc./CANADA) *6*(6):211
Reinhard, Martin (Stanford University/USA) *6*(7):95
Reisinger, H. James (Integrated Science & Technology Inc/USA) *6*(1):183
Rek, Dorota (IT Corporation/USA) *6*(2):73
Reynolds, Charles M. (U.S. Army Engineer Research & Development Center/USA) *6*(5):9
Reynolds, Daniel E. (Air Force Institute of Technology/USA) *6*(2):173
Rice, John M. (RMT, Inc./USA) *6*(7):181
Richard, Don E. (Barr Engineering Company/USA) *6*(3):219; *6*(5):105
Richardson, Ian (Conestoga-Rovers & Associates/USA) *6*(10):131
Richnow, Hans H. (UFZ-Centre for Environmental Research/GERMANY) *6*(4):99

Rijnaarts, Huub H.M. (TNO Institute of Environmental Science/THE NETHERLANDS) *6*(2):231
Ringelberg, David B. (U.S. Army Corps of Engineers/USA) *6*(5):9; *6*(6):43; *6*(10):115
Ríos-Leal, E. (CINVESTAV-IPN/MEXICO) *6*(3):243
Ripp, Steven (University of Tennessee/USA) *6*(4):45
Ritter, Michael (URS Corporation/USA) *6*(2):239
Ritter, William F. (University of Delaware/USA) *6*(6):9
Riva, Vanessa (Parsons Engineering Science, Inc./USA) *6*(2):39
Rivas-Lucero, B.A. (Centro de Investigacion en Materiales Avanzados/MEXICO) *6*(6):361
Rivetta, A. (Universita degli Studi di Milano/ITALY) *6*(5):157
Robb, Joseph (ENSR International/USA) *6*(1):75
Robertiello, Andrea (EniTecnologie S.p.A./ITALY) *6*(6):173
Robertson, K. (Queen's University/CANADA) *6*(2):121
Robinson, David (ERM, Inc./USA) *6*(7):45
Robinson, Sandra L. (Virginia Polytechnic Institute & State University/USA) *6*(5):1
Rockne, Karl J. (University of Illinois-Chicago/USA) *6*(5):297
Rodríguez-Vázquez, Refugio (CINVESTAV-IPN/MEXICO) *6*(3):243; *6*(6):219, 369
Römkens, Paul (Alterra/THE NETHERLANDS) *6*(5):137
Rongo, Rocco (University della Calabria/ITALY) *6*(4):131
Roorda, Marcus L. (Rowan University/USA) *6*(5):215
Rosser, Susan J. (University of Cambridge/UK) *6*(5):69
Rowland, Martin A. (Lockheed-Martin Michoud Space Systems/USA) *6*(7):1
Royer, Richard (The Pennsylvania State University/USA) *6*(8):201
Ruggeri, Bernardo (Politecnico di Torino/ITALY) *6*(3):211
Ruiz, Graciela M. (University of Iowa/USA) *6*(1):195
Rupassara, S. Indumathie (University of Illinois at Urbana-Champaign/USA) *6*(5):181

Sacchi, G.A. (Universita degli Studi di Milano/ITALY) *6*(5):157
Sahagun, Tracy (U.S. Marine Corps./USA) *6*(10):1
Sakakibara, Yutaka (Waseda University/JAPAN) *6*(8):113; *6*(9):201
Sakamoto, T. (Queen's University/CANADA) *6*(10):171
Salam, Munazza (Crescent Heights High School/CANADA) *6*(5):53
Salanitro, Joseph P. (Equilon Enterprises, LLC/USA) *6*(1):11; *6*(8):61
Salvador, Maria Cristina (CSD-GEOKLOCK/BRAZIL) *6*(1):27
Samson, Réjean (École Polytechnique de Montréal/CANADA) *6*(3):115; *6*(4):139; *6*(9):27
San Felipe, Zenaida (Monash University/AUSTRALIA) *6*(4):1
Sánchez, F.N. (ECOPETROL-ICP/COLOMBIA) *6*(6):319
Sánchez, Gisela (PDVSA Intevep/VENEZUELA) *6*(6):257
Sánchez, Luis (PDVSA Intevep/VENEZUELA) *6*(6):257
Sanchez, M. (Universidad Complutense de Madrid/SPAIN) *6*(4):29; *6*(6):377
Sandefur, Craig A. (Regenesis/USA) *6*(7):257; *6*(10):87
Sanford, Robert A. (University of Illinois at Urbana-Champaign/USA) *6*(9):179
Santangelo-Dreiling, Theresa (Colorado Dept. of Transportation/USA) *6*(10):231
Saran, Jennifer (Kennecott Utah Copper Corp./USA) *6*(9):323
Sarpietro, M.G. (Universita di Catania/ITALY) *6*(3):149

Sartoros, Catherine (Université du Québec à Montréal/CANADA) *6*(3):165
Saucedo-Terán, R.A. (Centro de Investigacion en Materiales Avanzados/MEXICO) *6*(6):361
Saunders, James A. (Auburn University/USA) *6*(9):105
Sayler, Gary S. (University of Tennessee/USA) *6*(4):45
Scalzi, Michael M. (Innovative Environmental Technologies, Inc./USA) *6*(10):23
Scarborough, Shirley (IT Corporation/USA) *6*(2):253
Schaffner, I. Richard (GZA GeoEnvironmental, Inc./USA) *6*(7):165
Scharp, Richard A. (U.S. EPA/USA) *6*(9):9
Schell, Heico (Water Technology Center/GERMANY) *6*(8):105
Scherer, Michelle M. (The University of Iowa/USA) *6*(3):1
Schipper, Mark (Groundwater Services) *6*(8):73
Schmelling, Stephen (U.S. EPA/USA) *6*(1):129
Schnoor, Jerald L. (University of Iowa/USA) *6*(8):147
Schoefs, Olivier (École Polytechnique de Montréal/CANADA) *6*(4):139
Schratzberger, Michaela (CEFAS Laboratory/UK) *6*(10):337
Schulze, Susanne (Water Technology Center/GERMANY) *6*(2):137
Schuur, Jessica H. (Lund University/SWEDEN) *6*(6):65
Scrocchi, Susan (Conestoga-Rovers & Associates/USA) *6*(1):99; *6*(10):131
Sczechowski, Jeff (California Polytechnic State University/USA) *6*(10):311
Seagren, Eric A. (University of Maryland/USA) *6*(10):57
Sedran, Marie A. (University of Cincinnati/USA) *6*(1):19
Seifert, Dorte (Technical University of Denmark/DENMARK) *6*(2):11
Semer, Robin (Harza Engineering Company, Inc./USA) *6*(7):157
Semprini, Lewis (Oregon State University/USA) *6*(10):145, 155, 179
Seracuse, Joe (Harding ESE/USA) *6*(7):205
Serra, Roberto (Centro Ricerche Ambientali/ITALY) *6*(4):131
Sewell, Guy W. (U.S. EPA/USA) *6*(1):167; *6*(7):125; *6*(8):167
Sharma, Pawan (Camp Dresser & McKee Inc./USA) *6*(7):305
Sharp, Robert R. (Manhattan College/USA) *6*(7):117
Shay, Devin T. (Groundwater & Environmental Services, Inc./USA) *6*(10):101
Shelley, Michael L. (Air Force Institute of Technology/USA) *6*(5):95
Shen, Hai (Dynamac Corporation/USA) *6*(1): 129, 167
Sherman, Neil (Louisiana-Pacific Corporation/USA) *6*(3):83
Sherwood Lollar, Barbara (University of Toronto/CANADA) *6*(4):91, 109
Shi, Jing (EFX Systems, Inc./USA) *6*(8):19
Shields, Adrian R.G. (Komex Europe/UK) *6*(10):123
Shiffer, Shawn (University of Illinois/USA) *6*(9):179
Shin, Won Sik (Lousiana State University/USA) *6*(5):313
Shiohara, Kei (Mississippi State University/USA) *6*(6):279
Shirazi, Fatemeh R. (Stratum Engineering Inc./USA) *6*(8):121
Shoemaker, Christine (Cornell University/USA) *6*(4):125
Sibbett, Bruce (IT Corporation/USA) *6*(2):73
Silver, Cannon F. (Parsons Engineering Science, Inc./USA) *6*(10):283
Silverman, Thomas S. (RMT, Inc./USA) *6*(10):201
Simon, Michelle A. (U.S. EPA/USA) *6*(10):293
Sims, Gerald K. (USDA-ARS/USA) *6*(5):181
Sims, Ronald C. (Utah State University/USA) *6*(4):67; *6*(6):1
Sincock, M. Jennifer (ENVIRON International Corp./USA) *6*(7):189

Sittler, Steven P. (Advanced Pollution Technologists, Ltd./USA) *6*(2):215
Skladany, George J. (ERM, Inc./USA) *6*(7):45, 213
Skubal, Karen L. (Case Western Reserve University/USA) *6*(8):193
Slenders, Hans (TNO-MEP/THE NETHERLANDS) *6*(7):289
Slomczynski, David J. (University of Cincinnati/USA) *6*(2):247
Slusser, Thomas J. (Wright State University/USA) *6*(5):95
Smallbeck, Donald R. (Harding Lawson/USA) *6*(10):231
Smets, Barth F. (University of Connecticut/USA) *6*(7):69
Smith, Christy (North Carolina State University/USA) *6*(1):145
Smith, Colin C. (University of Sheffield/UK) *6*(2):207
Smith, John R. (Alcoa Inc./USA) *6*(6):89
Smith, Jonathan (The Environment Agency/UK) *6*(4):17
Smith, Steve (King's College London/UK) *6*(2):223; *6*(3):173; *6*(10):123
Smyth, David J.A. (University of Waterloo/CANADA) *6*(9):71
Sobecky, Patricia (Georgia Institute of Technology/USA) *6*(9):165
Sola, Adrianna (Spelman College/USA) *6*(9):53
Sordini, E. (EniTechnologie/ITALY) *6*(6):173
Sorensen, James A. (University of North Dakota/USA) *6*(6):141
Sorenson, Kent S. (Idaho National Engineering and Environmental Laboratory./USA) *6*(7):265
South, Daniel (Harding ESE/USA) *6*(7):205
Spain, Jim (U.S. Air Force/USA) *6*(3):59; *6*(7):125
Spasova, Irena Ilieva (University of Mining & Geology/BULGARIA) *6*(9):97
Spataro, William (University della Calabria/ITALY) *6*(4):131
Spinnler, Gerard E. (Equilon Enterprises, LLC/USA) *6*(1):11; *6*(8):61
Springael, Dirk (VITO/BELGIUM) *6*(4):35
Srinivasan, P. (GeoTrans, Inc./USA) *6*(2):163
Stansbery, Anita (California Polytechnic State University/USA) *6*(10):311
Starr, Mark G. (DuPont Co./USA) *6*(8):43
Stehmeier, Lester G. (NOVA Research Technology Centre/CANADA) *6*(4):117; *6*(5):53
Stensel, H. David (University of Washington/USA) *6*(10):211
Stordahl, Darrel M. (Camp Dresser & McKee Inc./USA) *6*(6):287
Stout, Scott (Battelle/USA) *6*(5):237
Strand, Stuart E. (University of Washington/USA) *6*(10):211
Stratton, Glenn (Nova Scotia Agricultural College/CANADA) *6*(3):197
Strybel, Dan (IT Corporation/USA) *6*(9):215
Stuetz, R.M. (Cranfield University/UK) *6*(6):329
Suarez, B. (ECOPETROL-ICP/COLOMBIA) *6*(6):319
Suidan, Makram T. (University of Cincinnati/USA) *6*(1):19; *6*(5):243, 253, 261; *6*(6):133,
Suthersan, Suthan S. (ARCADIS Geraghty & Miller/USA) *6*(7):53
Suzuki, Masahiro (Nippon Institute of Technology/JAPAN) *6*(5):321
Sveum, Per (Deconterra AS/NORWAY) *6*(6):157
Swallow, Ian (BAE Systems Properties Ltd./UK) *6*(6):119
Swann, Benjamin M. (Camp Dresser & McKee Inc./USA) *6*(7):305
Swannell, Richard P.J. (AEA Technology Environment/UK) *6*(10):337

Tabak, Henry H. (U.S. EPA/USA) *6*(5):243, 253, 261, 269; *6*(9):1, 17
Takai, Koji (Fuji Packing/JAPAN) *6*(10):345

Talley, Jeffrey W. (University of Notre Dame/USA) *6*(3):189; *6*(4):59; *6*(6):43; *6*(7):125; *6*(10):115
Tao, Shu (Peking University/CHINA) *6*(7):61
Taylor, Christine D. (North Carolina State University/USA) *6*(1):83
Ter Meer, Jeroen (TNO Institute of Environmental Science/THE NETHERLANDS) *6*(2):231; *6*(7):289
Tétreault, Michel (Royal Military College of Canada/CANADA) *6*(8):95
Tharpe, D.L. (Equilon Enterprises LLC/USA) *6*(8):61
Theeuwen, J. (Grontmij BV/THE NETHERLANDS) *6*(7):289
Thomas, Hartmut (WASAG DECON GMbH/GERMANY) *6*(3):67
Thomas, Mark (EG&G Technical Services, Inc./USA) *6*(10):49
Thomas, Paul R. (Thomas Consultants, Inc./USA) *6*(5):189
Thomas, Robert C. (University of Georgia/USA) *6*(9):105
Thomson, Michelle M. (URS Corporation/USA) *6*(2):81
Thornton, Steven F. (University of Sheffield/UK) *6*(1):59, *6*(2):207
Tian, C. (University of Cincinnati/USA) *6*(8):35
Tiedje, James M. (Michigan State University/USA) *6*(7):125; *6*(8):19
Tiehm, Andreas (Water Technology Center/GERMANY) *6*(2):137; *6*(8):105
Tietje, David (Foster Wheeler Environmental Corportation/USA) *6*(9):249
Timmins, Brian (Oregon State University/USA) *6*(10):179
Togna, A. Paul (Envirogen Inc/USA) *6*(9):281
Tolbert, David E.(U.S. Army/USA) *6*(9):281
Tonnaer, Haimo (TAUW BV/THE NETHERLANDS) *6*(7):297; *6*(10):253
Toth, Brad (Harding ESE/USA) *6*(10):231
Tovanabootr, Adisorn (Oregon State University/USA) *6*(10):145
Travis, Bryan (Los Alamos National Laboratory/USA) *6*(10):163
Trudnowski, John M. (MSE Technology Applications, Inc./USA) *6*(9):35
Truax, Dennis D. (Mississippi State University/USA) *6*(9):241
Trute, Mary M. (Camp Dresser & McKee, Inc./USA) *6*(2):113
Tsuji, Hirokazu (Obayashi Corporation Ltd./JAPAN) *6*(6):111, 249; *6*(10):239
Tsutsumi, Hiroaki (Prefectural University of Kumamoto/JAPAN) *6*(10):345
Turner, Tim (CDM Federal Programs Corp./USA) *6*(6):81
Turner, Xandra (International Biochemicals Group/USA) *6*(10):23
Tyner, Larry (IT Corporation/USA) *6*(1):51; *6*(2):73

Ugolini, Nick (U.S. Navy/USA) *6*(10):65
Uhler, Richard (Battelle/USA) *6*(5):237
Unz, Richard F. (The Pennsylvania State University/USA) *6*(8):201
Utgikar, Vivek P. (U.S. EPA/USA) *6*(9):17

Valderrama, Brenda (Universidad Nacional Autónoma de México/MEXICO) *6*(6):17
Vallini, Giovanni (Universita degli Studi di Verona/ITALY) *6*(3):267
van Bavel, Bert (Umeå University/SWEDEN) *6*(3):181
van Breukelen, Boris M. (Vrije University/THE NETHERLANDS) *6*(4):91
VanBroekhoven, K. (Catholic University of Leuven/BELGIUM) *6*(4):35
Vandecasteele, Jean-Paul (Institut Français du Pétrole/FRANCE) *6*(3):227
VanDelft, Frank (NOVA Chemicals/CANADA) *6*(5):53
van der Gun, Johan (BodemBeheer bv/THE NETHERLANDS) *6*(5):289

van der Werf, A. W. (Bioclear Environmental Technology/THE NETHERLANDS) *6*(8):11
van Eekert, Miriam (TNO Environmental Sciences /THE NETHERLANDS) *6*(2):231; *6*(7):289
Van Hout, Amy H. (IT Corporation/USA) *6*(3):35
Van Keulen, E. (DHV Environment and Infrastructure/THE NETHERLANDS) *6*(8):11
Vargas, M.C. (ECOPETROL-ICP/COLOMBIA) *6*(6):319
Vazquez-Duhalt, Rafael (Universidad Nacional Autónoma de México/MEXICO) *6*(6):17
Venosa, Albert (U.S. EPA/USA) *6*(1):19
Verhaagen, P. (Grontmij BV/THE NETHERLANDS) *6*(7):289
Verheij, T. (DAF/THE NETHERLANDS) *6*(7):289
Vidumsky, John E. (E.I. du Pont de Nemours & Company/USA) *6*(2):81; *6*(8):185
Villani, Marco (Centro Ricerche Ambientali/ITALY) *6*(4):131
Vinnai, Louise (Investigative Science Inc./CANADA) *6*(2):27
Visscher, Gerolf (Province of Groningen/THE NETHERLANDS) *6*(7):141
Voegeli, Vincent (TranSystems Corporation/USA) *6*(7):229
Vogt, Bob (Louisiana-Pacific Corporation/USA) *6*(3):83
Volkering, Frank (TAUW bv/THE NETHERLANDS) *6*(4):91
von Arb, Michelle (University of Iowa) *6*(3):1
Vondracek, James E. (Ashland Inc./USA) *6*(5):121
Vos, Johan (VITO/BELGIUM) *6*(9):87
Voscott, Hoa T. (Camp Dresser & McKee, Inc./USA) *6*(7):305
Vough, Lester R. (University of Maryland/USA) *6*(5):77

Waisner, Scott A. (TA Environmental, Inc./USA) *6*(4):59; *6*(10):115
Walecka-Hutchison, Claudia M. (University of Arizona/USA) *6*(9):231
Wall, Caroline (CEFAS Laboratory/UK) *6*(10):337
Wallace, Steve (Lattice Property Holdings Plc./UK) *6*(4):17
Wallis, F.M. (University of Natal/REP OF SOUTH AFRICA) *6*(6):101; *6*(9):79
Walton, Michelle R. (Idaho National Engineering & Environmental Laboratory/USA) *6*(7):77
Walworth, James L. (University of Arizona/USA) *6*(9):231
Wan, C.K. (Hong Kong Baptist University/CHINA) *6*(6):73
Wang, Chuanyue (Rice University/USA) *6*(5):85
Wang, Qingren (Chinese Academy of Sciences/CHINA [PRC]) *6*(9):113
Wani, Altaf (Applied Research Associates, Inc./USA) *6*(10):115
Wanty, Duane A. (The Gillette Company/USA) *6*(7):87
Warburton, Joseph M. (Parsons Engineering Science/USA) *6*(7):173
Watanabe, Masataka (National Institute for Environmental Studies/JAPAN) *6*(5):321
Watson, James H.P. (University of Southampton/UK) *6*(9):61
Wealthall, Gary P. (University of Sheffield/UK) *6*(1):59
Weathers, Lenly J. (Tennessee Technological University/USA) *6*(8):139
Weaver, Dallas E. (Scientific Hatcheries/USA) *6*(1):91
Weaverling, Paul (Harding ESE/USA) *6*(10):231
Weber, A. Scott (State University of New York at Buffalo/USA) *6*(6):89
Weeber, Philip A. (Geotrans/USA) *6*(10):163
Wendt-Potthoff, Katrin (UFZ Centre for Environmental Research/GERMANY) *6*(9):43
Werner, Peter (Technical University of Dresden/GERMANY) *6*(3):227; *6*(8):105

West, Robert J. (The Dow Chemical Company/USA) *6*(2):89
Westerberg, Karolina (Stockholm University/SWEDEN) *6*(3):133
Weston, Alan F. (Conestoga-Rovers & Associates/USA) *6*(1):99; *6*(10):131
Westray, Mark (ThermoRetec Corp/USA) *6*(7):1
Wheater, H.S. (Imperial College of Science and Technology/UK) *6*(10):123
White, David C. (University of Tennessee/USA) *6*(4):73; *6*(5):305
White, Richard (EarthFax Engineering Inc/USA) *6*(6):263
Whitmer, Jill M. (GeoSyntec Consultants/USA) *6*(9):105
Wick, Lukas Y. (Swiss Federal Institute of Technology/SWITZERLAND) *6*(3):251
Wickramanayake, Godage B. (Battelle/USA) *6*(10):1
Widada, Jaka (The University of Tokyo/JAPAN) *6*(4):51
Widdowson, Mark A. (Virginia Polytechnic Institute & State University/USA) *6*(2):105; *6*(5):1
Wieck, James M. (GZA GeoEnvironmental, Inc./USA) *6*(7):165
Wiedemeier, Todd H. (Parsons Engineering Science, Inc./USA) *6*(7):241
Wiessner, Arndt (UFZ - Centre for Environmental Research/GERMANY) *6*(5):337
Wilken, Jon (Harding ESE/USA) *6*(10):231
Williams, Lakesha (Southern University at New Orleans/USA) *6*(5):145
Williamson, Travis (Battelle/USA) *6*(10):245
Willis, Matthew B. (Cornell University/USA) *6*(4):125
Willumsen, Pia Arentsen (National Environmental Research Institute/DENMARK) *6*(3):141
Wilson, Barbara H. (Dynamac Corporation/USA) *6*(1):129
Wilson, Gregory J. (University of Cincinnati/USA) *6*(1):19
Wilson, John T. (U.S. EPA/USA) *6*(1):43, 167
Wiseman, Lee (Camp Dresser & McKee Inc./USA) *6*(7):133
Wisniewski, H.L. (Equilon Enterprises LLC/USA) *6*(8):61
Witt, Michael E. (The Dow Chemical Company/USA) *6*(2):89
Wong, Edwina K. (University of Guelph/CANADA) *6*(6):185
Wong, J.W.C. (Hong Kong Baptist University/CHINA) *6*(6):73
Wood, Thomas K. (University of Connecticut/USA) *6*(5):199
Wrobel, John (U.S. Army/USA) *6*(5):207

Xella, Claudio (Water & Soil Remediation S.r.l./ITALY) *6*(6):179
Xing, Jian (Global Remediation Technologies, Inc./USA) *6*(6):311

Yamamoto, Isao (Sumitomo Marine Research Institute/JAPAN) *6*(10):345
Yamazaki, Fumio (Hyogo Prefectural Institute of Environmental Science/JAPAN) *6*(5):321
Yang, Jeff (URS Corporation/USA) *6*(2):239
Yerushalmi, Laleh (Biotechnology Research Institute/CANADA) *6*(3):165
Yoon, Woong-Sang (Sam) (Battelle/USA) *6*(7):13
Yoshida, Takako (The University of Tokyo/JAPAN) *6*(4):51; *6*(6):111
Yotsumoto, Mizuyo (Obayashi Corporation Ltd./JAPAN) *6*(6):111
Young, Harold C. (Air Force Institute of Technology/USA) *6*(2):173

Zagury, Gérald J. (École Polytechnique de Montréal/CANADA) *6*(9): 27, 129
Zahiraleslamzadeh, Zahra (FMC Corporation/USA) *6*(7):221
Zaluski, Marek H. (MSE Technology Applications/USA) *6*(9):35
Zappi, Mark E. (Mississippi State University/USA) *6*(9):241

Zelennikova, Olga (University of Connecticut/USA) *6*(7):69
Zhang, Chuanlun L. (University of Missouri/USA) *6*(9):165
Zhang, Wei (Cornell University/USA) *6*(4):125
Zhang, Zhong (University of Nevada Las Vegas/USA) *6*(9):257
Zheng, Zuoping (University of Oslo/NORWAY) *6*(2):181
Zocca, Chiara (Universita degli Studi di Verona/ITALY) *6*(3):267
Zwick, Thomas C. (Battelle/USA) *6*(10):1

KEYWORD INDEX

This index contains keyword terms assigned to the articles in the ten-volume proceedings of the Sixth International In Situ and On-Site Bioremediation Symposium (San Diego, California, June 4-7, 2001). Ordering information is provided on the back cover of this book.

In assigning the terms that appear in this index, no attempt was made to reference all subjects addressed. Instead, terms were assigned to each article to reflect the primary topics covered by that article. Authors' suggestions were taken into consideration and expanded or revised as necessary. The citations reference the ten volumes as follows:

6(1): Magar, V.S., J.T. Gibbs, K.T. O'Reilly, M.R. Hyman, and A. Leeson (Eds.), *Bioremediation of MTBE, Alcohols, and Ethers*. Battelle Press, Columbus, OH, 2001. 249 pp.

6(2): Leeson, A., M.E. Kelley, H.S. Rifai, and V.S. Magar (Eds.), *Natural Attenuation of Environmental Contaminants*. Battelle Press, Columbus, OH, 2001. 307 pp.

6(3): Magar, V.S., G. Johnson, S.K. Ong, and A. Leeson (Eds.), *Bioremediation of Energetics, Phenolics, and Polycyclic Aromatic Hydrocarbons*. Battelle Press, Columbus, OH, 2001. 313 pp.

6(4): Magar, V.S., T.M. Vogel, C.M. Aelion, and A. Leeson (Eds.), *Innovative Methods in Support of Bioremediation*. Battelle Press, Columbus, OH, 2001. 197 pp.

6(5): Leeson, A., E.A. Foote, M.K. Banks, and V.S. Magar (Eds.), *Phytoremediation, Wetlands, and Sediments*. Battelle Press, Columbus, OH, 2001. 383 pp.

6(6): Magar, V.S., F.M. von Fahnestock, and A. Leeson (Eds.), *Ex Situ Biological Treatment Technologies*. Battelle Press, Columbus, OH, 2001. 423 pp.

6(7): Magar, V.S., D.E. Fennell, J.J. Morse, B.C. Alleman, and A. Leeson (Eds.), *Anaerobic Degradation of Chlorinated Solvents*. Battelle Press, Columbus, OH, 2001. 387 pp.

6(8): Leeson, A., B.C. Alleman, P.J. Alvarez, and V.S. Magar (Eds.), *Bioaugmentation, Biobarriers, and Biogeochemistry*. Battelle Press, Columbus, OH, 2001. 255 pp.

6(9): Leeson, A., B.M. Peyton, J.L. Means, and V.S. Magar (Eds.), *Bioremediation of Inorganic Compounds*. Battelle Press, Columbus, OH, 2001. 377 pp.

6(10): Leeson, A., P.C. Johnson, R.E. Hinchee, L. Semprini, and V.S. Magar (Eds.), *In Situ Aeration and Aerobic Remediation*. Battelle Press, Columbus, OH, 2001. 391 pp.

A

abiotic/biotic dechlorination ***6*(8)**:193
acenaphthene ***6*(5)**:253
acetate as electron donor ***6*(3)**:51; ***6*(9)**:297
acetone ***6*(2)**:49
acid mine drainage, (*see also* mine tailings) ***6*(9)**:1, 9, 27, 35, 43, 53
acrylic vessel ***6*(5)**:321
actinomycetes ***6*(10)**:211
activated carbon biomass carrier ***6*(6)**:311; ***6*(8)**:113

activated carbon **6(8)**:105
adsorption **6(3)**:243; **6(5)**:253; **6(6)**:377; **6(7)**:77; **6(8)**:131; **6(9)**:86
advanced oxidation **6(1)**:121; **6(10)**:33
aerated submerged **6(10)**:329
aeration **6(6)**:203
anaerobic/aerobic treatment **6(6)**:361; **6(7)**:229
age dating **6(5)**:231, 237
air sparging **6(1)**:115, 175; **6(2)**:239; **6(9)**:215; **6(10)**:1, 9, 41, 49, 65, 101, 115, 123, 163, 223
alachlor **6(6)**:9
algae **6(5)**:181
alkaline phosphatase **6(9)**:165
alkane degradation **6(5)**:313
alkylaromatic compounds **6(6)**:173
alkylbenzene **6(2)**:19
alkylphenolethoxylate **6(5)**:215
Amaranthaceae **6(5)**:165
Ames test **6(6)**:249
ammonia **6(1)**:175; **6(5)**:337
amphipod toxicity test **6(5)**:321
anaerobic **6(1)**:35, 43; **6(3)**:91; 205; **6(5)**:17, 25, 261, 297, 313; **6(6)**:133; **6(7)**:249, 297; **6(9)**:147, 303
anaerobic biodegradation **6(1)**:137; **6(5)**:1; **6(8)**:167
anaerobic bioventing **6(3)**:9
anaerobic petroleum degradation **6(5)**:25
anaerobic sparging **6(7)**:297
aniline **6(6)**:149
Antarctica **6(2)**:57
anthracene **6(3)**:165, 251; **6(6)**:73
aquatic plants **6(5)**:181
arid-region soils **6(9)**:231
aromatic dyes **6(6)**:369
arsenic **6(2)**:239, 261; **6(5)**:173; **6(9)**: 97, 129
atrazine **6(5)**:181; **6(6)**:9
azoaromatic compounds **6(6)**:149
Azomonas **6(6)**:219

B

bacterial transport **6(8)**:1
barrier technologies **6(1)**:11; **6(3)**:165; **6(7)**:289; **6(8)**:61, 79, 87, 105, 121; **6(9)**:27, 71, 195, 209, 309
basidiomycete **6(6)**:101
benthic **6(10)**:337
benzene **6(1)**:1, 67, 75, 145, 167, 203; **6(4)**:91,117; **6(8)**:87; **6(10)**:123
benzene, toluene, ethylbenzene, and xylenes (BTEX) **6(1)**:43, 51, 59, 107, 129, 167, 195; **6(2)**:11, 19, 137, 215, 223, 270; **6(4)**:99; **6(5)**:33; **6(7)**:133; **6(8)**:105; **6(10)**: 1, 23, 49, 65, 95, 123, 131
benzo(a)pyrene **6(3)**:149; **6(6)**:101
benzo(e)pyrene **6(3)**:149
BER, *see* biofilm-electrode reactor
bioassays **6(3)**:219
bioaugmentation **6(1)**:11; **6(3)**:133; **6(4)**:59; **6(6)**:9, 43, 111; **6(7)**:125; **6(8)**:1, 11, 19, 27, 43, 53, 61, 147, 175
bioavailability **6(3)**:115, 157, 173, 189, 51; **6(4)**:7; **6(5)**:253, 279, 289; **6(6)**:1
bioavailable FeIII assay **6(8)**:209
biobarrier **6(1)**:11; **6(3)**:165; **6(7)**:289; **6(8)**:61, 79, 105, 121; **6(9)**:27, 71, 209, 309
BIOCHLOR model **6(2)**:155
biocide **6(7)**:321, 333
biodegradability **6(6)**:193
biodegradation **6(1)**:19,153; **6(3)**:165, 181, 205, 235; **6(10)**:187
biofilm **6(3)**:251; **6(4)**:149; **6(8)**:79; **6(9)**:201, 303
biofilm-electrode reactor (BER) **6(9)**:201
biofiltration **6(4)**:149
biofouling **6(7)**:321, 333
bioindicators **6(1)**:1; **6(3)**:173; **6(5)**:223
biological carbon regeneration **6(8)**:105
bioluminescence **6(1)**:1; **6(3)**:173; **6(4)**:45
biopile **6(6)**:81, 127, 141, 227, 249, 287
bioreactors **6(1)**:91; **6(6)**:361; **6(8)**:11, 35; **6(9)**:1, 265, 281, 303, 315; **6(10)**:171, 211
biorecovery of metals **6(9)**:9
bioreporters **6(4)**:45
biosensors **6(1)**:1
bioslurping **6(10)**:245, 253, 267, 275
bioslurry and bioslurry reactors **6(3)**:189; **6(6)**:51, 65
biosparging **6(10)**:115, 163
biostabilization **6(6)**:89
biostimulation **6(6)**:43
biosurfactant **6(3)**:243; **6(7)**:53
bioventing **6(10)**:109, 115, 131
biphasic reactor **6(3)**:181

biological oxygen demand (BOD) **6(10)**:311
BTEX, *see* benzene, toluene, ethylbenzene, and xylenes
Burkholderia cepacia **6(1)**:153; **6(7)**:117; **6(8)**:53
butane **6(1)**:137, 161
butyrate **6(7)**:289

C

cadmium **6(3)**:91; **6(9)**:79, 147
carAa, see carbazole 1,9a-dioxygenase gene
carbazole-degrading bacterium **6(6)**:111
carbazole 1,9a-dioxygenase gene (*carAa*) **6(4)**:51
Carbokalk **6(9)**:43
carbon isotope **6(4)**:91, 99, 109, 117; **6(10)**:115
carbon tetrachloride (CT) **6(2)**:81, 89; **6(5)**:113; **6(7)**:241; **6(8)**:185, 193
cesium-137 **6(5)**:231
CF, *see* chloroform
charged coupled device camera **6(2)**:207
chelators addition (EDGA, EDTA) **6(5)**:129, 137, 145, 151; **6(9)**:123, 147
chemical oxidation **6(7)**:45
chicken manure **6(9)**:289
chlorinated ethenes **6(7)**:27, 61, 69, 109; **6(10)**:163, 201, 231
chlorinated solvents **6(2)**:145; **6(7)**:all; **6(8)**:19; **6(10)**:231
chlorobenzene **6(8)**:105
chloroethane **6(2)**:113; **6(7)**:133, 249
chloroform (CF) **6(2)**:81; **6(8)**:193
chloromethanes **6(8)**:185
chlorophenol **6(3)**:75, 133
chlorophyll fluorescence **6(5)**:223
chromated copper arsenate **6(9)**:129
chromium (Cr[VI]) **6(8)**:139, 147; **6(9)**:129, 139, 315
chrysene **6(6)**:101
citrate and citric acid **6(5)**:137; **6(7)**:289
cleanup levels **6(6)**:1
coextraction method **6(4)**:51
Coke Facility waste **6(2)**:129
combined chemical toxicity (*see also* toxicity) **6(5)**:305
cometabolic air sparging **6(10)**:145, 155, 223
cometabolism **6(1)**:137, 145, 153, 161; **6(2)**:19; **6(6)**:81, 141; **6(7)**:117; **6(10)**:145, 155, 163, 171, 179, 193, 201, 211, 217, 223, 231; 239
competitive inhibition **6(2)**:19
composting **6(3)**:83; **6(5)**:129, **6(6)**:73, 119, 165, 257; **6(7)**:141
constructed wetlands **6(5)**:173, 329
contaminant aging **6(3)**:157, 197
contaminant transport **6(3)**:115
copper **6(9)**:79, 129
cosolvent effects **6(1)**:175, 195, 203, 243
cosolvent extraction **6(7)**:125
cost analyses and economics of environmental restoration **6(1)**:129; **6(4)**:17; **6(8)**:121; **6(9)**:331; **6(10)**:65, 211
Cr(VI), *see* chromium
creosote **6(3)**:259; **6(4)**:59; **6(5)**:1, 237, 329; **6(6)**:81, 101, 141, 295
cresols **6(10)**:123
crude oil **6(5)**:313; **6(6)**:193, 249; **6(10)**:329
CT, *see* carbon tetrachloride
cyanide **6(9)**:331
cytochrome P-450 **6(6)**:17

D

2,4-DAT, *see* diaminotoluene
DCA, *see* dichloroethane
1,1-DCA, *see* 1,1-dichloroethane
1,2-DCA, *see* 1,2-dichloroethane
DCE, *see* dichloroethene
1,1-DCE, *see* 1,1-dichloroethene
1,2-DCE, *see* 1,2-dichloroethene
c-DCE, *see* cis-dichloroethene
DCM, *see* dichloromethane
DDT, *see also* dioxins *and* pesticides **6(6)**:157
2,4-DNT, *see* dinitrotoluene
dechlorination kinetics **6(2)**:105; **6(7)**:61
dechlorination **6(2)**:231; **6(3)**:125; **6(5)**:95; **6(7)**:13, 61, 165, 173, 333; **6(8)**:19, 27, 43
DEE, *see* diethyl ether
Dehalococcoides ethenogenes **6(8)**:19, 43
dehalogenation **6(8)**:167
denaturing gradient gel electrophoresis (DGGE) **6(1)**:19; **6(4)**:35

denitrification **6(2)**:19; **6(4)**:149; **6(5)**:17, 261; **6(8)**:95; **6(9)**:179, 187, 195, 201, 209, 223, 309
dense, nonaqueous-phase liquid (DNAPL) **6(7)**:13, 19, 35, 181; **6(10)**:319
depletion rate **6(1)**:67
desorption **6(3)**:235, 243; **6(5)**:253; **6(6)**:377; **6(7)**:53, 77; **6(8)**:131
DGGE, *see* denaturing gradient gel electrophoresis
DHPA, *see* dihydroxyphenylacetate
dialysis sampler **6(5)**:207
diaminotoluene (2,4-DAT) **6(6)**:149
dibenzofuran-degrading bacterium **6(6)**:111
dibenzo-p-dioxin **6(6)**:111
dibenzothiophene **6(3)**:267
dichlorodiethyl ether **6(10)**:301
dichloroethane (DCA) **6(2)**:39; **6(7)**:289
1,1-dichloroethane (1,1-DCA; 1,2-DCA) **6(2)**:113; **6(5)**:207; **6(7)**:133, 165
1,2-dichloroethane (1,2-DCA) **6(5)**:207
dichloroethene, dichloroethylene **6(2)**:97, 155; **6(4)**:125; **6(5)**:105,113; **6(7)**:157, 197
cis-dichloroethene, *cis*-dichloroethylene (*c*-DCE) **6(2)**:39, 65, 73; 105, 173; **6(5)**:33, 95, 207; **6(7)**:1, 13, 61, 133, 141, 149, 165, 173, 181, 189, 205, 213, 221, 249, 273, 281, 289, 297, 305; **6(8)**:11, 19, 27, 43, 73, 105, 157, 209; **6(10)**:41, 145, 155, 179, 201
1,1-dichloroethene, 1,1-dichloroethylene (1,1-DCE) **6(2)**:39; **6(7)**:165, 229; **6(8)**:157; **6(10)**:231
1,2-dichloroethene and 1,2-dichloroethylene (1,2-DCE) **6(2)**:113
dichloromethane (DCM) **6(2)**:81; **6(8)**:185
diesel fuel **6(1)**:175; **6(2)**:57; **6(5)**:305; **6(6)**:81, 141, 165; **6(10)**:9
diesel-range organics (DRO) **6(10)**:9
diethyl ether (DEE) **6(1)**:19
dihydroxyphenylacetate (DHPA) **6(4)**:29
diisopropyl ether (DIPE) **6(1)**:19, 161
1,3-dinitro-5-nitroso-1,3,5-triazacyclohexane (MNX) (*see also* explosives *and* energetics) **6(3)**:51; **6(8)**:175
dinitrotoluene (2,4-DNT) **6(3)**:25, 59; **6(6)**:127, 149
dioxins **6(6)**:111
DIPE, *see* diisopropyl ether
dissolved oxygen **6(2)**:189, 207
16S rDNA sequencing **6(8)**:19
DNAPL, *see* dense, nonaqueous-phase liquid
DNX, *see* explosives and energetics
DRO, *see* diesel-range organics
dual porosity aquifer **6(1)**:59
dyes **6(6)**:369

E

ecological risk assessment **6(4)**:1
ecotoxicity, (*see also* toxicity) **6(1)**:1; **6(4)**:7
ethylenedibromide (EDB) **6(10)**:65
EDGA, *see* chelate addition
EDTA, *see* chelate addition
effluent **6(4)**:1
electrokinetics **6(9)**:241, 273
electron acceptors and electron acceptor processes **6(2)**:1, 137, 163, 231; **6(5)**:17, 25, 297; **6(7)**:19
electron donor amendment **6(3)**:25, 35, 51, 125; **6(7)**:69, 103,109, 141, 181, 249, 289, 297; **6(8)**:73; **6(9)**:297, 315
electron donor delivery **6(7)**:19, 27, 133, 173, 213, 221, 265, 273, 281, 305
electron donor mass balance **6(2)**:163
electron donor transport **6(4)**:125; **6(7)**:133; **6(9)**:241
embedded carrier **6(9)**:187
encapsulated bacteria **6(5)**:269
enhanced aeration **6(10)**:57
enhanced desorption **6(7)**:197
environmental stressors **6(4)**:1
enzyme induction **6(6)**:9; **6(10)**:211
ERIC sequences **6(4)**:29
ethane **6(2)**:113; **6(7)**:149
ethanol 6(1):19,167,175, 195, 203; **6(5)**:243; **6(6)**:133; **6(9)**:289
ethene and ethylene **6(2)**:105,113; **6(5)**:95; **6(7)**:1, 95, 133, 141, 205, 281, 297, 305; **6(8)**:11, 43, 167, 175, 209
ethylene dibromide **6(10)**:193
explosives and energetics **6(3)**:9, 17, 25, 35, 43, 51, 67; **6(5)**:69; **6(6)**:119, 127, 133; **6(7)**:125

F

fatty acids ***6*(5)**:41
Fe(II), *see* iron
Fenton's reagent ***6*(6)**:157
fertilizer ***6*(5)**:321; ***6*(6)**:35; ***6*(10)**:337
fixed-bed and fixed-film reactors ***6*(5)**:221, 337; ***6*(6)**:361; ***6*(9)**:303
flocculants ***6*(6)**:279
flow sensor ***6*(10)**:293
fluidized-bed reactor ***6*(1)**:91; ***6*(6)**:133, 311; ***6*(9)**:281
fluoranthene ***6*(3)**:141; ***6*(6)**:101
fluorogenic probes ***6*(4)**:51
food safety ***6*(9)**:113
formaldehyde ***6*(6)**:329
fractured shale ***6*(10)**:49
free-product recovery ***6*(6)**:211
Freon ***6*(2)**:49
fuel oil ***6*(5)**:321
fungal remediation ***6*(3)**:75, 99; ***6*(5)**:61, 279; ***6*(6)**:17, 101, 157, 263, 319, 329, 369
Funnel-and-Gate™ ***6*(8)**:95

G

gas flux ***6*(6)**:185
gasoline ***6*(1)**:35, 75, 161, 167, 195; ***6*(10)**:115
gasoline-range organics (GRO) ***6*(10)**:9
manufactured gas plants and gasworks ***6*(2)**:137; ***6*(10)**:123
GCW, *see* groundwater circulating well
gel-encapsulated biomass ***6*(8)**:35
GEM, see genetically engineered microorganisms
genetically engineered microorganisms (GEM) ***6*(4)**:45; ***6*(5)**:199; ***6*(7)**:125
genotoxicity, (*see also* toxicity) ***6*(3)**:227
Geobacter ***6*(3)**:1
geochemical characterization ***6*(4)**:91
geographic information system (GIS) ***6*(2)**:163
geologic heterogeneity ***6*(2)**:11
germination index 6(3):219; ***6*(6)**:73
GFP, *see* green fluorescent protein
GIS, *see* geographic information system
glutaric dialdehyde dehydrogenase ***6*(4)**:81
Gordonia terrae ***6*(1)**:153
green fluorescent protein (GFP) ***6*(5)**:199
GRO, see gasoline-range organics
groundwater ***6*(3)**:35; ***6*(8)**: 35, 87, 121; ***6*(10)**:231
groundwater circulating well (GCW) ***6*(7)**:229, 321; ***6*(10)**:283, 293

H

H_2 gas, *see* hydrogen
H_2S, *see* hydrogen sulfide
halogenated hydrocarbons ***6*(9)**:61
halorespiration ***6*(8)**:19
heavy metal ***6*(2)**:239; ***6*(5)**:137, 145, 157, 165, 173; ***6*(6)**:51; ***6*(9)**:53, 61, 71, 79, 86, 97, 113, 129, 147
herbicides ***6*(5)**:223; ***6*(6)**:35
hexachlorobenzene ***6*(3)**:99
hexane ***6*(3)**:181, ***6*(6)**:329
HMX, *see* explosives and energetics
hollow fiber membranes ***6*(5)**:269
hopane ***6*(6)**:193; ***6*(10)**:337
hornwort ***6*(5)**:181
HRC® (a proprietary hydrogen-release compound) ***6*(3)**:17, 25, 107; ***6*(7)**:27, 103, 157, 189, 197, 205, 221, 257, 305, ***6*(8)**:157, 209
^{2}H-tetradecane (*see also* tetradecane) ***6*(2)**:27
humates ***6*(1)**:99
hybrid treatment ***6*(10)**:311
hydraulic containment ***6*(8)**:79
hydraulically facilitated remediation ***6*(2)**:239
hydrocarbon ***6*(6)**:235; ***6*(10)**:329
hydrogen (H_2 gas) 6(2):199; ***6*(9)**:201
hydrogen injection, in situ ***6*(7)**:19
hydrogen isotope ***6*(4)**:91
hydrogen peroxide ***6*(1)**:121; ***6*(6)**:353; ***6*(10)**:33
hydrogen release compound, *see* HRC®
hydrogen sulfide (H_2S) ***6*(9)**:123
hydrogen ***6*(2)**:231, ***6*(7)**:61, 305
hydrolysis ***6*(1)**:83
hydrophobicity ***6*(3)**:141
hydroxyl radical ***6*(1)**:121
hydroxylamino TNT intermediates ***6*(5)**:85

I

immobilization ***6*(8)**:53
immobilized cells ***6*(8)**:121
immobilized soil bioreactor ***6*(10)**:171

in situ oxidation ***6*(7)**:1
industrial effluents ***6*(6)**:303, 361
inhibition ***6*(9)**:17
injection strategies, in situ ***6*(7)**:19, 133, 173, 213, 221, 265, 273, 305, 313; ***6*(9)**:223; ***6*(10)**:23, 163
insecticides ***6*(6)**:27
intrinsic biodegradation ***6*(2)**:89, 121
intrinsic remediation, *see* natural attenuation
ion migration ***6*(9)**:241
iron (Fe[II]) ***6*(5)**:1
iron barrier ***6*(8)**:139, 147, 157, 167
iron oxide ***6*(3)**:1
iron precipitation ***6*(3)**:211
iron-reducing processes ***6*(2)**:121; ***6*(3)**:1; ***6*(5)**:1, 17, 25; ***6*(6)**:149; ***6*(8)**:193, 201, 209; ***6*(9)**:43, 323
IR-spectroscopy ***6*(4)**:67
isotope analyses ***6*(2)**:27; ***6*(4)**:91; ***6*(8)**:27
isotope fractionation ***6*(4)**:99, 109, 117

J

jet fuel ***6*(10)**:95, 139

K

KB-1 strain ***6*(8)**:27
kerosene ***6*(6)**:219
kinetics ***6*(8)**:131, ***6*(1)**:1, 19, 27, 167; ***6*(2)**:11, 19, 105; ***6*(3)**:173; ***6*(4)**:131; ***6*(7)**:61
Klebsiella oxytoca ***6*(7)**:117
Kuwait ***6*(6)**:249

L

laccase ***6*(3)**:75; ***6*(6)**:319
lactate and lactic acid ***6*(7)**:103, 109, 165, 181, 213, 265, 281, 289; ***6*(8)**:139; ***6*(9)**:155, 273
lagoons ***6*(6)**:303
land treatment units (LTU) ***6*(6)**:1; ***6*(6)**:81, 141, 287, 295
landfarming ***6*(3)**:259; ***6*(4)**:59; ***6*(5)**:53, 279; ***6*(6)**:1, 43, 59, 179, 203, 211, 235
landfills ***6*(2)**:145, 247; ***6*(4)**:91; ***6*(8)**:113
leaching ***6*(9)**:187
lead ***6*(5)**:129, 145, 151, 157
lead-210 ***6*(5)**:231
light, nonaqueous-phase liquids (LNAPL) ***6*(1)**:59; ***6*(4)**:35; ***6*(10)**:57, 109, 245, 253, 275
lindane, (*see also* pesticides) ***6*(5)**:189
linuron (*see also* herbicides) ***6*(5)**:223
LNAPL, *see* light, nonaqueous-phase liquids
Lolium multiflorum ***6*(5)**:9
LTU, *see* land treatment units
lubricating oil ***6*(6)**:173
luciferase ***6*(3)**:133
lux ***6*(4)**:45

M

mackinawite ***6*(9)**:155
macrofauna ***6*(10)**:337
magnetic separation ***6*(9)**:61
magnetite ***6*(3)**:1; ***6*(8)**:193
manganese ***6*(2)**:261
manufactured gas plant (MGP) ***6*(2)**:19; ***6*(3)**:211, 227; ***6*(10)**:123
mass balance ***6*(2)**:163
mass transfer limitation ***6*(3)**:157
mass transfer ***6*(1)**:67
MC-100, see mixed culture
media development ***6*(9)**:147
Meiofauna ***6*(5)**:305; ***6*(10)**:337
membrane ***6*(5)**:269; ***6*(9)**:1, 265
metabolites ***6*(3)**:227
metal reduction ***6*(8)**:1
metal precipitation ***6*(9)**:9, 165
metals, biorecovery of ***6*(9)**:9
metals speciation ***6*(9)**:129
metal toxicity (*see also* toxicity) ***6*(9)**:17, 129
metals ***6*(5)**:129, 305; ***6*(8)**:1; ***6*(9)**:9, 17, 27, 105, 123, 129, 155, 165
methane oxidation ***6*(10)**:171, 187, 193, 201, 223, 231
methane ***6*(1)**:183; ***6*(8)**:113
methanogenesis ***6*(1)**:35, 43, 183; ***6*(3)**:205; ***6*(9)**:147
methanogens ***6*(3)**:91
methanol ***6*(1)**:183; ***6*(7)**:141, 289, 297
methanotrophs ***6*(10)**:171, 187, 201
methylene chloride ***6*(2)**:39; ***6*(10)**:231
Methylosinus trichosporium ***6*(10)**:187
methyl *tert*-butyl ether *or* methyl *tertiary*-butyl ether (MTBE) ***6*(1)**:1, 11, 19, 27, 35, 43, 51, 59, 67, 75, 83, 91, 107,

115, 121, 129, 137, 145, 153,161, 195, ***6*(2)**:215; ***6*(8)**:61; ***6*(10)**:1, 65
MGP, *see* manufactured gas plant
microbial heterogeneity ***6*(4)**:73
microbial isolation ***6*(3)**:267
microbial population dynamics ***6*(4)**:35
microbial regrowth ***6*(2)**:253; ***6*(7)**:1, 13; ***6*(10)**:319
microcosm studies ***6*(7)**:109; ***6*(10)**:179
microencapsulation ***6*(8)**:53
microfiltration ***6*(9)**:201
microporous membrane ***6*(9)**:265
microtox assay ***6*(3)**:227
mine tailings (*see also* acid mine drainage) ***6*(5)**:173; ***6*(9)**:27, 71
mineral oil ***6*(5)**:279, 289; ***6*(6)**:59
mineralization ***6*(2)**:121; ***6*(3)**:165; ***6*(6)**:165; ***6*(8)**:175; ***6*(9)**:139, 155
MIP, *see* membrane interface probe
mixed culture ***6*(8)**:61
mixed wastes ***6*(3)**:91; ***6*(7)**:133; ***6*(9)**:139
MNX, *see* 1,3-dinitro-5-nitroso-1,3,5-triazacyclohexane
modeling ***6*(1)**:51; ***6*(2)**:105, 155, 181, ***6*(4)**:125, 131, 139, 149; ***6*(6)**:339, 377; ***6*(8)**:185; ***6*(9)**:27, 105; ***6*(10)**:163
moisture content ***6*(2)**:247
molasses as electron donor ***6*(3)**:35; ***6*(7)**:53, 103, 149, 173; ***6*(9)**:315
monitored natural attenuation *(see also* natural attenuation) ***6*(1)**:183, ***6*(2)**:11, 163, 199, 223, 253, 261
monitoring techniques ***6*(2)**:27,189, 199, 207; ***6*(4)**:59
motor oil ***6*(5)**:53
MPE, *see* multiphase extraction
multiphase extraction (MPE) well design ***6*(10)**:245, 259
MTBE, *see* methyl *tert*-butyl ether
multiphase extraction ***6*(10)**:245, 253, 259, 267, 275
municipal solid waste ***6*(2)**:247
Mycobacterium sp. IFP 2012 ***6*(1)**:153
Mycobacterium adhesion ***6*(3)**:251
mycoremediation ***6*(6)**:263

N

naphthalene ***6*(1)**:1; ***6*(2)**:121; ***6*(3)**:173, 227; ***6*(5)**:1, 253; ***6*(6)**:51; ***6*(8)**:95, ***6*(9)**:139; ***6*(10)**:123
NAPL, *see* nonaqueous-phase liquid
natural attenuation ***6*(1)**:27, 35, 43, 51, 59, 75, 83, 183, 195; ***6*(2)**:1,39, 73, 81, 89, 97, 105, 137, 145, 173, 181, 215; ***6*(4)**:91, 99, 117; ***6*(5)**:33, 189, 321; ***6*(8)**:185, 209; ***6*(9)**:179; ***6*(10)**:115, 163
natural gas ***6*(10)**:193
natural organic carbon ***6*(2)**:261
natural organic matter ***6*(2)**:81, 97; ***6*(8)**:201
natural recovery ***6*(5)**:132, 231
nitrate contamination ***6*(9)**:173
nitrate reduction ***6*(3)**:51; ***6*(5)**:25; ***6*(9)**:331
nitrate utilization efficiency ***6*(6)**:353
nitrate ***6*(2)**:1; ***6*(3)**:17, 43; ***6*(6)**:353; ***6*(8)**:95, 147; ***6*(9)**:179, 187, 195, 209, 223, 257
nitrification ***6*(4)**:149; ***6*(5)**:337; ***6*(9)**:215
nitroaromatic compounds (*see also* explosives and energetics) ***6*(3)**:59, 67; ***6*(6)**:149
nitrobenzene, *see also* explosives and energetics ***6*(6)**:149
nitrocellulose, *see also* explosives and energetics ***6*(6)**:119
nitrogen fixation ***6*(6)**:219
nitrogen utilization ***6*(9)**:231
nitrogenase ***6*(6)**:219
nitroglycerin, *see also* explosives and energetics ***6*(5)**:69
nitrotoluenes, *see also* explosives and energetics ***6*(6)**:127
nitrous oxide ***6*(8)**:113
^{13}C-NMR, *see* nuclear magnetic resonance spectroscopy
nonaqueous-phase liquids (NAPLs) ***6*(1)**:67, 203; ***6*(3)**:141; ***6*(7)**:249
nonylphenolethoxylates ***6*(5)**:215
nuclear magnetic resonance spectroscopy (^{13}C-NMR) ***6*(4)**:67
nutrient augmentation ***6*(3)**:59; ***6*(5)**:329; ***6*(6)**:257; ***6*(7)**:313; ***6*(9)**:331; ***6*(10)**:23
nutrient injection ***6*(10)**:101
nutrient transport ***6*(9)**:241

O

oily waste ***6*(4)**:35; 6(6):257; ***6*(10)**:337, 345
oil-coated stones ***6*(10)**:329

optimization **6(5)**:279
ORC® (a proprietary oxygen-release compound) **6(1)**:99,107; **6(2)**:215; **6(3)**:107; **6(7)**:229; **6(10)**:9, 15, 87, 95, 139
organic acids **6(2)**:39
organophosphorus **6(6)**:17, 27
advanced oxidation **6(6)**:157, **6(10)**:311
oxygen-release compound, *see* ORC®
oxygen-release material **6(10)**:73
oxygen respiration **6(9)**:231; **6(10)**:57
oxygenation **6(1)**:107, 145
ozonation **6(1)**:121; **6(10)**:33, 149, 301

P

packed-bed reactors **6(9)**:249; **6(10)**:329
PAHs, *see* polycyclic aromatic hydrocarbons
paper mill waste **6(4)**:1
paraffins **6(3)**:141
partitioning **6(9)**:129
PCBs, *see* polychlorinated biphenyls
PCP toxicity (*see also* toxicity) **6(3)**:125
PCP, *see* pentachlorophenol
PCR analysis, *see* polymerase chain reaction
pentachlorophenol (PCP) **6(3)**:83, 91, 99, 107, 115, 125; **6(5)**:329; **6(6)**:279, 287, 295, 329
percarbonate **6(10)**:73
perchlorate **6(9)**:249, 257, 265, 273, 281, 289, 297, 303, 309, 315
perchloroethene, perchloroethylene **6(7)**:53
permeable reactive barriers **6(3)**:1; **6(8)**: 73, 87, 95, 121, 139, 147, 157, 167, 175, 185; **6(9)**:71, 309, 323; **6(10)**:95
pesticides **6(5)**:189; **6(6)**:9, 17, 35
PETN reductase **6(5)**:69
petroleum hydrocarbon degradation **6(4)**:7; **6(5)**:9, 17, 25; **6(8)**:131; **6(10)**: 65, 101, 245, 345
phenanthrene **6(2)**:121; **6(3)**:227, 235, 243; **6(6)**:51, 65, 73
phenol **6(6)**:303, 319, 329
phenolic waste **6(6)**:311
phenol-oxidizing cultures **6(10)**:211, 217, 239
phenyldodecane **6(2)**:27
phosphate precipitation **6(9)**:165
PHOSter **6(10)**:65
photocatalysis **6(10)**:311
physical/chemical pretreatment **6(1)**:1, 51; **6(2)**:253; **6(3)**:149; **6(5)**:9, 33, 41, 53, 61, 69, 77, 85,105, 113, 121, 129,137, 145, 151, 157, 165, 189, 199, 207, 279, 337; **6(6)**:59, 157, 241; **6(7)**:1, 13; **6(9)**:113, 173; **6(10)**:239, 311, 319
phytotoxicity (*see also* toxicity) **6(5)**:41, 223
phytotransformation **6(5)**:85
pile-turner **6(6)**:249
PLFA, *see* phospholipid fatty acid analysis
polychlorinated biphenyls (PCBs) **6(2)**:39,105,173; **6(5)**:33, 61, 95, 113, 231, 289; **6(6)**:89, **6(7)**:13, 61, 69, 95, 109, 125, 133, 141, 149, 165, 181, 189, 197, 205, 213, 241, 249, 273, 297, 305; **6(8)**:11,19, 27, 43, 157, 167, 193, 209; **6(10)**:33, 41, 231, 283
polycyclic aromatic hydrocarbons (PAHs) **6(2)**:19, 121, 129, 137; **6(3)**:141, 149, 157, 165, 173, 181, 189, 197, 205, 211, 219, 227, 235, 243; **6(4)**:35, 45, 59, 67; **6(5)**:1, 9, 17, 41, 237, 243, 251, 253, 261, 269, 279, 289, 305, 329; **6(6)**:43, 51, 59, 65, 73, 81, 89, 101, 279, 295, 297; **6(7)**:125; **6(8)**:95; **6(9)**:139; **6(10)**:33, 123
polymerase chain reaction (PCR) analysis **6(4)**:29, 35, 51; **6(8)**:43
polynuclear aromatic hydrocarbons, *see* polycyclic aromatic hydrocarbons
poplar lipid fatty acid analysis (PLFA) **6(3)**:189
poplar trees **6(5)**:113, 121, 189
potassium permanganate **6(2)**:253; **6(7)**:1
precipitation **6(9)**:105; **6(10)**:301
pressurized-bed reactor **6(6)**:311
propane utilization **6(1)**:137; **6(10)**:145, 155, 179, 193
propionate **6(7)**:265, 289
Pseudomonas fluorescens **6(3)**:173
pyrene **6(3)**:165, 235; **6(4)**:67; **6(6)**: 65, 73, 101
pyridine **6(4)**:81

R

RABITT, *see* reductive anaerobic biological in situ treatment technology
radium ***6*(5)**:173
rapeseed oil ***6*(6)**:65
RDX, *see* research development explosive
rebound ***6*(10)**:1
recirculation well ***6*(7)**:333, 341; ***6*(10)**:283
redox measurement and control ***6*(1)**:35; ***6*(2)**:11, 231; ***6*(5)**:1; ***6*(9)**:53
reductive anaerobic biological in situ treatment technology (RABITT) ***6*(7)**:109
reductive dechlorination ***6*(2)**:39, 65, 97, 105, 145, 173; ***6*(4)**:125; ***6*(7)**:45, 53, 87, 103, 109, 133, 141, 149, 157,181, 197, 205, 213, 221, 249, 257, 265, 273, 289, 297; ***6*(8)**:11, 73, 105, 157, 209
reductive dehalogenation ***6*(7)**:69
reed canary grass ***6*(5)**:181
research development explosive (RDX) ***6*(3)**:1, 9, 17, 25, 35, 43, 51; ***6*(6)**:133; ***6*(8)**:175
respiration and respiration rates ***6*(2)**:129; ***6*(4)**:59; ***6*(6)**:185, 227
respirometry ***6*(6)**:127; ***6*(10)**:217
rhizoremediation ***6*(5)**:9, 61, 199
Rhodococcus opacus ***6*(4)**:81
risk assessment ***6*(2)**:215; ***6*(4)**:1
16S rRNA sequencing ***6*(8)**:43; ***6*(9)**:147
rock-bed biofiltration ***6*(4)**:149
rotating biological contactor ***6*(9)**:79
rototiller ***6*(6)**:203
RT3D ***6*(10)**:163

S

salinity ***6*(9)**:257
salt marsh ***6*(5)**:313
SC-100, *see* single culture
Sea of Japan ***6*(5)**:321
sediments ***6*(3)**:91; ***6*(5)**:231, 237, 253, 261, 269, 279, 289, 297, 305; ***6*(6)**:51, 59; ***6*(9)**:61
selenium ***6*(9)**:323, 331
semivolatile organic carbon (SVOC) ***6*(2)**:113
sheep dip ***6*(6)**:27
Shewanella putrefaciens ***6*(8)**:201
silicon oil ***6*(3)**:141, 181
single culture ***6*(8)**:61
site characterization ***6*(10)**:139
site closure ***6*(2)**:215
slow-release fertilizer ***6*(2)**:57
sodium glycine ***6*(9)**:273
soil treatment ***6*(3)**:181; ***6*(6)**:1
soil washing ***6*(5)**:243; ***6*(6)**:241
soil-vapor extraction (SVE) ***6*(1)**:183; ***6*(10)**:1, 41, 131, 223
solids residence time ***6*(10)**:211
sorption ***6*(5)**:215, 253; ***6*(6)**:377; ***6*(8)**:131; ***6*(9)**:79, 105
source zone ***6*(7)**:13, 19, 27, 181; ***6*(10)**:267
soybean oil ***6*(7)**:213
sparging ***6*(10)**:33, 145, 155
stabilization ***6*(6)**:89
substrate delivery ***6*(7)**:281
sulfate reduction ***6*(1)**:35; ***6*(3)**:43, 91; ***6*(5)**:261, 313; ***6*(6)**:339; ***6*(7)**:69, 95; ***6*(8)**:139, 147, 193; ***6*(9)**:1, 9, 17, 27, 35, 43, 61, 71, 86, 105, 123, 147
sulfide precipitation ***6*(9)**:123
surfactants ***6*(5)**:215; ***6*(6)**:73; ***6*(7)**:213, 321, 333; ***6*(8)**:131
sustainability ***6*(6)**:1
SVE, *see* soil vapor extraction
SVOC, *see* semivolatile organic carbon
synthetic pyrethroid ***6*(6)**:27

T

TCA, see trichlorethane
1,1,1-TCA, *see* 1,1,1-trichloroethane
1,1,2-TCA, *see* 1,1,2-trichloroethane
2,4,6-TCP, *see* 2,4,6-trichlorophenol
1,1,1,2-TeCA,*see* tetrachloroethane
1,1,2,2-TeCA, *see* tetrachloroethane
1,3,5-TNB, *see* 1,3,5-trinitrobenzene
TAME, *see* tertiary methyl-amyl ether
TBA, *see* tertiary butyl alcohol
TBF, *see* tertiary butyl formate
TCE oxidation, *see* trichloroethene, trichloroethylene
TCE, *see* trichloroethene
TCP, *see* trichlorophenol
t-DCE, *see* trans-dichloroethene, trans-dichloroethylene
technology comparisons ***6*(7)**:45; ***6*(9)**:323
terrazyme ***6*(10)**:345

tertiary butyl alcohol (TBA) ***6*(1)**:19, 27, 35, 51, 59, 91, 145, 153, 161
tertiary butyl formate (TBF) ***6*(1)**:145, 161
tertiary methyl-amyl ether (TAME) ***6*(1)**:59, 161
tetrachloroethane (1,1,1,2-TeCA, 1,1,2,2-TeCA) ***6*(5)**:207; ***6*(7)**:321, 341; ***6*(8)**:193
tetradecane (see also ^{2}H-tetradecane) ***6*(3)**:181
thermal desorption ***6*(3)**:189, ***6*(6)**:35
TNB, *see* trinitrobenzene
TNT, see trinitrotoluene
TNX, *see* 1,3,5-trinitroso-1,3,5-triazacyclohexane
tobacco plant ***6*(5)**:69
toluene ***6*(1)**:145; ***6*(2)**:181; ***6*(7)**:95; ***6*(8)**:35, 131
total petroleum hydrocarbons (TPH) ***6*(2)**:1; ***6*(5)**:9; ***6*(6)**:127, 173, 179, 193, 227, 241, 249; ***6*(10)**:15, 73, 115, 337
toxicity ***6*(1)**:1; ***6*(3)**:67, 189, 227; ***6*(4)**:7; ***6*(5)**:41, 61, 223, 305; ***6*(9)**:17, 129
TPH, *see* total petroleum hydrocarbons
trace gas emissions ***6*(6)**:185
trans-dichloroethene, trans-dichloroethylene ***6*(5)**:95, 207; ***6*(7)**:165
transgenic plants ***6*(5)**:69
transpiration ***6*(5)**:189
Trecate oil spill ***6*(6)**:241; ***6*(10)**:109
trichloroethane (TCA) ***6*(7)**:241, 281
1,1,1-trichloroethane (1,1,1-TCA; 1,1,2-TCA) ***6*(2)**:39, 113, 464; ***6*(5)**:207; ***6*(7)**:87,165, 281
1,1,2-trichloroethane (1,1,2-TCA) ***6*(5)**:207
trichloroethene, trichloroethylene (TCE) ***6*(2)**:39, 65, 73, 97, 105, 113, 155, 173, 253; ***6*(4)**:125; ***6*(5)**:33, 95, 105, 113, 207; ***6*(7)**:1, 13, 53, 61, 69, 77, 87, 109, 117, 133, 141, 149, 157, 181, 189, 197, 205, 213, 221, 241, 249, 265, 273, 281, 297, 305, ***6*(8)**:11, 19, 27, 35, 43, 53, 73, 105,147, 157, 193, 209; ***6*(10)**:41, 131, 145, 155, 163, 171, 179, 187, 201, 211, 217, 223, 231, 239, 283, 319
2,4,6-trichlorophenol (2,4,6-TCP) ***6*(3)**:75; ***6*(8)**:121
trichlorotrifluoroethane ***6*(2)**:49
trinitrobenzene (TNB) ***6*(3)**:9, 25
1,3,5-trinitroso-1,3,5-triazacyclohexane (TNX) ***6*(8)**:175
trinitrotoluene (TNT) ***6*(3)**:35, 67; ***6*(5)**:69, 77, 85; ***6*(6)**:133

U

underground storage tank (UST) ***6*(1)**:67, 129
uranium ***6*(5)**:173; ***6*(7)**:77; ***6*(9)**:155, 165
UST, *see* underground storage tank

V

vacuum extraction ***6*(1)**:115
vadose zone ***6*(1)**:183; ***6*(2)**:39, 65, 97, 105, 113, 155, 173; ***6*(3)**:9; ***6*(5)**:33, 105; ***6*(7)**:1,13, 61, 133, 141, 197, 205, 213, 249, 273, 281, 305; ***6*(8)**:11,19, 43, 73, 157, 209; ***6*(10)**:41, 163
vegetable oil ***6*(6)**:65; ***6*(7)**103, 213, 241, 249
vinyl chloride ***6*(2)**:73; ***6*(4)**:109; ***6*(5)**:95; ***6*(7)**:95,149, 157, 165, 173, 289, 297, ***6*(10)**:231
vitamin B_{12} ***6*(7)**:321, 333, 341
VOCs, *see* volatile organic carbons
volatile fatty acid ***6*(7)**:61
volatile organic carbons (VOCs) ***6*(2)**:113, 189; ***6*(5)**:113, 121

W

wastewater treatment ***6*(5)**:215; ***6*(6)**:149; ***6*(9)**:173
water potential ***6*(9)**:231
weathering ***6*(4)**:7
wetlands ***6*(5)**:33, 95, 105, 313, 329; ***6*(9)**:97
white rot fungi, (*see also* fungal remediation) ***6*(3)**:75, 99; ***6*(6)**:17, 157, 263
windrow ***6*(6)**:81, 119, 141
wood preservatives ***6*(3)**:83, 259; ***6*(4)**:59; ***6*(6)**:279

X

xylene ***6*(1)**:67

Y

Z